Development of Novel Functional Materials for the Manufacture of Electronic and Optoelectronic Devices

Development of Novel Functional Materials for the Manufacture of Electronic and Optoelectronic Devices

Guest Editors

Jijun Feng
Shengli Pu

Basel • Beijing • Wuhan • Barcelona • Belgrade • Novi Sad • Cluj • Manchester

Guest Editors

Jijun Feng
School of Optical-Electrical
and Computer Engineering
University of Shanghai for
Science and Technology
Shanghai
China

Shengli Pu
College of Science
University of Shanghai for
Science and Technology
Shanghai
China

Editorial Office
MDPI AG
Grosspeteranlage 5
4052 Basel, Switzerland

This is a reprint of the Special Issue, published open access by the journal *Materials* (ISSN 1996-1944), freely accessible at: www.mdpi.com/journal/materials/special_issues/Electron_Optoelectron_Devices18.

For citation purposes, cite each article independently as indicated on the article page online and using the guide below:

Lastname, A.A.; Lastname, B.B. Article Title. *Journal Name* **Year**, *Volume Number*, Page Range.

ISBN 978-3-7258-3486-0 (Hbk)
ISBN 978-3-7258-3485-3 (PDF)
https://doi.org/10.3390/books978-3-7258-3485-3

© 2025 by the authors. Articles in this book are Open Access and distributed under the Creative Commons Attribution (CC BY) license. The book as a whole is distributed by MDPI under the terms and conditions of the Creative Commons Attribution-NonCommercial-NoDerivs (CC BY-NC-ND) license (https://creativecommons.org/licenses/by-nc-nd/4.0/).

Contents

Preface vii

Jun Zhu, Zhihao Xu, Sihua Ha, Dongke Li, Kexiong Zhang and Hai Zhang et al.
Gallium Oxide for Gas Sensor Applications: A Comprehensive Review
Reprinted from: *Materials* **2022**, *15*, 7339, https://doi.org/10.3390/ma15207339 1

Jian Chen, Ji-Jun Feng, Hai-Peng Liu, Wen-Bin Chen, Jia-Hao Guo and Yang Liao et al.
Femtosecond Laser Modification of Silica Optical Waveguides for Potential Bragg Gratings Sensing
Reprinted from: *Materials* **2022**, *15*, 6220, https://doi.org/10.3390/ma15186220 37

Joong-Mok Park, Zhi Xiang Chong, Richard H. J. Kim, Samuel Haeuser, Randy Chan and Akshay A. Murthy et al.
Probing Non-Equilibrium Pair-Breaking and Quasiparticle Dynamics in Nb Superconducting Resonators Under Magnetic Fields
Reprinted from: *Materials* **2025**, *18*, 569, https://doi.org/10.3390/ma18030569 47

Yu-Shyan Lin and Wei-Hung Chen
Dye-Sensitized Solar Cells with Modified TiO_2 Scattering Layer Produced by Hydrothermal Method
Reprinted from: *Materials* **2025**, *18*, 278, https://doi.org/10.3390/ma18020278 60

Julia Talecka, Janusz Kluczyński, Katarzyna Jasik, Ireneusz Szachogłuchowicz and Janusz Torzewski
Strength and Electrostatic Discharge Resistance Analysis of Additively Manufactured Polyethylene Terephthalate Glycol (PET-G) Parts for Potential Electronic Application
Reprinted from: *Materials* **2024**, *17*, 4095, https://doi.org/10.3390/ma17164095 74

Liga Avotina, Liga Bikse, Yuri Dekhtyar, Annija Elizabete Goldmane, Gunta Kizane and Aleksei Muhin et al.
Tungsten–SiO_2–Based Planar Field Emission Microtriodes with Different Electrode Topologies
Reprinted from: *Materials* **2023**, *16*, 5781, https://doi.org/10.3390/ma16175781 89

Xueyan Hou, Xiaohan Duan, Mengnan Liang, Zixuan Wang and Dong Yan
Application of Bis-Adducts of Phenyl-C_{61} Butyric Acid Methyl Ester in Promoting the Open-Circuit Voltage of Indoor Organic Photovoltaics
Reprinted from: *Materials* **2023**, *16*, 2613, https://doi.org/10.3390/ma16072613 100

Ronghui Xu, Yipu Xue, Minmin Xue, Chengran Ke, Jingfu Ye and Ming Chen et al.
Simultaneous Measurement of Magnetic Field and Temperature Utilizing Magnetofluid-Coated SMF-UHCF-SMF Fiber Structure
Reprinted from: *Materials* **2022**, *15*, 7966, https://doi.org/10.3390/ma15227966 110

Weinan Liu, Shengli Pu, Zijian Hao, Jia Wang, Yuanyuan Fan and Chencheng Zhang et al.
Fiber-Optic Vector-Magnetic-Field Sensor Based on Gold-Clad Bent Multimode Fiber and Magnetic Fluid Materials
Reprinted from: *Materials* **2022**, *15*, 7208, https://doi.org/10.3390/ma15207208 123

Qiupeng Wu, Zhiheng Yu, Fengli Huang and Jinmei Gu
Electrospun PA66/Graphene Fiber Films and Application on Flexible Triboelectric Nanogenerators
Reprinted from: *Materials* **2022**, *15*, 5191, https://doi.org/10.3390/ma15155191 133

Weikan Jin, Zhiheng Yu, Guohong Hu, Hui Zhang, Fengli Huang and Jinmei Gu
Effects of Three-Dimensional Circular Truncated Cone Microstructures on the Performance of Flexible Pressure Sensors
Reprinted from: *Materials* **2022**, *15*, 4708, https://doi.org/10.3390/ma15134708 . . . . **144**

Congmeng Li, Haitian Luo, Hongwei Gu and Hui Li
BTO-Coupled CIGS Solar Cells with High Performances
Reprinted from: *Materials* **2022**, *15*, 5883, https://doi.org/10.3390/ma15175883 . . . . **153**

Preface

With the rapid advancement of modern technology, the innovation and application of functional materials have become a central research focus in the manufacturing of electronic and optoelectronic devices. This Special Issue, entitled "Development of Novel Functional Materials for the Manufacture of Electronic and Optoelectronic Devices", brings together a collection of cutting-edge research covering the design, fabrication, and application of novel functional materials in microelectronics, solar cells, sensors, and beyond.

The articles included in this Special Issue highlight the latest advancements in functional materials, emphasizing their role in enhancing device performance, improving energy conversion efficiency, and expanding new application scenarios. Some studies focus on electronic materials and sensor development, while others explore the optimization of nanostructures to improve the efficiency of photovoltaic and microelectronic devices. Additionally, the Special Issue features research on microelectronics and nanotechnology, which play a crucial role in the development of high-performance optoelectronic devices.

As new functional materials continue to emerge, their significance in electronic and optoelectronic device manufacturing will only grow. This Special Issue not only summarizes the latest developments in the field but also provides researchers and engineers with new insights to drive future technological breakthroughs and industrial applications.

We sincerely appreciate the hard work of all the authors and reviewers, whose contributions have made this Special Issue a high-quality academic exchange platform. We hope that the content of this Special Issue will inspire researchers and provide strong support for the advancement of functional materials in electronic and optoelectronic device manufacturing.

Jijun Feng and Shengli Pu
Guest Editors

Review

Gallium Oxide for Gas Sensor Applications: A Comprehensive Review

Jun Zhu [1], Zhihao Xu [2], Sihua Ha [3], Dongke Li [4,*], Kexiong Zhang [5], Hai Zhang [3,*] and Jijun Feng [6,*]

1. School of Physical Science and Technology, Inner Mongolia University, Hohhot 010021, China
2. Global Zero Emission Research Center (GZR), National Institute of Advanced Industrial Science and Technology (AIST), Tsukuba 3058560, Japan
3. College of Sciences, Inner Mongolia University of Technology, Hohhot 010051, China
4. ZJU-Hangzhou Global Scientific and Technological Innovation Center, School of Materials Science and Engineering, Zhejiang University, Hangzhou 311200, China
5. School of Microelectronics, Dalian University of Technology, Dalian 116602, China
6. Shanghai Key Laboratory of Modern Optical System, Engineering Research Center of Optical Instrument and System (Ministry of Education), School of Optical-Electrical and Computer Engineering, University of Shanghai for Science and Technology, Shanghai 200093, China

* Correspondence: ldkest@zju.edu.cn (D.L.); z_hai@imut.edu.cn (H.Z.); fjijun@usst.edu.cn (J.F.)

Citation: Zhu, J.; Xu, Z.; Ha, S.; Li, D.; Zhang, K.; Zhang, H.; Feng, J. Gallium Oxide for Gas Sensor Applications: A Comprehensive Review. *Materials* **2022**, *15*, 7339. https://doi.org/10.3390/ma15207339

Academic Editor: Marina N. Rumyantseva

Received: 19 September 2022
Accepted: 12 October 2022
Published: 20 October 2022

Publisher's Note: MDPI stays neutral with regard to jurisdictional claims in published maps and institutional affiliations.

Copyright: © 2022 by the authors. Licensee MDPI, Basel, Switzerland. This article is an open access article distributed under the terms and conditions of the Creative Commons Attribution (CC BY) license (https://creativecommons.org/licenses/by/4.0/).

Abstract: Ga_2O_3 has emerged as a promising ultrawide bandgap semiconductor for numerous device applications owing to its excellent material properties. In this paper, we present a comprehensive review on major advances achieved over the past thirty years in the field of Ga_2O_3-based gas sensors. We begin with a brief introduction of the polymorphs and basic electric properties of Ga_2O_3. Next, we provide an overview of the typical preparation methods for the fabrication of Ga_2O_3-sensing material developed so far. Then, we will concentrate our discussion on the state-of-the-art Ga_2O_3-based gas sensor devices and put an emphasis on seven sophisticated strategies to improve their gas-sensing performance in terms of material engineering and device optimization. Finally, we give some concluding remarks and put forward some suggestions, including (i) construction of hybrid structures with two-dimensional materials and organic polymers, (ii) combination with density functional theoretical calculations and machine learning, and (iii) development of optical sensors using the characteristic optical spectra for the future development of novel Ga_2O_3-based gas sensors.

Keywords: Ga_2O_3; electric properties; preparation methods; gas sensors; enhancement strategies

1. Introduction

Gallium oxide (Ga_2O_3), as one type of ultrawide bandgap (UWBG) semiconducting material [1], has received tremendous attention ever since 2012 when Higashiwaki et al. successfully developed the first single-crystal Ga_2O_3 field-effect transistors (FETs) [2]. Over the past decade, Ga_2O_3 has found main applications in power electronics, solar-blind ultraviolet (UV) photodetectors, and radiation detectors, as well as gas sensors [3,4]. A variety of electronic and optoelectronic devices such as Schottky barrier diodes (SBDs) [5,6] and FETs, including MESFETs, MOSFETs, MODFETs, and HEMTs [6–9] based on Ga_2O_3 bulk single crystals, thin films, and nanostructured materials, have been achieved thanks to the advance in growth and characterization technologies and the unique properties of Ga_2O_3. There have been tens of review articles [5–40] concerning Ga_2O_3 that cover the growth techniques, the physical and chemical properties, and the state-of-the-art device fabrications. Although initial studies on the gas-sensing properties of Ga_2O_3 thin films were launched by Fleischer and Meixner [41,42] in the early 1990s, few review papers on Ga_2O_3-based gas sensors exist in the literature. Except for a complete review [26,27] that focuses on the gas sensors made by β-Ga_2O_3 nanowires and thin films, the related review can be only found in several works [4,14,15], which are usually not specialized in the domain of sensors.

It is known that Ga_2O_3 is a very important gas-sensing material widely used for monitoring exhaust gases of automobiles, flue gases of incinerators, pollutant gases of refinery plants, and explosive gases from military applications [43]. The exploration of Ga_2O_3-based gas sensors has never broken off in the last three decades, and more than five publications on average were present every year, as evident from Figure 1. In this article, we give a comprehensive overview on the distinctive aspects of Ga_2O_3 for gas sensor applications. The rest of the article is organized as follows. The polymorphs and crystal structures of Ga_2O_3 are presented in Section 2, followed by an introduction of the basic electrical properties of Ga_2O_3 in Section 3. The preparation methods used for fabricating Ga_2O_3-sensing material are provided in Section 4. Four key aspects in Ga_2O_3-based gas sensors, i.e., sensing mechanisms, evaluation criteria, classification of typical sensors, and performance enhancement strategies, are discussed in Section 5. Finally, a brief conclusion and outlook will be described in the last section.

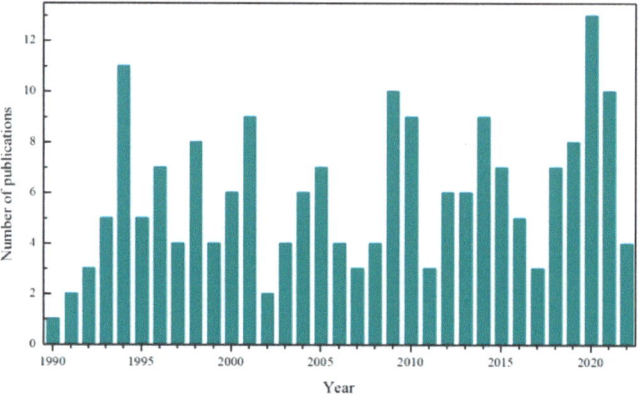

Figure 1. Number of publications on Ga_2O_3-based gas sensors from 1990 to 2022. Data extracted from Web of Science with keywords "sensor" or "sensing" and "Ga_2O_3" or "gallium oxide".

2. Polymorphs and Crystal Structures of Ga_2O_3

In the context of crystallography, polymorphism is the occurrence of different crystal structures for the same chemical entity [44]. Roy et al. [45] first identified the polymorphs of Ga_2O_3 by X-ray diffraction (XRD) when studying the phase equilibriums of the Al_2O_3-Ga_2O_3-H_2O system in 1952. They reported that there were five polymorphs of Ga_2O_3, labeled as α, β, γ, δ, and ε. The polymorphs were further confirmed by Yoshioka et al. [46] via first-principle calculations and by Zinkevich and Aldinger [47] via thermodynamic theoretical calculations. The Ga-O binary phase diagram has been also established for the first time in the work of Zinkevich and Aldinger [47]. Playford et al. [48] and Cora et al. [49] discovered the new κ-phase of Ga_2O_3 with a mixture of the β- or ε-phase. The electronic structures and polar properties of κ-Ga_2O_3 were demonstrated using density functional theory (DFT) [50]. Single-domain κ-Ga_2O_3 thin films have been successfully grown on the ε-$GaFeO_3$ substrate by Nishinaka et al. in 2020 [51]. XRD φ-scan and transmission electron microscopy (TEM) revealed that the as-grown κ-Ga_2O_3 thin films comprised a single-domain, and none of the in-plane rotational domains were present in the films.

A summary of these six polymorphs observed in Ga_2O_3 until now is presented in Table 1. Among these polymorphs, β-phase is the most stable thermodynamically. All the other polymorphs are metastable and will transform into β-phase over a certain temperature, as shown in Figure 2. The interconversion of other Ga_2O_3 polymorphs possibly occurs under different temperatures and pressures. The formation-free energies of all the phases except κ follow the $β < ε < α < δ < γ$ order at a low temperature [46]. Figure 3a depicts the schematic crystal structure of each polymorph. Almost all the phases demonstrate anisotropy. Taking β-Ga_2O_3 as an example, it is clearly seen from Figure 3b that the atomic

Review

Gallium Oxide for Gas Sensor Applications: A Comprehensive Review

Jun Zhu [1], Zhihao Xu [2], Sihua Ha [3], Dongke Li [4,*], Kexiong Zhang [5], Hai Zhang [3,*] and Jijun Feng [6,*]

1. School of Physical Science and Technology, Inner Mongolia University, Hohhot 010021, China
2. Global Zero Emission Research Center (GZR), National Institute of Advanced Industrial Science and Technology (AIST), Tsukuba 3058560, Japan
3. College of Sciences, Inner Mongolia University of Technology, Hohhot 010051, China
4. ZJU-Hangzhou Global Scientific and Technological Innovation Center, School of Materials Science and Engineering, Zhejiang University, Hangzhou 311200, China
5. School of Microelectronics, Dalian University of Technology, Dalian 116602, China
6. Shanghai Key Laboratory of Modern Optical System, Engineering Research Center of Optical Instrument and System (Ministry of Education), School of Optical-Electrical and Computer Engineering, University of Shanghai for Science and Technology, Shanghai 200093, China
* Correspondence: ldkest@zju.edu.cn (D.L.); z_hai@imut.edu.cn (H.Z.); fjjun@usst.edu.cn (J.F.)

Citation: Zhu, J.; Xu, Z.; Ha, S.; Li, D.; Zhang, K.; Zhang, H.; Feng, J. Gallium Oxide for Gas Sensor Applications: A Comprehensive Review. *Materials* **2022**, *15*, 7339. https://doi.org/10.3390/ma15207339

Academic Editor: Marina N. Rumyantseva

Received: 19 September 2022
Accepted: 12 October 2022
Published: 20 October 2022

Publisher's Note: MDPI stays neutral with regard to jurisdictional claims in published maps and institutional affiliations.

Copyright: © 2022 by the authors. Licensee MDPI, Basel, Switzerland. This article is an open access article distributed under the terms and conditions of the Creative Commons Attribution (CC BY) license (https://creativecommons.org/licenses/by/4.0/).

Abstract: Ga_2O_3 has emerged as a promising ultrawide bandgap semiconductor for numerous device applications owing to its excellent material properties. In this paper, we present a comprehensive review on major advances achieved over the past thirty years in the field of Ga_2O_3-based gas sensors. We begin with a brief introduction of the polymorphs and basic electric properties of Ga_2O_3. Next, we provide an overview of the typical preparation methods for the fabrication of Ga_2O_3-sensing material developed so far. Then, we will concentrate our discussion on the state-of-the-art Ga_2O_3-based gas sensor devices and put an emphasis on seven sophisticated strategies to improve their gas-sensing performance in terms of material engineering and device optimization. Finally, we give some concluding remarks and put forward some suggestions, including (i) construction of hybrid structures with two-dimensional materials and organic polymers, (ii) combination with density functional theoretical calculations and machine learning, and (iii) development of optical sensors using the characteristic optical spectra for the future development of novel Ga_2O_3-based gas sensors.

Keywords: Ga_2O_3; electric properties; preparation methods; gas sensors; enhancement strategies

1. Introduction

Gallium oxide (Ga_2O_3), as one type of ultrawide bandgap (UWBG) semiconducting material [1], has received tremendous attention ever since 2012 when Higashiwaki et al. successfully developed the first single-crystal Ga_2O_3 field-effect transistors (FETs) [2]. Over the past decade, Ga_2O_3 has found main applications in power electronics, solar-blind ultraviolet (UV) photodetectors, and radiation detectors, as well as gas sensors [3,4]. A variety of electronic and optoelectronic devices such as Schottky barrier diodes (SBDs) [5,6] and FETs, including MESFETs, MOSFETs, MODFETs, and HEMTs [6–9] based on Ga_2O_3 bulk single crystals, thin films, and nanostructured materials, have been achieved thanks to the advance in growth and characterization technologies and the unique properties of Ga_2O_3. There have been tens of review articles [5–40] concerning Ga_2O_3 that cover the growth techniques, the physical and chemical properties, and the state-of-the-art device fabrications. Although initial studies on the gas-sensing properties of Ga_2O_3 thin films were launched by Fleischer and Meixner [41,42] in the early 1990s, few review papers on Ga_2O_3-based gas sensors exist in the literature. Except for a complete review [26,27] that focuses on the gas sensors made by β-Ga_2O_3 nanowires and thin films, the related review can be only found in several works [4,14,15], which are usually not specialized in the domain of sensors.

It is known that Ga_2O_3 is a very important gas-sensing material widely used for monitoring exhaust gases of automobiles, flue gases of incinerators, pollutant gases of refinery plants, and explosive gases from military applications [43]. The exploration of Ga_2O_3-based gas sensors has never broken off in the last three decades, and more than five publications on average were present every year, as evident from Figure 1. In this article, we give a comprehensive overview on the distinctive aspects of Ga_2O_3 for gas sensor applications. The rest of the article is organized as follows. The polymorphs and crystal structures of Ga_2O_3 are presented in Section 2, followed by an introduction of the basic electrical properties of Ga_2O_3 in Section 3. The preparation methods used for fabricating Ga_2O_3-sensing material are provided in Section 4. Four key aspects in Ga_2O_3-based gas sensors, i.e., sensing mechanisms, evaluation criteria, classification of typical sensors, and performance enhancement strategies, are discussed in Section 5. Finally, a brief conclusion and outlook will be described in the last section.

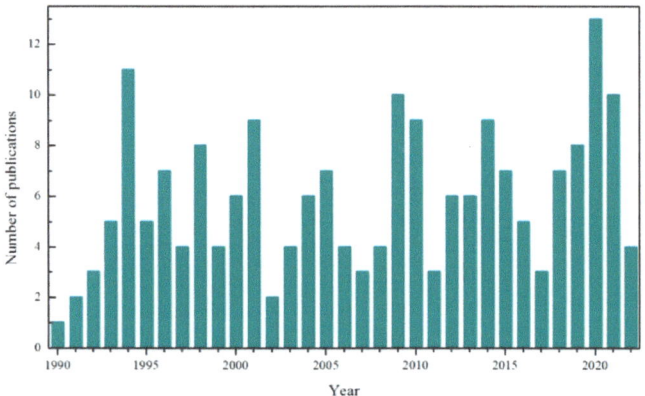

Figure 1. Number of publications on Ga_2O_3-based gas sensors from 1990 to 2022. Data extracted from Web of Science with keywords "sensor" or "sensing" and "Ga_2O_3" or "gallium oxide".

2. Polymorphs and Crystal Structures of Ga_2O_3

In the context of crystallography, polymorphism is the occurrence of different crystal structures for the same chemical entity [44]. Roy et al. [45] first identified the polymorphs of Ga_2O_3 by X-ray diffraction (XRD) when studying the phase equilibriums of the Al_2O_3-Ga_2O_3-H_2O system in 1952. They reported that there were five polymorphs of Ga_2O_3, labeled as α, β, γ, δ, and ε. The polymorphs were further confirmed by Yoshioka et al. [46] via first-principle calculations and by Zinkevich and Aldinger [47] via thermodynamic theoretical calculations. The Ga-O binary phase diagram has been also established for the first time in the work of Zinkevich and Aldinger [47]. Playford et al. [48] and Cora et al. [49] discovered the new κ-phase of Ga_2O_3 with a mixture of the β- or ε-phase. The electronic structures and polar properties of κ-Ga_2O_3 were demonstrated using density functional theory (DFT) [50]. Single-domain κ-Ga_2O_3 thin films have been successfully grown on the ε-$GaFeO_3$ substrate by Nishinaka et al. in 2020 [51]. XRD φ-scan and transmission electron microscopy (TEM) revealed that the as-grown κ-Ga_2O_3 thin films comprised a single-domain, and none of the in-plane rotational domains were present in the films.

A summary of these six polymorphs observed in Ga_2O_3 until now is presented in Table 1. Among these polymorphs, β-phase is the most stable thermodynamically. All the other polymorphs are metastable and will transform into β-phase over a certain temperature, as shown in Figure 2. The interconversion of other Ga_2O_3 polymorphs possibly occurs under different temperatures and pressures. The formation-free energies of all the phases except κ follow the $\beta < \varepsilon < \alpha < \delta < \gamma$ order at a low temperature [46]. Figure 3a depicts the schematic crystal structure of each polymorph. Almost all the phases demonstrate anisotropy. Taking β-Ga_2O_3 as an example, it is clearly seen from Figure 3b that the atomic arrangements for each crystal plane are different, which leads to different atomic configurations and dangling bond densities and therefore nonequivalent sensing properties along different crystal orientations [52].

Table 1. Basic properties of Ga_2O_3 polymorphs [22,32].

Polymorph	System	Space Group	Lattice Parameters
α	hexagonal	$R\bar{3}c$	$a = b = 4.98–5.04$ Å, $c = 13.4–13.6$ Å, $\alpha = \beta = 90°, \gamma = 120°$
β	monoclinic	$C2/m$	$a = 12.12–12.34$ Å, $b = 3.03–3.04$ Å, $c = 5.80–5.87$ Å, $\alpha = \beta = 90°, \gamma = 103.8°$
γ	cubic	$Fd\bar{3}m$	$a = b = c = 8.24–8.30$ Å, $\alpha = \beta = \gamma = 90°$
δ	cubic	$Ia3$	$a = b = c = 9.40–10.1$ Å, $\alpha = \beta = \gamma = 90°$
ε	hexagonal	$P6_3mc$	$a = 5.06–5.12$ Å, $b = 8.69–8.79$ Å, $c = 9.3–9.4$ Å, $\alpha = \beta = 90°, \gamma = 120°$
κ	orthorhombic	$Pna2_1$	$a = 5.05$ Å, $b = 8.69$ Å, $c = 9.27$ Å, $\alpha = \beta = \gamma = 90°$

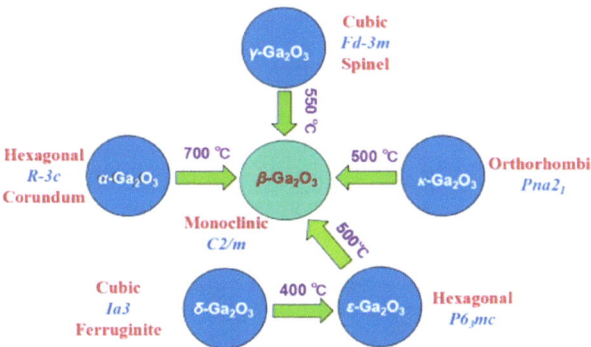

Figure 2. Interconversion relation of Ga_2O_3 polymorphs [22]. Copyright 2019 Elsevier.

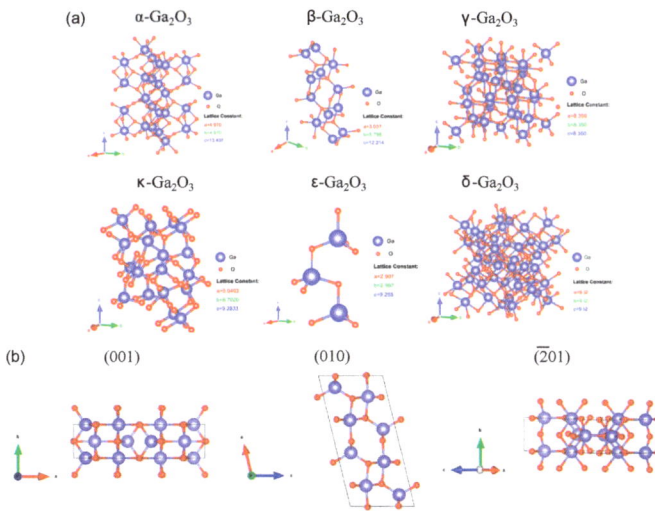

Figure 3. (**a**) Crystal structures of six polymorphs of Ga_2O_3. (**b**) The (001), (010), and ($\bar{2}$01) planes for β-Ga_2O_3. Original drawing using VESTA was done by us.

3. Electrical Properties of Ga$_2$O$_3$

It is of interest to understand the electrical properties of Ga$_2$O$_3$ that play a critical role in determining the operation and functionality of a gas sensor.

As known, the electrical conductivity (σ) of a semiconductor has a relationship with the carrier concentration (n) and carrier mobility (μ), which are written as

$$\sigma = en\mu, \tag{1}$$

where e is the elementary charge.

Theoretically, undoped stoichiometric Ga$_2$O$_3$ is a transparent insulator because of its ultrawide bandgap of ~4.9 eV. However, the as-prepared Ga$_2$O$_3$ exhibits intrinsic n-type conductivity, and the reason for the unintentional doping is still under debate. Oxygen vacancies have been considered as a source of the intrinsic electrical conductivity for a long time [53,54]. Varley et al. [55] suggested that oxygen vacancies should act as deep donors that cannot directly account for the intrinsic electrical conductivity, because the ionization energy of such vacancies was calculated as more than 1 eV. It has been widely accepted that residual impurities such as Si and H with lower formation energies in the growth process have possibly given rise to the underlying conductivity of Ga$_2$O$_3$ since then [14,15,29]. By doping impurities such as Sn, Si, and Ge in Ga$_2$O$_3$, the free electron concentration could reach the magnitude of 10^{19} cm^{-3} in the bulk crystals and 10^{20} cm^{-3} in the case of thin films at room temperature [25,28,29,37]. In general, the ionization energy of the shallow donors has a value of less than 70 meV, which decreases with the dopant concentration. For instance, the ionization energy for Sn ranges from 7.4 meV to 60 meV; for Si, ranges from 16 meV to 50 meV; and for Ge, ranges from 17.5 meV to 30 meV [29]. The free electron concentration can be also increased by the growth conditions and post-thermal treatment [15]. However, p-type doping of Ga$_2$O$_3$ is still a huge challenge.

So far, the measured room temperature Hall mobility in β-Ga$_2$O$_3$ reaches 184 cm^2/Vs [56], lower than the theoretical predicted value of 300 cm^2/Vs [2]. The electron mobility varies as functions of the temperature and carrier density. Fleischer and Meixner [57], Irmscher et al. [58], Ma et al. [59], and Gato et al. [60] studied the temperature dependence of the electron mobility of single-crystal and polycrystalline Ga$_2$O$_3$ through Hall effect measurements. As shown in Figure 4, both the electron density and the Hall mobility increase with the increasing temperature. At an elevated temperature, the electron mobility in both cases is not determined by grain boundaries but the crystal lattice itself [57]. The intrinsic mobility is limited by optical phonon scattering at a high temperature and by ionized impurity scattering at a low temperature [59]. Moreover, the electron mobility falls when the electron density increases. The electronic effective mass at low and moderate free electron concentrations is estimated at about 0.28 of the free electronic mass, and the static dielectric constant is near 10 [61]. Table 2 highlights the basic electric properties of β-Ga$_2$O$_3$ reviewed in the literature. It should be pointed out that these material parameters, including electronic mobility, electronic effective mass, and dielectric constant, are greatly modified by the crystal anisotropy effect.

Table 2. Basic electric properties of β-Ga$_2$O$_3$ [32,61].

Electric Properties	Value
Electronic effective mass (m_0)	0.28
Static dielectric constant	10
High frequency dielectric constant	3.9
Electron mobility (cm$^2 \cdot$V$^{-1} \cdot$s^{-1})	200
Range of free electron concentration (cm^{-3})	$10^{16} \sim 10^{20}$
Range of doping concentration (cm^{-3})	$10^{17} \sim 10^{20}$
Electron affinity (eV)	4.0
Break down field (eV/cm)	8.0
Typical types of shallow donors	Sn, Ge, Si

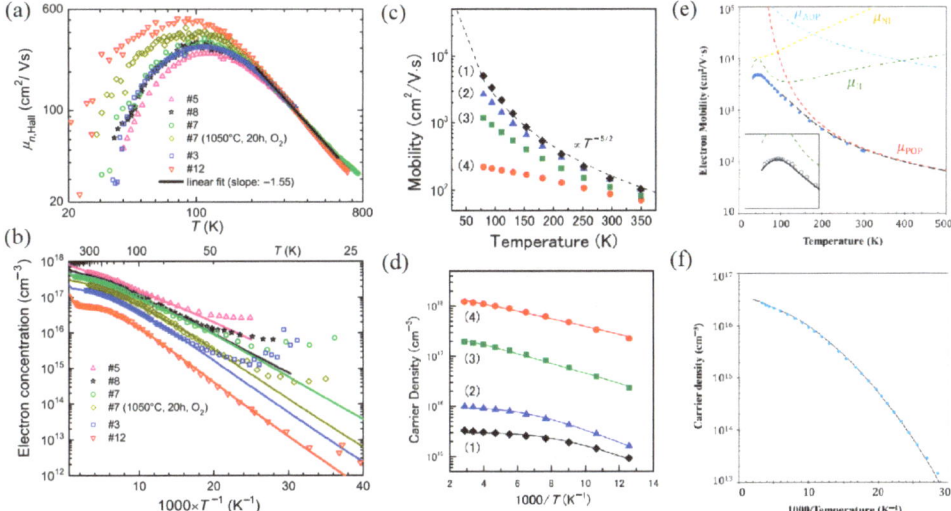

Figure 4. (**a**) The Hall mobility and (**b**) electron density as functions as the temperature for β-Ga$_2$O$_3$ single crystals grown by the Czochralski method [58]. Copyright 2011 American Institute of Physics. (**c**) The Hall mobility and (**d**) electron density as functions as the temperature for Si-doped homoepitaxial films grown on β-Ga$_2$O$_3$ (001) substrate by halide vapor-phase epitaxy [60]. Copyright 2018 Elsevier. (**e**) The Hall mobility and (**f**) electron density as functions as the temperature for silicon-doped β-Ga$_2$O$_3$ homoepitaxial films grown on β-Ga$_2$O$_3$ (010)-oriented substrates via metalorganic chemical vapor deposition [56]. Copyright 2019 American Institute of Physics.

4. Preparation Methods of Ga$_2$O$_3$ Gas-Sensing Material

Owing to its high chemical and thermal stability, Ga$_2$O$_3$ is regarded as an important n-type gas-sensing material [43]. Amidst six polymorphs, β-Ga$_2$O$_3$ is the most favorable modification for constructing gas sensors. Sometimes, the amorphous [62–64] and metastable phases [65,66] of Ga$_2$O$_3$ have also been proposed for the purposes of detecting oxygen, ozone, and so on. Nonstoichiometric amorphous film with certain degree of oxygen vacancy leads to a lower electric conductivity and, thus, a faster rising response and a higher sensitivity to O$_2$ [62,63]. The amorphous Ga$_2$O$_3$ thin films, which were decorated with InGaZnO nanoparticles, showed a boosted room temperature-sensing capability of ozone [64]. α- and ε (κ)-Ga$_2$O$_3$ thin films demonstrated sensing properties towards a large amount of oxidizing and reducing gases [65,66]. Usually, bulk crystals; thin films; and nanostructures (e.g., nanowires, nanorods, nanoparticles, etc.) have been prepared for various types of gas sensors. In this section, particular emphasis is placed on the commonly used growth techniques to prepare Ga$_2$O$_3$-sensing material.

4.1. Bulk Single-Crystal Growth

As for gas sensor applications, a few melt growth techniques have been used for preparing bulk β-Ga$_2$O$_3$ single crystals. These techniques contain the floating zone (FZ) [67–69], Czochralski (CZ) [70,71], and edge-defined film-fed (EFG) [52] growths. At high temperatures, β-Ga$_2$O$_3$ single crystal shows an oxygen-sensing property by the electrical resistance measurement. Oxygen diffusion in bulk takes place more likely by interstitial migration rather than by vacancy migration. The hydrogen-sensing characteristics of Pt Schottky diodes have been investigated using ($\bar{2}$01) and (010) β-Ga$_2$O$_3$ single crystals grown by the FZ and EFG methods. Different gallium and oxygen atomic configurations of Ga$_2$O$_3$ surfaces result in somewhat distinctive hydrogen responses of ($\bar{2}$01) and (010) facets. However, it remains unknown which is the chemical active crystal face of β-Ga$_2$O$_3$. The carrier mobility in single-crystal and polycrystalline ceramics was shown to be identical by the high-temperature Hall measurements [57]. Therefore, Ga_2O_3 polycrystalline thin films and nanostructures have become more popular than bulk crystals for advanced gas sensors. Since there have been quite a bit of research on gas-sensing characteristics of β-Ga_2O_3 single crystals, the bulk crystal growth methods of Ga_2O_3 are not discussed in detail here and can be found in other reviews [3,14–16,36]. Numerous approaches have been employed for the preparation of Ga_2O_3-sensing materials in the form of thin film and nanostructures. Later on, we will focus our discussion on the four most prevalent synthesis routes in the literature: magnetron sputtering, chemical vapor deposition (CVD), sol–gel synthesis, and hydrothermal synthesis.

4.2. Magnetron Sputtering

One of the most popular techniques for the growth of Ga_2O_3-sensing material is magnetron sputtering because of its low cost, simplicity, and low operating temperature. It is a physical vapor deposition technique to deposit thin films in which atoms are removed from a sintered target by bombardment with positive gas ions such as Ar^+. A magnetic field is added beneath the target to create plasma at lower working pressures and to deflect and confine electrons. Since gallium has a low melting point of 29.8 °C, Ga_2O_3 thin films can be sputtered only from the metal oxide target. A radio frequency (RF) voltage is usually applied in order to prevent the target from charging up due to the bombardment from positively charged ions, as the Ga_2O_3 target is an insulating ceramic one. Saikumar et al. [24] provided a review of RF-sputtered films of Ga_2O_3 and clarified the influences on the crystallinity and film properties of Ga_2O_3 from various sputtering parameters, such as substrate temperature, sputtering power, annealing temperatures, and oxygen partial pressure in the reactive sputtering.

Different substrates such as sapphire [72–88], BeO [41,42,89–97], silicon [98–101], and quartz glass [101–104] frequently equipped with an interdigital electrode have been chosen to deposit Ga_2O_3-sensing layers using the sputtering technique. The obtained films are polycrystalline or amorphous Ga_2O_3 thin films. After post annealing, the sputtering β-Ga_2O_3 polycrystalline films are very suitable for making high-temperature sensors owing to their low-term chemical stability. Ga_2O_3 with incomplete crystallinity or the amorphous phase has also aroused interest in a room temperature ozone sensor due to the large base resistance and oxygen vacancies [64]. Figure 5 shows the AFM images and XRD patterns of the Cr-doped β-Ga_2O_3 thin films synthesized on the sapphire substrates by means of RF magnetron sputtering of the Ga_2O_3 target in oxygen–argon plasma. After annealing, the Ga_2O_3 thin films with and without the addition of Cr dopant had a β-phase polycrystalline structure, and the average grain size was estimated to be 90–100 nm for pure β-Ga_2O_3 and 25–50 nm for Cr-doped β-Ga_2O_3. However, the main drawback for gas sensor applications is that the sputtered films are too compact to increase the specific surface area.

4.3. Chemical Vapor Deposition

Chemical vapor deposition (CVD) is particularly interesting not only because it gives rise to high-quality Ga_2O_3 thin films and nanostructures but also because it is applicable to large-scale production [105]. CVD, also referred to as chemical vapor transport or vapor phase epitaxy, is a conventional method often using a multitemperature zone tubular furnace (see Figure 6a) or a stainless steel chamber as a reactor. As metal organic precursors are used, the CVD technique is called MOCVD, while, in the case of hydride or halide precursors, the technique is named HVPE. Strictly speaking, thermal oxidation should not belong to the CVD technique. It will be discussed together in this section due to the high growth temperature and chemical reaction in an activated gaseous environment similar to CVD. In the CVD process, the reactant gases are diluted in carrier gases and introduced into the reaction chamber or region. When the reactant gases approach the surface of the heated substrate, chemical reactions occur on the surface to form the deposited material. The energy necessary to start the desired chemical reaction can be supplied as thermal energy or photon energy or glow discharge plasma. Gaseous reaction by-products are then evacuated with the carrier gas by a rotary pump.

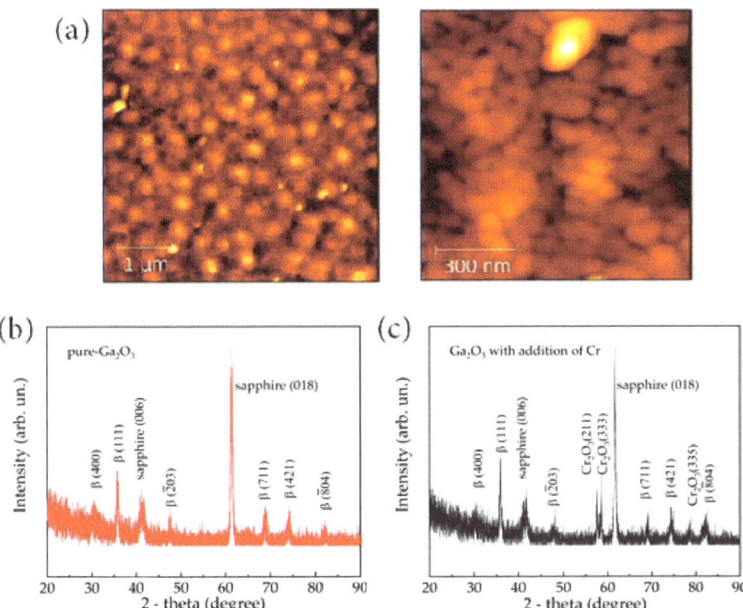

Figure 5. (**a**) AFM images of the Cr-doped Ga$_2$O$_3$ sputtering thin film [87]. XRD patterns of (**b**) pure Ga$_2$O$_3$ thin films and (**c**) Cr-doped Ga$_2$O$_3$ thin films [87]. Copyright 2020 Elsevier.

For the growth of Ga$_2$O$_3$-sensing material, liquid gallium [106–123], gallium nitride [124–132], gallium sulfides, including Ga$_2$S$_3$ [133] and GaS [134], gallium acetylacetonate [135], and gallium chloride [65,66,136–138] are often used as the gallium source, while pure oxygen or water vapor is selected as the oxygen source. Argon or nitrogen is used as the carrier gas. Sometimes, a mixture of Ga$_2$O$_3$ and carbon powders was also used as the precursor material to fabricate Ga$_2$O$_3$ nanostructures [139–141]. The key growth parameters, including precursors, growth temperature, growth pressure, growth duration, and gas flow rate, as well as separation distance between the metal source and substrate, all play important roles in depositing Ga$_2$O$_3$. Mostly, β-Ga$_2$O$_3$ nanostructures in the forms of nanowires, nanobelts, and nanorods will be possibly obtained, determined by the vapor–liquid–solid (VLS) or vapor–solid (VS) growth mechanisms [11]. Figure 6b–e shows some newly CVD-deposited Ga$_2$O$_3$ nanostructures, such as nanowires, nanorods, nanobelts, nanoribbons, and core–shell nanowires in recent years. Although a large variety of Ga$_2$O$_3$ nanostructures can be grown epitaxially on substrates using CVD, its disadvantage is that a high operating temperature is required, and the by-product gas may be hazardous and harmful.

4.4. Sol–Gel Synthesis

Of all the solution synthesis approaches, the sol–gel method is the most extensively used method to synthesize metal oxide-sensing material, since it allows for exquisite control over the size, shape, and crystal phase of the resultant material. The sol–gel process usually involves several steps: (i) dispersion of colloidal particles in a liquid to form a sol, (ii) deposition of the sol solution on a substrate by spraying or dipping or spinning, (iii) polymerization of the particles in the sol to become a gel by stabilizing the removal of the component, and (iv) pyrolysis of the remaining organic or inorganic components, thus forming the final film. Inorganic metal salts or metal organic compounds, e.g., gallium nitrate [142–145] and gallium isopropoxide [146–152], were commonly available precursor solutions used in the sol–gel process of Ga_2O_3-sensing material. The sol–gel synthesis of Ga_2O_3 films depends on the solvent, pH value, viscosity, temperature, and so on. The surface morphology and crystallinity can be modified by subsequent heat treatment. However, disadvantages such as weak adhesion and low wear resistance limit its application in sensor fabrication.

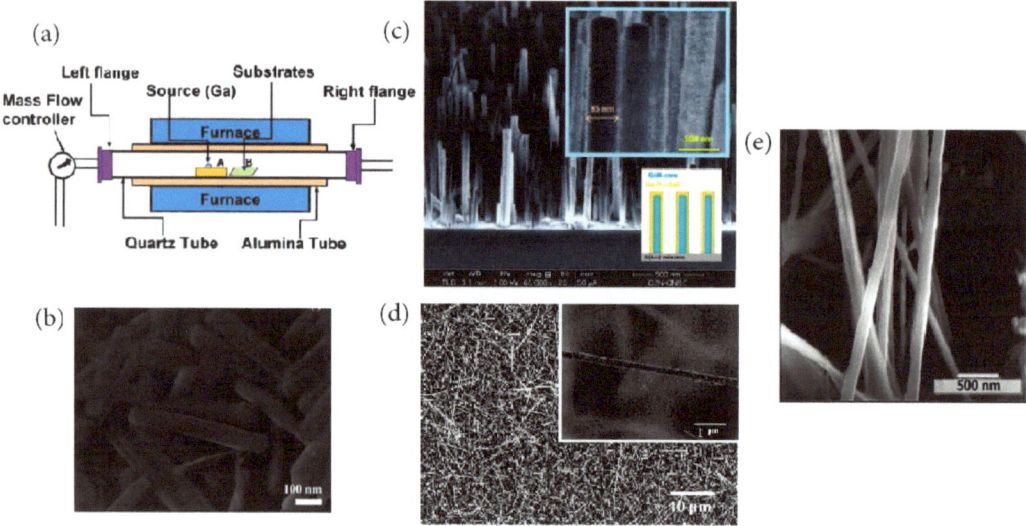

Figure 6. (**a**) Schematic experimental setup of the CVD furnace for the growth of Ga_2O_3 nanostructures [11]. Copyright 2013 Wiley. (**b**) SEM image of β-Ga_2O_3 nanorods [135]. Copyright 2016 The Royal Society of Chemistry. (**c**) SEM image of GaN/Ga_2O_3 core–shell nanowires [131]. Copyright 2019 MDPI. (**d**) SEM and HR-TEM images of β-Ga_2O_3 nanowires [140]. Copyright 2020 Elsevier. (**e**) SEM image of SnO_2-coated β-Ga_2O_3 nanobelts [132]. Copyright 2021 Elsevier.

4.5. Hydrothermal Synthesis

Due to easy operation and tunable growth parameters, hydrothermal synthesis is an important approach for the preparation of nanocomposites for gas sensor application. The hydrothermal process begins with an aqueous mixture of soluble metal salt precursors. Then, the solution is placed in an autoclave for reaction under relatively high pressure and moderate temperature conditions. In most cases, gallium nitrate hydrate $Ga(NO_3)_3 \bullet xH_2O$ [153–162] was used as the precursor for synthetizing Ga_2O_3 nanomaterials. The chosen solution has been always distilled water, while other organic solvent such as alcohol [156,157] was also used. After hydrothermal growth, the precipitates were calcinated for a couple of hours to improve the crystallinity. To obtain Ga_2O_3-sensing material with a certain size and morphology, it needs to precisely modulate the PH value and concentration of the solution, temperature, pressure, and reaction time. However, it is difficult to control the tailored phases and exact morphologies. Pilliadugula and Krishnan [159] studied the effect of PH on the surface morphology of hydrothermal synthesized β-Ga_2O_3 powders. Figure 7 shows the morphology evolutions of the as-prepared samples at different PH values. It was demonstrated that an extremely alkaline solution (PH = 14) caused higher-order hierarchical structures, whereas the acidic solution (PH = 5) facilitated nanorod structures. Cocoon-shaped morphology was formed as the PH value was increased from 7 to 11.

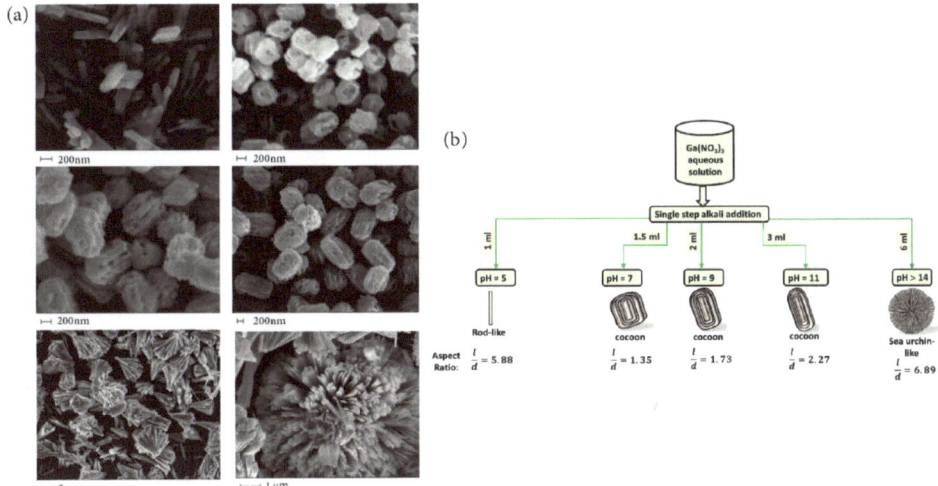

Figure 7. (**a**) SEM images of the morphological evolution of β-Ga$_2$O$_3$ nanostructures [159]. (**b**) Schematic representation of morphological evolutions at different pH values [159]. Copyright 2020 Elsevier.

4.6. Other Methods

Besides the above-discussed methods, there have been other vacuum and nonvacuum growth techniques, such as pulsed laser deposition (PLD) [163–167], atomic layer deposition (ALD) [168], coprecipitation [169,170], and spray pyrolysis [171], for the fabrication of Ga$_2$O$_3$-sensing material. Photoelectrochemical oxidation [172–174] was also used to grow Ga$_2$O$_3$ thin films on the GaN surface. In this process, the GaN epitaxial film was dipped into a phosphoric acid solution and then oxidized under the illumination of a He-Cd laser source with wavelength of 325 nm. An amorphous Ga$_2$O$_3$ film was directly grown and could be converted to the β-Ga$_2$O$_3$ phase by annealing in O$_2$ or N$_2$ ambiance. Recently, a novel liquid gallium-based sonication approach [175–177] was developed to prepare Ga$_2$O$_3$ thin films with lots of microparticles. Using this approach, ternary alloy oxides with tuning compositions were achieved by the probe sonication of liquid gallium with traces of In, Sn, and Zn in the water medium or other solvents. After annealing, the as-prepared samples can be applied for sensing NO$_2$ at a low temperature.

5. Ga$_2$O$_3$-Based Gas Sensors

Over the past three decades, several types of gas sensors using Ga$_2$O$_3$ bulk crystals, thin films, and nanomaterials have been developed to detect diverse oxidizing and reducing gases during a wide operational temperature range from room temperature to more than 1000 °C. In this section, we will focus on the advances in Ga$_2$O$_3$-based gas sensors, which cover the fundamentals of semiconductor metal oxide gas sensors and the prevailing strategies for how to improve the performance of Ga$_2$O$_3$ gas sensors.

5.1. Sensing Mechanisms

Gas-sensing mechanisms explain why the gas can cause changes in the electrical properties of a sensor. Several common gas-sensing mechanisms of metal oxide semiconductor gas sensors were reviewed by Ji et al. [178] in detail. However, the mechanisms governing Ga$_2$O$_3$-based gas sensors seem quite different. Most gas sensors using Ga$_2$O$_3$ are based on the resistive or conductive change upon target gas exposure. The sensing mechanisms of conductivity for Ga$_2$O$_3$-based gas sensors can be found in Refs. [3,26,179–181]. As mentioned above, the conductance of n-type Ga$_2$O$_3$ is determined by the carrier concentration and electron mobility. Both of them are temperature-dependent. When detecting gas, the change in conductance of Ga_2O_3 is dominated by the variation of electron density at a high temperature, while the influence of the conductance of Ga_2O_3 from electron mobility at a low temperature is more distinctive. As far as the mechanism that describes the interaction between the target gas and the sensitive metal oxide is concerned, three temperature-dependent regimes can be distinguished for Ga_2O_3-based resistive gas sensors [4,26,179–182] from Figure 8a.

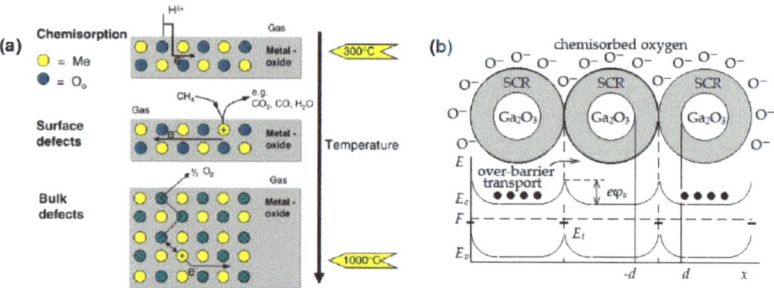

Figure 8. (a) The model of the three temperature-dependent regimes of the gas reaction of Ga_2O_3 [182]. Copyright 2008 IOP. (b) The model and corresponding energy diagram of Ga_2O_3 grain contact [87]. Copyright 2020 Elsevier.

In a high operating temperature range above 800 °C, the oxygen in the surrounding atmosphere and in the crystal lattice is in dynamic equilibrium, which means the oxygen exchange undergoes constantly between the bulk lattice of Ga_2O_3 and the ambient. If there is a reduction in the proportion of oxygen in the surrounding atmosphere, Ga_2O_3 crystal will experience an increase in the concentration of positively ionized oxygen defects in the lattice. Using the Kröger–Vink notation, the sum of the processes can be described as [180]

$$Ga_2O_3 \leftrightarrow 2Ga_{Ga}^x + 2O_O^x + V_O^* + e^- + 1/2O_2^{(g)}. \qquad (2)$$

As a consequence, the conductivity of Ga_2O_3 increases because of the delocalized electrons in crystal lattice. It can be said that this sensing regime is dominated by bulk oxygen defects. One can use it to implement high temperature oxygen sensors, which obey a typical power law [26,99,183]

$$\sigma \propto p_{O_2}^m e^{-E_A/kT}, \qquad (3)$$

where p_{O2} is the oxygen partial pressure and E_A is thermal activation energy of a dopant, k is the Boltzmann constant, and T is the temperature. The first part in above equation represents the contribution from the oxygen partial pressure, and the last part reveals the temperature dependence of the conductivity on the doped specimens.

In the intermediate operation temperature range below 800 °C, only an exchange of oxygen near the surface of Ga_2O_3 with the surrounding gas takes place. If a certain reducing gas such as CH_4 approaches the surface, it will be oxidized by oxygen from the near-surface region, leading to the formation of oxygen vacancy donors at surface and a greater conductivity due to the release of conduction electrons. The process then will be recovered in the absence of reducing gas, since the loss of oxygen of the Ga_2O_3 surface is compensated by the oxygen from the ambient atmosphere. In this case, the sensing regime is mainly caused by surface oxygen defects.

At even lower temperatures, the oxygen defect equilibrium disappears. The gas-sensing behavior is dominated by the change in electron mobility controlled by grain boundaries, rather than the change in oxygen defect-affected charged carrier density. As shown in Figure 8b, on the one hand, similar to other metal oxides, a potential barrier is formed at the grain boundary of Ga_2O_3 by pre-adsorbed oxygen ions with negative charge in the air, leading to a charge depletion region between grains. The adsorbed oxygen ions are O_2^- below 200 °C, O^- in the range of 200–550 °C, and O^{2-} above 550 °C. The width of charge depletion region is determined by the Debye length [26]:

$$L_D = \sqrt{\varepsilon k T / n e^2}, \quad (4)$$

where ε is the permittivity, n is the electron density, and e is the electronic charge.

When exposed to a reducing gas such as CO, the gas molecules react with chemisorbed oxygen ions at the Ga_2O_3 surface and return the captured electrons to the conduction band, resulting in the decrease of the depletion width. Hence, the conductance is increased due to the easier electronic transport across the grain boundary. In the case of the presence of an oxidizing gas such as NO_2, a competitive adsorption process may take place, and the overall results are a wider depletion region, smaller electron mobility, and a decrease of conductivity. However, the sensitivity toward NO_2 could be very low, since the Debye length of Ga_2O_3 is as large as several μm [181].

On the other hand, reducing a gas such as H_2 will be chemisorbed at the surface via covalent bonds to form adsorbed molecules with positive charges. To satisfy the electric neutrality, more conduction electrons are released, yielding a conductive increase in n-type Ga_2O_3. The chemisorption process happens even with no oxygen in the ambient. However, the overall concentration of the surface adsorbed species is subject to the Weisz limit [184]. Clearly, gas chemisorption and reaction at the surface play an important role in this low-temperature-sensing regime.

It should be noted that there are no upper and lower temperature limits for the three gas-sensing regimes. The modulation of the conductance of Ga_2O_3 by the target gas can be a consequence of one or a combination of grain boundaries, gas absorption and reaction, and oxygen vacancies. There have also been various Ga_2O_3-based nonresistive gas sensors using different operating principles. For example, the gas-induced changes in work function, in capacitance, in the Schottky potential barrier, and in the ionic conductivity can be used for sensing a wide range of target gases. These gas-sensing mechanisms are similar to those of other semiconductor metal oxide sensors, which can be easily found in a lot of the literatures such as [185].

5.2. Evaluation Criteria

Usually, the performance of Ga_2O_3-based gas sensors can be evaluated with respect to the '4s'-criteria: sensitivity, speed, selectivity, and stability [43].

The sensitivity S, also named as response, in the presence of gases is usually defined in several different forms. It is generally calculated from the ratio of the electrical readout (resistance R or conductance G and current I or volt V) of the sensor in background gas (usually air) Y_a to that upon exposure to certain concentrations of the target analyte Y_g:

$$S = Y_a / Y_g, \quad (5)$$

for oxidizing the target gas and

$$S = Y_g / Y_a, \quad (6)$$

for reducing the target gas.

Other expression forms of sensitivity that describes the variation degree of the output signal are also in use:

$$S = \Delta Y_a / Y_g. \quad (7)$$

The detection speed is evaluated by the response and recovery time. The response time is defined as the time required for a sensor to reach 90% of the total response upon exposure to the target gas. Recovery time is defined as the time required for a sensor to return to 90% of the original baseline signal upon removal of the target gas.

The selectivity of a sensor describes how much the sensor is disturbed by interfering gases from the target gas. The selectivity Q, which reflects the ability of a sensor to

differentiate between the target gas x and the other components in the gaseous environment x', can be expresses as:
$$Q = S_x/S_{x'}. \tag{8}$$

Obviously, larger Q means better selection of target gas and stronger resistive to interfering gas.

The stability (reproducibility) describes the endurance of a gas sensor to maintain its output signal over a long period of time and/or to the analyte gas of varying concentrations. The stability is greatly affected by thermal aging and gaseous poisoning of the sensor layer, especially when operating in a harsh environment.

Additionally, the operating temperature, power consumption, size, and cost are the other concerns that should be considered, depending on particular applications of gas sensors. It is difficult to achieve all the optimal results of the above performance parameters at the same time. Therefore, to keep the balance according to the specific situation and requirements has become a major aim for the development of Ga_2O_3 gas sensors.

5.3. Classification of Ga_2O_3-Based Gas Sensors

According to the famous Yamazoe model [186], a gas sensor can be considered as an integration of a receptor and a transducer. The former is provided with a material or a material system that interacts with a target gas and thus induces a change in its own properties or emits heat or light. The latter is a device to transform such an effect into an electrical signal. There are various approaches used for gas sensor classification. In terms of transduction principles, the gas sensors that many groups have always developed using Ga_2O_3 as sensing materials can be classified into three categories: electrical gas sensors, electrochemical gas sensors or solid electrolyte-based gas sensors, and optical gas sensors [187].

5.3.1. Electrical Gas Sensors

Electrical gas sensors, operating due to electronic conduction induced by a surface interaction with target gas, contain resistor-type gas sensors and nonresistive sensors such as SBD-, FET-, and capacitor-type gas sensors.

(1) Resistor-type gas sensors are the conductometric sensors that measure the change in resistance caused by the interaction between the sensing element and analyte gas. They possess advantages such as simple configuration, easy fabrication, and cost effectiveness and can be easily miniaturized and integrated on a microelectronic mechanical system platform. A typical resistor-type gas sensor based on Ga_2O_3 is depicted in Figure 9a. Pt and Au are commonly used as measuring electrodes, since the electron affinity of Ga_2O_3 is as large as about 4 eV [40]. Similar to solar-blind ultraviolet photodetectors, the interdigitated electrode geometry, which enables a wide contact area, is the most widely accepted geometry for a resistor sensor. In most cases, it forms the electrodes first and then deposits the Ga_2O_3-sensing layer on them, thereby causing no damage to the sensing material. Additionally, a Pt heater is placed on the back side of the substrate to heat the sensor.

In early years, Fleischer and Meixner in Siemens AG and their cooperators adopted the RF magnetron sputtering technique to fabricate β-Ga_2O_3 polycrystalline thin films and constructed resistor sensors to detect a variety of gases, such as O_2 [41,42,74,79,96,102, 188–190]; O_3 [81]; H_2 [74,78,79,82,89,90,95,191–194]; CO [75,79,82,89,91,97,188,189,191,192]; NO [78,79,96]; NH_3 [79,96]; and hydrocarbons (HCs) such as CH_4 [73,77–79,82,94,97,102, 188,189,195], C_3H_8 [79,82,196], and C_4H_8 [79,188], as well as volatile organic compounds (VOCs) such as C_2H_6O [78,79,97,195] and C_3H_6O [78,80,82]. The operating temperature ranges from 1100 °C to 400 °C. Typically, polycrystalline Ga_2O_3 thin films can be used either for sensing oxygen (>900 °C) or reducing gases (<900 °C) [14,91]. The co-adsorption of H_2O and other coexisting gases of the sputtered films were considered by Giber et al. [197], Reti et al. [198–200], and Pohle et al. [201]. Due to a high operating temperature, a self-cleaning effect was observed on the sensor surface, and the unwanted species could be eliminated to a large extent. Varhegyi et al. [202] investigated the influence from corrosive gases such as Cl_2 and SO_2 on the Ga_2O_3-sputtered layer and found that Ga_2O_3 was more resistant against SO_2 but had a very fast reaction with Cl_2 at 800 °C. A screen-printing technique was developed by Frank et al. [188], Pole et al. [201], Wiesner et al. [195], and Biskupski et al. [196] to prepare porous Ga_2O_3 thick films for detecting oxidizing gases such as CO_2 and O_3 and inflammable gases such as C_4H_{10} and C_3H_8, as well as VOCs. Additionally, many other groups such as Macri et al. [72], Ogita et al. [62,63,97,103], Hovhannisyan et al. [203], Dyndal et al. [204], Almaev et al. [86,87], Manandhar et al. [101], and Sui et al. [64] studied the gas sensitivities of resistor sensors with sputtering films for O_2, H_2, CO, C_7H_8, C_2H_6O, and C_3H_6O. However, it is very hard to reduce the operating temperature of these resistor sensors using the Ga_2O_3 compact films to a lower value.

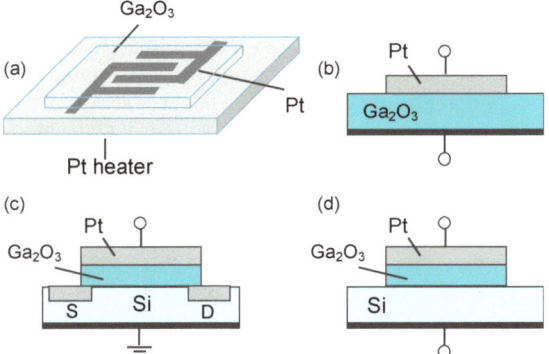

Figure 9. Schematic device structures for (**a**) resistor-type (**b**) SBD-type, (**c**) FET-type, and (**d**) capacitor-type Ga_2O_3-based gas sensors. Redrawn with permission from G. Korotcenkov [187]. Copyright 2014 Springer.

With the progress of advanced synthesis technologies, many novel Ga_2O_3 nanomaterials with different compositions and microstructures have been prepared for resistor sensors. The morphology of these Ga_2O_3 functional nanomaterials varies from one dimensional to three-dimensional nanostructures such as nanospheres, nanoflowers, nanowires, nanorods, nanobelts, and so on. These sensing materials showed high resistive response towards O_2 [109], CO [109,124,129,135,140,154,155], CO_2 [158], H_2 [113,132,176], NH_3 [159,162,177], NO_2 [125, 126,128,141,160,161,176], VOCs such as C_2H_6O [117,123,141,168], C_3H_6O [123,133,157], C_3H_8O [120], and water vapor [127,130,134], as reported by many authors. The operating temperature can be low to room temperature due to the large specific surface area and robust properties of nanostructures.

(2) Since Lundstrom et al. [205] first achieved a H_2-sensitive Pd-gate gasFET device, there has been of great interest in semiconductor metal oxide nonresistive-type gas sensors due to the simple electrical circuit required to operate them. In contrast to the above-discussed resistor-type gas sensors, the gas-sensing properties of nonresistive ones will be less affected by the morphology of Ga_2O_3-sensing material. In generally, nonresistive type gas sensors consist of two types of structures, i.e., metal/semiconductor (MS) is metal/insulator/semiconductor (MIS) structures. Comparatively, the MIS structure is always adopted in designing Ga_2O_3 sensors with Pd or Pt as catalytic metals and Ga_2O_3 as reactive insulator material. There are usually three types of nonresistive-type gas sensors based on various operating principles, namely, SBD-type, FET-type, and capacitor-type gas sensors. Figure 9b–d depict the schematic device structures of these three typical Ga_2O_3-based nonresistive-type gas sensors. The basis with respect to SBD- or FET-type (including MIS capacitor) gas sensors with a catalytic metal is modulation in the Schottky barrier height or the flat band potential by target gas [187]. Catalytic decomposition of hydrogen on the surface of noble metal and subsequent diffusion of hydrogen atoms to

the interface between metal and insulator make these devices respond to H_2 and many hydrogen containing gases.

SBD-type sensors operate on the change in Schottky barrier height affected either by formation of a dipole layer or by modification of work function once gaseous species interact with the metal surface. Then, the gas-induced rectifying property can be recognized by I–V characteristics under a forward, as well as reverse, bias voltage. $Pt/Ga_2O_3/SiC$ [146,149–152] and $Pt/Ga_2O_3/GaN$ [172–174] Schottky diode gas sensors were built to detect H_2 and C_3H_6 at different operating temperatures. Notably, Jang et al. [52] and Nakagomi et al. [68,69] reported Schottky diode gas sensors based on β-Ga_2O_3 single crystals, which showed an enhanced response to H_2 and stable operation at elevated temperatures. Almaev et al. [136] studied the gas-sensing properties of $Pt/α-Ga_2O_3$: Sn/Pt Schottky metal–semiconductor–metal structures when exposed to H_2, O_2, CO, NO, CH_4, and NH_3 in the temperature range of 25–500 °C.

FET-type sensors are based on the readout of work function change of the sensing material. The gas response of FET-type sensors is measured as a shift in either gate-source voltage or drain-source current, and it has been shown that this response is related to a shift in the threshold voltage. This kind of gas sensors can detect many reducing gases, such as H_2 and VOCs. A hybrid FET-type sensor was designed using Ga_2O_3 as the sensitive layer and the measured variation of the work function indicated a multiple response to H_2, NH_3, and NO_2 [206–208]. In their work, an analytic theoretical model was proposed to explain the inner sensing mechanism. Lampe et al. [84] made a gas FET sensor in which a sputtered Ga_2O_3 thin film activated with Pd was used to detect CO. Stegmeier et al. [85,86] used sputtered Ga_2O_3 film to construct a FET-type sensor for detecting VOCs. Nakagomi et al. [118,119] reported a field–effect hydrogen sensor using Ga_2O_3 thin film prepared by CVD. Shin et al. [209] investigated the channel length scaling effects on the signal-to-noise the ratio of a FET-type NO_2 sensor. It was demonstrated that the FET gas sensors had an advantage of working at room temperature.

MIS capacitor-type sensors are changes that can be made to the relative permittivity of the dielectric, the area of the electrode, or the distance between the two electrodes and, therefore, by measuring the change in the capacitance. Arnold et al. [110] reported an interdigital comb-finger structural capacitor gas sensor with β-Ga_2O_3 nanowires as dielectric. By capacitance measurement using a balanced AC bridge circuit, the sensor showed a rapid and reversible response to VOCs such as C_2H_6O and C_3H_6O and a more limited response to some HC_s, including C_7H_8 at room temperature. Mazeina et al. [114,115] compared pure and functionalized Ga_2O_3 nanowires as active material in room temperature capacitance-based gas sensors. It was found that the functionalization of Ga_2O_3 nanowires with acetic acids showed a significant decrease in response to C_3H_8O and CH_3NO_2 as well as $C_6H_{15}N$. In the case of pyruvic acid-functionalized nanowires, no response was observed to CH_3NO_2, but one order of magnitude increased response to $C_6H_{15}N$ was obtained compared to the pure nanowires.

5.3.2. Electrochemical Gas Sensors

An electrochemical gas sensor is a device that yields an output as a result of an electrical charge exchange process at the interface between ionic or electronic conductors. Solid electrolytes exhibit high ionic conductivity resulting from the migration of ions through the point defect sites in their lattices. In solid electrolyte-based sensors, electronic conduction only makes typically less than 1% contribution to the total conductivity, while the ionic conductivity contributes to 99% of the rest. Schematic device structures for two kinds of electrochemical gas sensors containing Ga_2O_3 are shown in Figure 10. In an early study, NH_4^+ ion conducting gallate solid electrolyte was prepared by mixing K_2CO_3, $RbCO_3$, and Ga_2O_3 [210]. Fabricated by the combination of NH_4^+-Ga_2O_3 and rare earth ammonium sulfate as a solid electrolyte and a solid reference electrode, the gas sensor showed outstanding NH_3 detection in good accordance with the Nernst relation. In a mixed potential gas sensor based on O^{2-} ion conducting yttria-stabilized zirconia (YSZ) solid electrolyte, Ga$_2$O$_3$ was used to stabilize the metal (Au, Pt) electrode in its morphology and to inhibit its catalytic activity [211–215]. The results indicated that these gas sensors had a high sensitivity and good selectivity of HCs, such as C$_3$H$_6$. An electrochemical Pt/YSZ/Au-doped Ga$_2$O$_3$ impedance metric sensor was fabricated by Wu et al. [216]. The impedance of the sensor originated from the Ohmic contact resistance, the electrolyte impedance, and the interfacial impedance between the electrolyte and the sensing electrode. Strong dependence of the interfacial impedance upon the CO concentration at 550 °C was found due to an electrochemical oxidation of CO. Yan et al. [217] synthesized a mesoporous β-Ga$_2$O$_3$ nanoplate to make an amperometric electrochemical sensors. It was summarized that more oxygen defects and action sites in β-Ga$_2$O$_3$ nanoplate account for enhanced gas sensitivity in detecting CO.

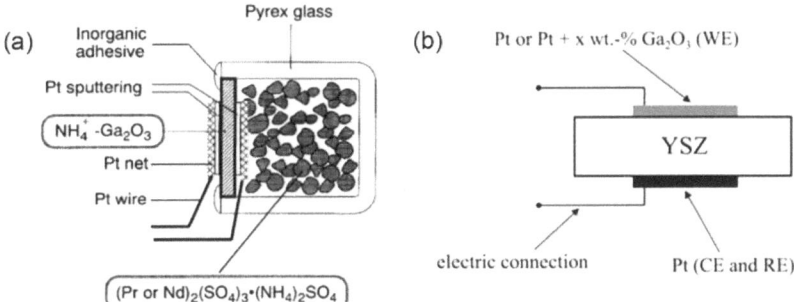

Figure 10. Schematic device structures for electrochemical gas sensors: (**a**) NH^{4+}-Ga$_2$O$_3$ solid electrolyte [210] Copyright 1998 The Electrochemical Society. (**b**) YSZ solid electrolyte [214]. Copyright 2004 Elsevier.

5.3.3. Optical Gas Sensors

Optical gas sensors detect changes in visible light or other electromagnetic waves during interactions with gaseous molecules. Reiprich et al. [122] used a corona discharge assisted growth morphology to prepare Sn-doped Ga$_2$O$_3$ for optical gas-sensing application. The Sn-doped Ga$_2$O$_3$ layer was shown to be capable of detecting small amounts of C$_2$H$_6$O, C$_3$H$_6$O, and C$_3$H$_8$O at room temperature. The reason was that the generation and quantity variance of negatively charged oxygen ions indirectly produced a change in the photoluminescence spectrum. It was also observed that the response toward C$_3$H$_6$O of a Ga$_2$O$_3$ layer-like structure would be increased by 30% relative to the nanowires. However, the study of Ga$_2$O$_3$-based optical gas sensors is in its infancy, further careful investigations are needed into this type of gas sensor.

To sum up, Ga$_2$O$_3$-based gas sensors exhibit broad-range sensitivity to a wide variety of analyte gases. Table 3 lists the available target gases of different types of Ga$_2$O$_3$-based gas sensors. The most preferable target gases are O$_2$, CO, H$_2$, and CH$_4$ [218].

Table 3. Target gases of Ga$_2$O$_3$-based gas sensors.

	Electrical Gas Sensors				Electrochemical Gas Sensor	Optical Gas Sensor
	Resistor	SBD	FET	Capacitor		
Environmental gases	O$_2$, CO$_2$, O$_3$, NH$_3$, SO$_2$	O$_2$, NH$_3$	NH$_3$		NH$_3$	
Highly toxic gases	CO, H$_2$S, NO, NO$_2$	CO, NO	CO, NO$_2$		CO	
Combustible gases	H$_2$, CH$_4$, C$_4$H$_{10}$, C$_3$H$_8$, C$_4$H$_8$, C$_7$H$_8$	H$_2$, CH$_4$, C$_3$H$_6$	H$_2$	C$_7$H$_8$	C$_3$H$_6$	
VOCs	C$_2$H$_6$O, C$_3$H$_6$O, C$_3$H$_8$O			C$_2$H$_6$O, C$_3$H$_6$O, C$_3$H$_8$O		C$_2$H$_6$O, C$_3$H$_6$O, C$_3$H$_8$O
Humidity	H$_2$O					
Other gases	C$_2$H$_6$S			CH$_3$NO$_2$, C$_6$H$_{15}$N		

5.4. Performance Enhancement Strategies

As is discussed in the former section, Ga_2O_3 can be used to make several types of gas sensors that respond to lots of gases, ensuring a wide application range. However, Ga_2O_3-based gas sensors still encounter some issues such as low selectivity, average sensitivity, and high operation temperature [26,182,187,218]. Multiple ways have been developed till now by many researchers to improve the gas-sensing performance of Ga_2O_3-based gas sensors through material engineering and device optimization.

5.4.1. Surface Modulation of Pure Ga_2O_3

It is well-known that the performance of Ga_2O_3 gas sensors greatly depends on the interaction between target gas and the Ga_2O_3 surface. Therefore, surface modulation, such as crystallinity, porosity, and oxygen vacancies, in the process of fabricating Ga_2O_3-sensing material plays an important role in determining the gas-sensing behavior. Ogita et al. [63,67,99,100,219] presented a series of studies on the influence of the annealing conditions on the high-temperature oxygen-sensing properties of the Ga_2O_3 thin film sensor. The results indicated that the annealing conditions affect not only the grain size, number of oxygen vacancies and surface roughness but also the oxygen-sensing properties of Ga_2O_3 thin films. The oxygen sensitivity of the sensors increases when the oxygen flow rate was increasing from 0 to 100% at 1000 °C. The response time of the Ga_2O_3 sensors decreased as the grain size increases with increasing the annealing temperature and annealing time. The authors also measured the oxygen-sensing characteristics at 1000 °C for β-Ga_2O_3 sputtered thin films and β-Ga_2O_3 single crystals and found that both sensors had response times in the same range. Grain boundaries played an important role in determining the value of the response time in sputtered films, while the surface processes primarily influenced the modifications of the resistance for single crystals. The influence of film thickness on sensing response at room temperature was investigated by Pandeeswari and Jeyaprakash [171]. The sensitivities towards 0.5 ppm and 50 ppm of NH_3 increased as the film thickness became larger.

Due to large surface to volume ratio and abundant surface active sites to the target gas, porous and nanostructured Ga_2O_3 was always used as the sensing layer. Cuong et al. [113] addressed the effect of growth temperature on the microstructural properties and H_2 sensing ability of Ga_2O_3 nanowires. Girija et al. [135] performed the gas-sensing analysis of morphology-dependent β-Ga_2O_3 nanostructure thin films. It can be seen from Figure 11 that nanorods of β-Ga_2O_3 had a higher surface area with a larger pore volume distribution than rectangular β-Ga_2O_3 structures. As a result, rod shaped β-Ga_2O_3 nanostructures showed better sensitivity, shorter response, and recovery time upon exposure to CO gas at 100 °C in comparison with rectangular shape β-Ga_2O_3 nanostructures. The improvement on sensor performance could be contributed from the high surface area, particle size, shape, numerous surface active sites and the oxygen vacancies. The effect of pH-dependent morphology evolutions on room temperature NH_3-sensing performances of β-Ga_2O_3 nanostructured films was investigated by Pilliadugula and Krishnan [159]. It could be seen that the sample with hierarchical morphology and relatively low crystallite size showed utmost sensing responses at all concentrations of NH_3 relative to the remaining samples.

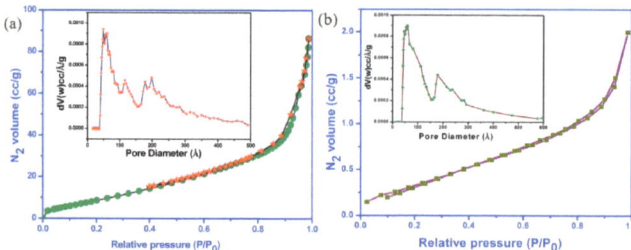

Figure 11. Cont.

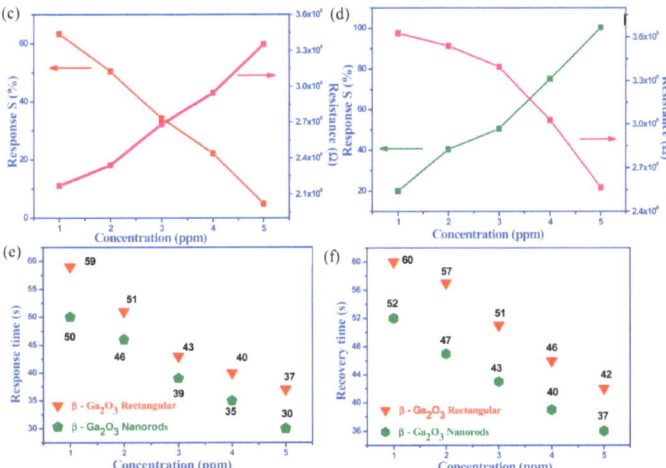

Figure 11. Nitrogen adsorption-desorption isotherm of the Ga_2O_3 (**a**) rectangular and (**b**) rod-shaped nanostructures (inset: the corresponding pore size distribution). Sensing response and electrical resistance as a function of CO concentration at 100 °C (**c**) β-Ga_2O_3 rectangular nanorods, (**d**) β-Ga_2O_3 nanorods, (**e**) response time, and (**f**) recovery time of different β-Ga_2O_3 nanostructure thin films [135]. Copyright 2016 The Royal Society of Chemistry.

High-energy crystal facet has a higher surface energy and more defect sites, thus exhibiting more active physicochemical properties and better gas-sensing performance. Jang et al. [52] demonstrated that the H_2 responses of the (010) facet were slightly higher than that of the ($\bar{2}$01) facet when studying the sensing characteristics of Schottky diode-type gas sensors using Ga_2O_3 single crystal. However, it is not clear which facet is the most active one for gas sensors so far.

5.4.2. Sensitizing by Noble Metal

In most of gas-sensing process, small noble metal nanoparticles or clusters distributed on the surface of Ga_2O_3 can improve the efficiency of catalytic reactions between target gases and the surface of sensing layers, thus remarkably enhancing the gas-sensing characteristics. At an early stage, Fleischer et al. [91], Bausewein et al. [74], and Schwebel et al. [75] studied the effects of Pt, Pd, and Au catalyst dispersions on the sensitivity to reducing gases of high temperature Ga_2O_3 thin film sensors. They found that Pt dispersion accelerated the response of CO, whereas significantly increased the sensitivity to H_2. Au clusters on the Ga_2O_3 surface could yield a high sensitivity to CO and a distinct reduction of the cross-sensitivity towards VOCs. However, although Pt accelerated the desorption of H_2 on Ga_2O_3, it had no impact on the sensitivity to H_2. Recently, Krawczyk et al. [138] reported the impregnation of Au nanoparticles of the HVPE-prepared n-type β-Ga_2O_3 layer would cause an inversion to p-type conductivity at the surface. The conductive sensitivity of the Au-modified β-Ga_2O_3 layer toward 16 ppm of dimethyl sulfide obviously was enhanced if compared to that of the unmodified β-Ga_2O_3 layer. Pt and Au-decorated Ga_2O_3 nanowires were synthesized by Kim et al. [124] and Weng et al. [140]. Figure 12 shows the SEM and TEM images and electrical response of the gas sensors fabricated by Pt- and Au-decorated Ga_2O_3 nanowires. As measuring at 100 °C, the responses of the Pt-functionalized nanowire were reported to improve 27.8-, 26.1-, 22.0-, and 16.9-fold at CO concentrations of 10, 25, 50, and 100 ppm, respectively, relative to the bare Ga_2O_3 nanowires. However, both the response and recovery times of this Pt-functionalized CO sensor were increased significantly. Through CO gas sensor measurements at room temperature, Au-decorated single β-Ga_2O_3 nanowire showed better results than single pure β-Ga_2O_3 nanowire and multiple networked Au-decorated β-Ga_2O_3 nanowires. The enhancement of the CO-sensing properties was resulted from a combination of the spillover effect and the enhanced chemisorption and dissociation of the target gas.

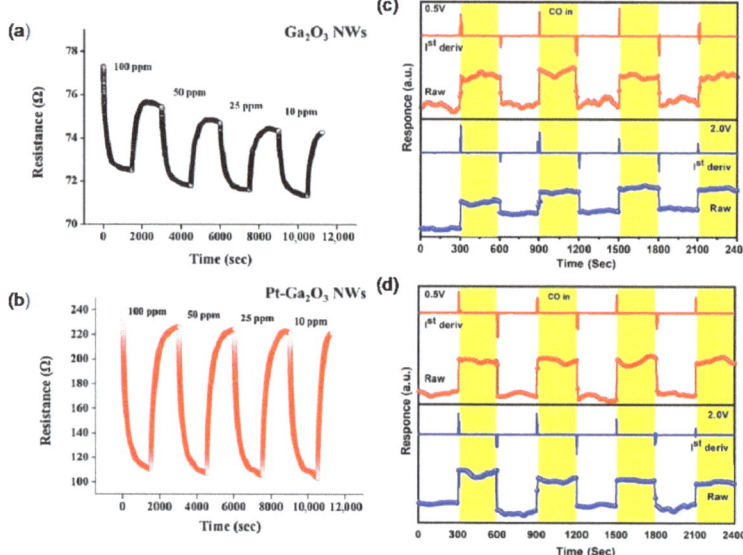

Figure 12. Gas responses of (**a**) bare Ga_2O_3 nanowires and (**b**) Pt-coated Ga_2O_3 nanowires to 10, 25, 50, and 100 ppm CO gas at 100 °C [124]. Copyright 2012 Elsevier. Room temperature CO gas sensor measurement results with different bias voltages for (**c**) pure β-Ga_2O_3 nanowires and (**d**) Au-decorated β-Ga_2O_3 nanowire with 2158 ppm CO concentration [140]. Copyright 2020 Elsevier.

So far, noble metals had also been used in FET-type and solid electrolyte type gas sensors to catalytically activate the sensing layer. Zosel et al. [212,213] achieved a high sensitivity for hydrocarbons in potentiometric zirconia-based gas sensors, attributed to the catalytic activity of the Au composites. Shuk et al. [215] reported that the mixed potential solid electrolyte sensor based on Au/Ga_2O_3 composite electrodes had a sensitivity limit of 5 ppm CO and showed good selectivity, reproducibility, and stability in the presence of other combustion gas species. The FET-type sensors thermally activated by Pd and Pt for the detection of reducing gases were studied by Lampe et al. [84] and Stegmeier et al. [85,86]. The results indicated that catalytic Pt dispersions on the micro- and nanoscale caused a remarkable gas response at room temperature to a large variety of hydrocarbons and VOCs with small concentrations (ppb-ppm). Overall, noble metals such as Pt, Pd, and Au play the roles of chemical sensitizers and electronic sensitizers in Ga_2O_3 sensors.

5.4.3. Doping Specific Element

Doping is an effective way to influence not only the electronic properties but also the structural properties of grains, such as size and shape, leading to the enhancement of sensing behavior. Frank et al. [102,188,189] found that the doping of sputter-deposited polycrystalline Ga_2O_3 thin films using donor type ions such as Zr^{4+}, Ti^{4+}, and especially Sn^{4+} caused a modulation of the base conductivity by two orders of magnitude and strong impact on the sensitivity to reducing gases. However, there was no influence on the sensitivity to reducing gases such as CH_4 and CO after doping Mg^{2+} in the Ga_2O_3 and no influence of whether donor-type or acceptor-type doping concentration on O_2 sensitivity at high temperature. However, in the work of Almaev et al. [87,88], the authors stated that addition of Cr stimulated dissociative adsorption of O_2 due to its high catalytic activity via a spillover mechanism, leading to a significant increase in the response of to O_2 over the temperature range 250–400 °C. The O_2 gas-sensing performance of Ga_2O_3

semiconducting thin films prepared by the sol–gel process and doped with Ce, Sb, W, and Zn was investigated by Li et al. [148]. The authors observed that the operating temperature of sensors doped with Zn reduced below 450 °C and Sensors doped with Ce had a very fast response time of typically 40 s. W-doped sensors showed the highest response to O_2 while Sb-doped films possessed the highest base resistance. Sn was proven to be the most effective dopant in Ga_2O_3 for gas detection at different operating temperatures [121,122,162,170,189]. As examples, Figure 13 exhibits the enhancement of the sensor sensitivity of Sn-doped Ga_2O_3 gas sensors when detecting different gases. An increase in conductivity of up to two orders of magnitude, as well as an enhancement of the gas sensitivity toward CO and CH_4, was found by employing SnO_2 as a doping material into sputtered polycrystalline Ga_2O_3 thin films [189]. Doping Sn could enhance the adsorption of NH_3 due to the higher Lewis acidity of Sn^{4+} cations than Ga^{3+} ones [170]. Moreover, the introduction of Sn causes a decrease in the average crystallite size and an increase in the specific surface area [162]. Therefore, Sn-doped Ga_2O_3 NH_3 sensor showed better performance than the pure Ga_2O_3 one regardless of the operating temperature. Manandhar et al. [101] demonstrated that the Ti-doped nanocrystalline β-Ga_2O_3 films significantly accelerated the oxygen response about 20 times while retaining the stability and repeatability.

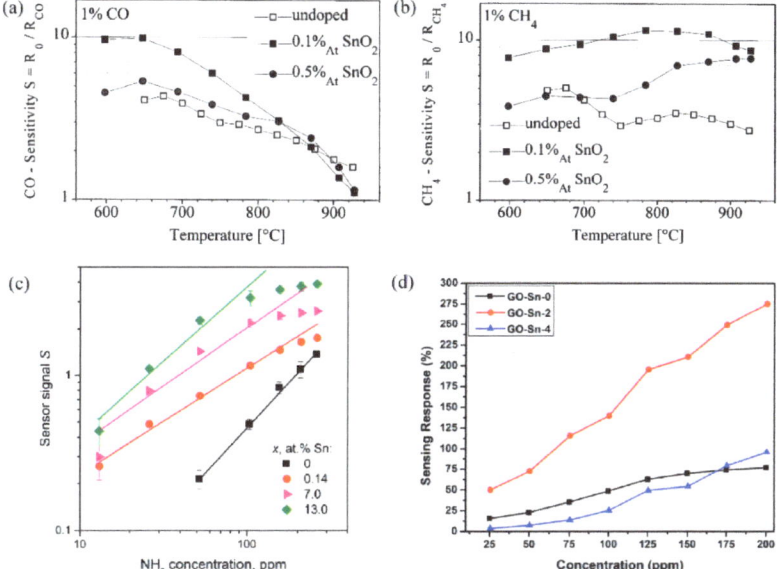

Figure 13. Enhancement of the sensitivity to (**a**) 1% CO and (**b**) 1% CH_4 of Sn-doped Ga_2O_3 thin films compared to undoped films [189]. Copyright 1998 Elsevier. (**c**) Sensor signal as a function of NH_3 concentration for Sn-doped Ga_2O_3 samples measured at 500 °C [170]. Copyright 2021 MDPI. (**d**) Room temperature-sensing response as a function of NH_3 concentration [162]. Copyright 2021 Elsevier.

It is noted that doping can not only enhance the sensing properties of resistor-type Ga_2O_3 sensors but also have a great impact on design of novel sensors. Reiprich et al. [122] used Sn-doped Ga_2O_3 nanostructure for optical gas-sensing at room temperature. The change in the photoluminescence spectra indicated that Sn-doped Ga_2O_3 was capable of detecting trace amounts of VOCs such as C_2H_6O, C_3H_6O, and C_3H_8O. Saidi et al. [176] developed a novel liquid metal-based ultrasonication process within which additional metallic elements (In, Sn, and Zn) were incorporated into liquid Ga and then sonicated in dimethyl sulfoxide (DMSO) and water. These new types of doped Ga_2O_3 sensors showed very tuning sensitivities to NO_2 and H_2. Impedance measurements were performed on the

Pt/YSZ/Au-doped Ga_2O_3 CO electrochemical sensor [216] and on the humidity sensor-based on Ga_2O_3 nanorods doped with Na^+ and K^+ [220]. Both types of Ga_2O_3 gas sensors showed fast response of target gases. Due to the synergistic effect of Ga_2O_3 nanorods and the doping of alkali metal ions, Ga_2O_3 nanorod gas sensor exhibited good linearity response and high stability over 25 days. So far, the dopants added in Ga_2O_3 were Zr, Ti, Mg, Sn, Ni, N, Na, K, Cr, In, Zn, Pt, and Au for the purpose of enhancing the sensing properties.

5.4.4. Constructing Ga_2O_3 Heterostructure

Semiconducting heterostructures show great potential in gas sensors because of a high surface-to-volume and synergistic effect [221]. Once heterointerface forms, Fermi level-mediated charge transfer and band bending occur, usually resulting in a higher sensitivity. According to the types of participating semiconductors, Ga_2O_3 heterostructures can be divided into the n–n heterojunction and p–n heterojunction. Figure 14 illustrates the schematic diagram of the energy band structures for such kinds of heterojunctions. Many heterostructures, including heterojunctions and hierarchical heterostructures, have been developed by many researchers to improve the performance of Ga_2O_3 sensors. In the early years, Fleischer et al. [79,81,96,190] discovered a dramatic influence on the sensitivity and selectivity of Ga_2O_3 thin film sensors by sputtering different metal oxide, i.e., WO_3, Ta_2O_5, NiO, $AlVO_4$, CeO_2, Sm_2O_3, RhO, RuO, Ir_2O_3, and In_2O_3, as the surface modification layer. These types of thin film sensors could be employed as selective oxygen sensors for gas atmospheres with an overall excess of oxygen.

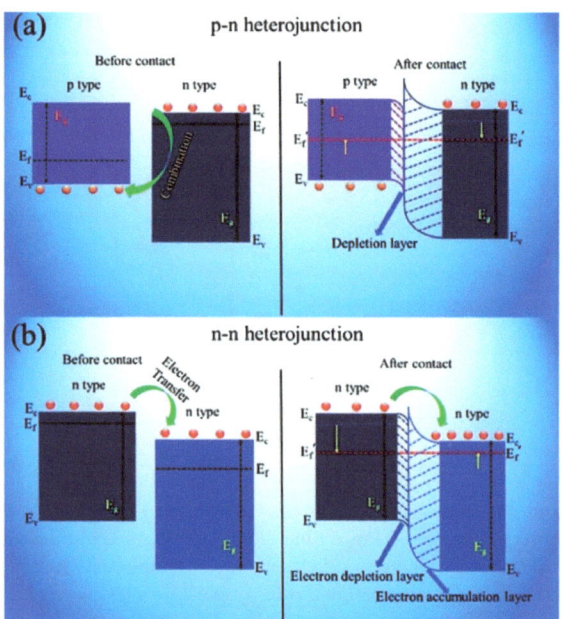

Figure 14. Schematic illustrations of the energy band structures at heterojunction interfaces of (**a**) p–n and (**b**) n–n of heterojunctions [178]. 2019 The Royal Society of Chemistry.

As for resistor-type of gas sensors, TiO_2, SnO_2, ZnO, WO_3, GaN, and GaS were used for constructing n–n heterostructures with Ga_2O_3 while NiO, CuO, LSFO, and LSCO of perovskite crystal structure for p–n heterostructures. Mohammadi and Fray [143] reported that mesoporous Ga_2O_3/TiO_2 thin film gas sensors had response values of 13.7 and 4.3 to 400 ppm CO and 10 ppm NO_2 gas at 200 °C, approximately 55% larger than pure TiO_2 sensors. Jang et al. [117], Liu et al. [139], and Abdullah et al. [132] studied the sensing properties of Ga_2O_3/SnO_2 core–shell nanowires and nanobelts. Compared with pure Ga_2O_3 nanowire sensor, the optimum sensing temperature was reduced by 200 °C for Ga_2O_3/SnO_2 core–shell nanowire sensors that showed the highest ethanol response at 400 °C. Due to the fast physisorption of water and the fast formation of depletion layers caused by the large surface area of the amorphous SnO_2 shell and Ga_2O_3/SnO_2 heterojunction, Ga_2O_3/SnO_2 core–shell nanoribbons had a very high sensitivity to humidity with quick response and recovery near room temperature. At 25 °C, the conductivity of this nanoribbon humidity sensor at 75% relative humidity was three orders of magnitude larger than that at 5% relative humidity. The response time and recovery time were approximately 28 s and 7 s, respectively, when the relative humidity was switched between 5 and 75%. The H_2 gas sensor based on Ga_2O_3/SnO_2 core–shell nanobelts exhibited significant enhanced performance at room temperature in terms of response, response/recovery time and repeatability. During the exposure of 100 ppm NO_2, multiple networked Ga_2O_3/ZnO core–shell nanorod sensors showed a super response of 32.778% at 300 °C, which was 692 and 1791 times larger than that of bare Ga_2O_3 and bare ZnO nanorod sensors [126]. The sensor made by a wafer-scale ultra-thin Ga_2O_3/WO_3 heterostructure [168] exhibited about 4- and 10-fold improvement in the response to C_2H_6O compared to that of pure WO_3 and Ga_2O_3 nanofilm sensors at 275 °C. Furthermore, the Ga_2O_3/WO_3 heterostructural sensor possessed a shorter response/recover time and excellent selectivity. Park et al. [129] developed a Ga_2O_3/GaN core–shell nanowire sensor by surface-nitridated Ga_2O_3 nanowire. It showed responses of 160–363% to CO concentrations of 10–200 ppm at 150 °C, which were 1.6–3.1-fold greater than those of pristine Ga_2O_3 nanowire sensors.

For n–n junction, an accumulation layer is usually created, whereas a depletion layer forms in p–n junction. Lin et al. [154] and Zhang et al. [161] fabricated p–n heterostructural gas sensors using perovskite-sensitized Ga_2O_3 nanorod arrays for CO and NO_2 detection at high temperature. Figure 15 gives the TEM images and measured sensing performances of gas sensors made by $Ga_2O_3/La_{0.8}Sr_{0.2}FeO_3$ (LSFO) and $Ga_2O_3/La_{0.8}Sr_{0.2}CoO_3$ (LSCO) nanorods. Compared with the pristine Ga_2O_3 nanorod array sensor, close to 10 times enhanced sensitivity to 100 ppm CO was discovered for $Ga_2O_3/LSFO$ sensors at 500 °C. In addition to the excellent CO sensitivity, the sensor based on $Ga_2O_3/LSFO$ p–n heterostructure has a faster response time than that of the pristine Ga_2O_3 sensor. $Ga_2O_3/LSCO$ sensors also showed nearly an order of magnitude enhanced sensitivity to 200 ppm NO_2 at 800 °C, along with much shorter response time. Ga_2O_3/CuO thin films were magnetron sputtered by Dyndal et al. [204] as a gas-sensitive material for C_3H_6O detection measured at 300 °C. Benefited from p–n heterostructure, Ga_2O_3/CuO thin film sensor exhibited 40% faster response time in comparison with pure CuO one. Sprincean et al. [134] made a $Ga_2O_3/GaS:Zn$ nanostructured room temperature humidity sensor, which demonstrated acceptable sensitivity on the air relative humidity in the range from 42 to 92% and stable static characteristics over 6 months.

Additionally, Wang et al. [160] investigated the gas-sensing behavior of Ga_2O_3/Al_2O_3 nanocomposite and found that the composite-based sensor had a 6.5 times higher response to 100 ppm NO_x than that of the pure Ga_2O_3 sensor at room temperature. Notably, Sivasankaran and Balaji [177] synthesized mesoporous Ga_2O_3/reduced graphene oxide (rGO) nanocomposites by hydrothermal method. The results indicated that the sensing response to 200 ppm NH_3 of Ga_2O_3/rGO nanocomposite was 3.7-fold larger than that of pure Ga_2O_3 at room temperature.

As known, Ga_2O_3-based Shottky diode H_2 sensors generally has the MIS-type heterostructure in which β-Ga_2O_3 is served as reactive insulator and SiC [146], GaN [172,174], and AlGaN [173] are chosen as semiconductors. Lee et al. [174] concluded that the MIS-type sensor diodes exhibited better forward response than the MS-type Schottky sensor diode, because the β-Ga_2O_3 insulator surface provided more trap sites for hydrogen atoms at the metal–insulator interface.

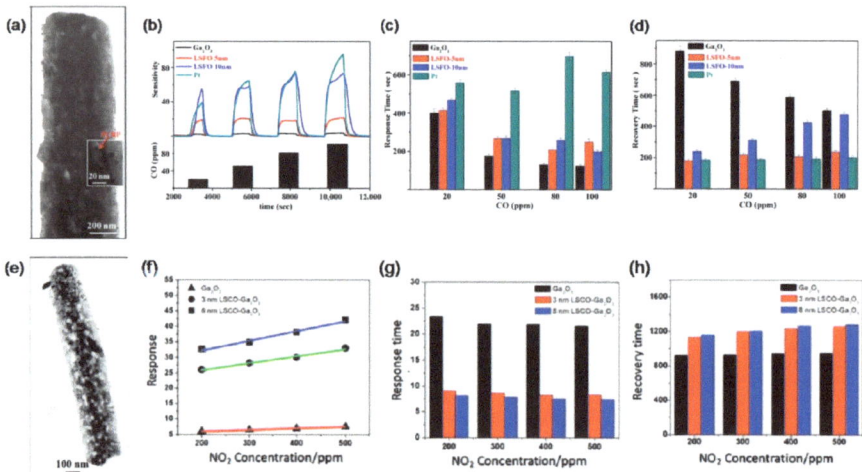

Figure 15. (a) TEM image of a β-Ga$_2$O$_3$ nanorod coated with 5 nm LSFO. (b) Sensitivity, (c) response time, and (d) recover time of gas sensors made by β-Ga$_2$O$_3$/LSFO nanorods upon exposure to CO at 500 °C [154]. Copyright 2016 American Chemical Society. (e) TEM image of a β-Ga$_2$O$_3$ nanorod coated with 8 nm LSCO. (f) Sensitivity, (g) response time, and (h) recover time of gas sensors made by β-Ga$_2$O$_3$/LSCO nanorods upon exposure to NO$_2$ at 800 °C [161]. Copyright 2020 The Royal Society of Chemistry.

5.4.5. Ga-Contained Metal Oxide

Aside from doping in Ga$_2$O$_3$, the gas-sensing properties can be optimized by intentional doping Ga^{3+} or solubility of Ga$_2$O$_3$ into other metal oxides. In general, host metal oxides are In$_2$O$_3$ [141,156,163–167,222], SnO$_2$ [223–225], and ZnO [145,226–229] for resistor gas sensors. Gas sensitive properties of Ga-doped In$_2$O$_3$ thin films and nanostructures were studied by Ratko et al. [222], Chen et al. [156], and Demin et al. [163–167]. Ga$_2$O$_3$ caused the formation of a porous or nanostructure in the In$_2$O$_3$-based ceramics, providing an active surface for reducing gases such as CH$_4$ [222] and CH$_2$O [156]. The response toward 100 ppm CH$_2$O of Ga$_x$In$_{2-x}$O$_3$ nanofiber was about four times higher than that of pure In$_2$O$_3$. Meanwhile, it has superior ability to selectively detect CH$_2$O against other interfering VOCs. Demin et al. [163,164] obtained a high selectivity to NH$_3$ against C$_2$H$_6$O, C$_3$H$_6$O and liquefied petroleum gas based on 50% In$_2$O$_3$-50% Ga$_2$O$_3$ thin film-sensitive layer. The sensitivities towards O$_2$, CO, C$_2$H$_6$O, and CH$_2$O of Ga-doped SnO$_2$ materials were investigated by Silver et al. [223], Bagheri et al. [224], and Du et al. [225]. The Ga-doped SnO$_2$ thin film O$_2$ sensor showed sensitivity up to 2.1 for a partial pressure of oxygen as low as 1 Torr [223]. Bagheri et al. [224] observed the highest responses of 315 and 119 for 300 ppm CO and C$_2$H$_6$O by the SnO$_2$ sensors containing 5 and 1 wt% Ga$_2$O$_3$, respectively. With adding more than 25 wt% Ga$_2$O$_3$, the sensors became selective to CO and showed negligible responses to C$_2$H$_6$O and CH$_4$. The sensitivity to 50 ppm CH$_2$O of Ga-doped SnO$_2$ sensor was 4.5 times greater than that of the pure SnO$_2$ sensor. Moreover, it exhibited a lower detection limit of 0.1 ppm CH$_2$O with sensitivity of 3 and a short response-recovery time (3/39 s) and good selectivity. Ga-doped ZnO nanocrystalline film sensors were explored for detecting C$_2$H$_6$O, H$_2$S, NO$_2$, and H$_2$. The results demonstrated that the gas response of Ga-doped ZnO sensors were greatly enhanced by compared to pristine ZnO sensor. For instance, Hou et al. [145] observed a 60% enhancement of sensing response to H$_2$ by optimizing Ga composition to 0.3 at% compared with undoped ZnO sensors when measured at 130 °C. Moreover, the 0.3 at% Ga-doped ZnO sensor had a shorter response time and a better selectivity to H$_2$ in a mixture of H$_2$, CH$_4$, and NH$_3$. Rashid et al. [228] developed a 3%-Ga modified ZnO H$_2$ sensor whose resistive response was improved six-fold compared with the pristine one at room temperature. It had a very low detection limit of 0.2 ppm. The enhanced H_2S-sensing properties of Ga-doped ZnO sensors were measured by Vorobyeva et al. [227] and by Girija et al. [229]. Figure 16 shows the H_2S response of Ga-doped ZnO gas sensors as functions of the gas concentration and temperature. In the presence of H_2S, chemisorbed oxygen interacts with the target gas as governed by the following reaction: $H_2S + 3O^- \rightarrow SO_2 (g) + H_2O (g) + 3e$. The enhanced sensitivity could be attributed to both excess of oxygen vacancies due to Ga^{3+} substitution and large adsorption energy of Ga for H_2S. Based on the liquid metal-based probe sonication route, Shafiei et al. [175] developed a quaternary $GaInSnO_x$ sensor and detection limits as low as 1 ppm and 20 ppm for NO_2 and NH_3 were obtained when operated at 100 °C. Table 4 summarizes the enhanced sensing performance of gas sensors using Ga-contained metal oxides toward different target gases for comparison.

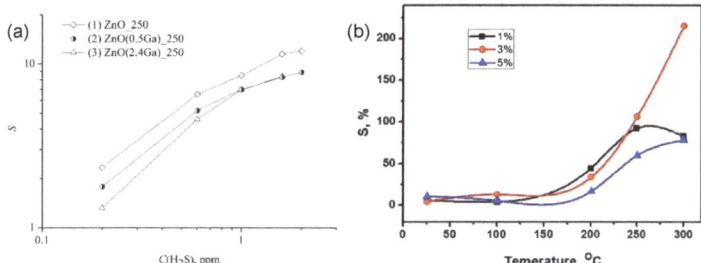

Figure 16. (a) H_2S Response of Ga-doped ZnO gas sensors as a function of the gas concentration [227]. Copyright 2013 Elsevier. (b) H_2S Response of Ga-doped ZnO gas sensors as a function of the temperature [229]. Copyright 2018 Elsevier.

5.4.6. Coating with Gas Filter

If working at high temperature, Ga_2O_3-based gas sensors (particularly resistor-type ones), similar to other metal oxide semiconductor sensors, suffer from an issue of poor selective gas detection, since they almost react to all reducing or oxidizing gases. One of the most efficient ways to improve the selectivity of gas sensors is the use of filters [182,230]. These filters, including physical and chemical filters, are highly permeable to the target gases and can hinder interfering gases from reaching the sensors surface. Figure 17 depicts the conceptual diagrams for physical and chemical filters of Ga_2O_3-based gas sensors. Fleischer et al. [95] designed a selective H_2 sensor by covering a SiO_2 physical gas-filtering layer on the Ga_2O_3 sputtered film for the first time. It was found that sensors with this surface layer structure had an extremely high specificity for H_2 upon exposure to a variety of interfering gases such as CO, CO_2, CH_4, C_4H_8, C_2H_6O, C_3H_6O, NO, and NH_3 at 700 °C. In another work of selective H_2, CH_4 and NO_x sensors operating at high temperature, Fleischer et al. [78] summarized that the SiO_2 physical filter can only permeate hydrogen and the porous Ga_2O_3 catalytic filter removes disturbing solvent vapors by oxidation, and the gas conversion filter composed of Pt supported on Al_2O_3 ensures a defined NO/NO_2 equilibrium. Flingelli et al. [77] fabricated a thin-film Ga_2O_3–gas sensor equipped with a screen printed porous Ga_2O_3 layer as catalytic filter for the selective detection of CH_4 even in the presence of C_2H_6O. It was observed that the cross-sensitivities eliminated for the organic solvents were oxidized while passing the filter, and only the quite stable methane was allowed to reach the sensor surface. Weh et al. [193,194] studied the optimization of physical filters for selective high-temperature H_2 sensors. A single SiO_2 filter, Ga_2O_3/SiO_2 and Al_2O_3/SiO_2 dual filter systems, and buried filter systems in which Ga_2O_3 or Cr-doped $SrTiO_3$ was buried between two SiO_2 layers were applied to improve the selectivity. The results indicated that optimizing filter systems could not only increase the selectivity but also the sensitivity of a given sensor. Furthermore, these filters can be additionally used to ensure the stability over the needed lifetime of the sensor.

Table 4. Enhanced sensing performances of gas sensors using Ga-contained metal oxides.

Sensing Material	Preparation	Sensitivity @Gas Concentration	Operating Temperature	Response/Recover Time (s)	Other Observations	Reference
Ga-doped In_2O_3 nanowire	CVD	△1.05 @ 80 ppm C_2H_6O	200	40/800		[141]
		△2.2 @ 4 ppm NO_2	200	1980/2780		
Ga-doped In_2O_3 nanofiber	Hydrothermal synthesis	◊52.5 @ 100 ppm CH_2O	150	1/70	The low limit of detection is 0.2 ppm	[156]
Ga-doped In_2O_3 film	PLD	□2.15 @ 25 ppm NH_3	623			[164]
		□20.5 @ 25 ppm C_2H_6O	504			
		□24.4 @ 25 ppm C_3H_6O	504			
		□7.47 @ 25 ppm CH_4	500			[167]
Ga-doped In_2O_3 ceramics	Coprecipitation	△0.85 @ 0.5% CH_4	380			[222]
Ga-doped SnO_2 film	Spray pyrolysis	◊3.1 @ 1 Torr O_2	350			[223]
Ga-doped SnO_2 nanocomposites	Coprecipitation	◊315 @ 300 ppm CO	300	36		[224]
Ga-doped SnO_2 nanocomposites		◊119 @ 300 ppm C_2H_6O	250	93		
Ga-doped SnO_2 microflowers	Hydrothermal synthesis	◊95.8 @ 50 ppm CH_2O	230	3	The low limit of detection is 0.1 ppm	[225]
Ga-doped ZnO film	Sol–gel synthesis	△0.5 @ 500 ppm H_2	130	475		[145]
Ga-doped ZnO Nanorod	Hydrothermal synthesis	△1.01% @ 250 ppm C_2H_6O	RT			[226]
Ga-doped ZnO nanoparticle	Spray pyrolysis	△56 @ 2 ppm NO_2	250			[227]
		△8 @ 1 ppm H_2S	250			
Ga-doped ZnO nanorod	Sol–gel synthesis	△0.91 @ 100 ppm H_2	RT	20		[228]
Ga-doped ZnO film	Magnetron sputtering	△2.4 @ 5 ppm H_2S	300			[229]

△ $S = (Ra - Rg)/Ra$; ◊ $S = Ra/Rg$; □ $S = (Ra - Rg)/Rg$

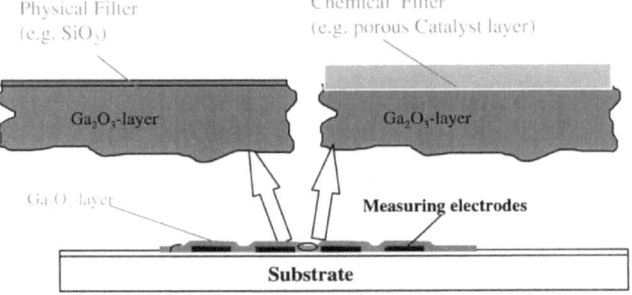

Figure 17. Conceptual diagrams for physical and chemical filters of Ga_2O_3-based gas sensors [230]. Copyright 1998 Elsevier.

5.4.7. Light Illumination

Light illumination rather than thermal activation is a promising strategy to enhance the gas-sensing of Ga_2O_3-based sensors. For clarity, the schematics and energy-level representations of the β-Ga_2O_3 nanowires before and after the 254 nm UV illumination are shown in Figure 18. When the sensors are illuminated under UV light with the photon energy equal or higher than the band gap of Ga_2O_3, electrons from the valence band can be rapidly excited to the conduction band, causing the desorption of oxygen from Ga_2O_3 surface and inducing the photosensitizing effect. Feng et al. [107] first reported a very fast room temperature oxygen response of the individual β-Ga_2O_3 nanowires synthesized by CVD under 254 nm UV illumination. This UV light-activated fast room temperature oxygen-sensing characteristic was demonstrated in the thermal evaporated β-Ga_2O_3 nanobelts by Ma and Fan [112]. Juan et al. [130] studied the effect on the humidity-sensing properties of a β-Ga2O3 nanowire sensor from UV light. However, they found that the humidity sensitivity with UV illumination was lower than that in the dark, since the water molecules captured the electrons and holes generated by UV light in an environment with high relative humidity. Lin et al. [155] compared the effect of UV radiation on the CO sensors made by pure Ga_2O_3 nanorod arrays, Pt-decorated Ga_2O_3 nanorod arrays, and LSFO/Ga_2O_3 nanorod arrays. The measured results are shown in Figure 19a,b. Under 254 nm UV illumination, the sensitivity to 100 ppm CO was enhanced by about 30%, 20%, and 50% for pristine β-Ga_2O_3 nanorod arrays sensors, LSFO/Ga_2O_3 nanorod arrays sensors and Pt decorated Ga_2O_3 nanorod array sensors at 500 °C, respectively. Additionally, the response times were reduced for all cases, and up to 30% reduction of response time was achieved for LSFO/Ga_2O_3 nanorod arrays sensors. An et al. [128] showed a significant enhancement in the response of the Pt-functionalized Ga_2O_3 nanorods to NO_2 gas by UV irradiation at room temperature. It can be clearly seen from Figure 19c that the response to 5 ppm NO_2 increases from 175% to 931% with increasing the UV light illumination intensity from 0.35 to 1.2 mW/cm^2. A combination of the spillover effect and the enhancement of chemisorption and dissociation of gas results in the enhanced electrical response of the Pt-functionalized Ga_2O_3 nanostructured sensors. Sui et al. [64] realized the room temperature ozone sensing capability of InGaZnO (IGZO)-decorated amorphous Ga_2O_3 films under UV illumination.

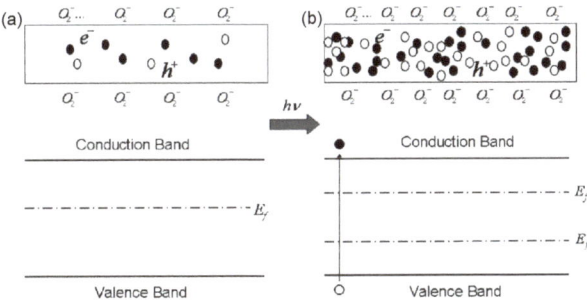

Figure 18. Schematics and energy-level representations of the β-Ga_2O_3 nanowires (**a**) before and (**b**) after the 254 nm UV illumination [107]. Copyright 2006 American Institute of Physics.

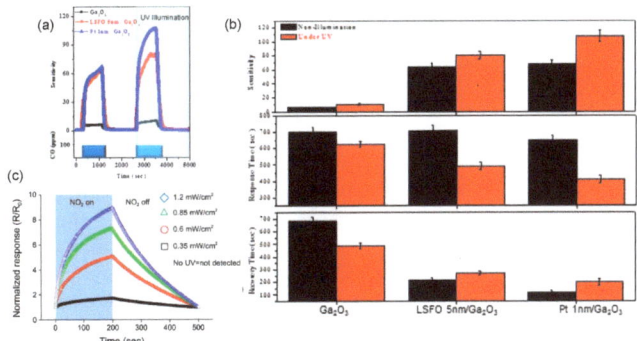

Figure 19. CO gas-sensing performance comparison of pristine, LSFO-decorated, and Pt-decorated β-Ga_2O_3 nanorod arrays: (**a**) normalized sensitivity time characteristics and (**b**) sensitivity, recovery time, and recovery time under dark or UV illumination tested at 500 °C [155]. Copyright 2017 American Institute of Physics. (**c**) Room temperature gas responses of Pt-functionalized Ga_2O_3 nanorod gas sensors to 5 ppm NO_2 under UV light illumination with varied intensities [128]. Copyright 2013 Korean Chemical Society.

6. Conclusions and Outlook

In conclusion, the past thirty years have witnessed very impressive progress in the field of Ga_2O_3-based gas sensors. It has been proven that β-Ga_2O_3 polycrystalline thin films and nanostructures have great potential in detecting oxygen and many reducing gases in a wide temperature range from room temperature to 1000 °C. Some growth techniques with low cost, simple process, and flexible operation are available to prepare Ga_2O_3-sensing material. Compared with the gas sensors made by other metal oxides such as SnO_2, Ga_2O_3-based gas sensors have the advantages of long-term stability, fast response and recovery times, good reproducibility, low cross-sensitivity to water vapor, and short preaging time [181]. Figure 20 displays the radar chart of seven performance enhancement strategies of gas sensors made by Ga_2O_3 bulk crystal, thin films, and nanostructures. Although the performance of Ga_2O_3-based gas sensors has been improved through various enhancement strategies, from a practical point of view, continuous efforts should be made to further increase the gas selectivity, to decrease the operating temperature, and to develop high-speed nonresistive types of sensors. In-depth understanding of the interaction between the target gas and Ga_2O_3 surface, as well as the gas-sensing behavior, are also needed using newly developed computational tools. Some suggestions on future research opportunities of Ga_2O_3-based gas sensors are listed below:

(i) Construction of hybrid structures with two-dimensional (2D) materials and organic polymers. Two-dimensional materials are regarded as having promising potential for gas-sensing devices owing to their large surface-to-volume ratio and high surface sensitivity [231]. The integration of 2D layers with Ga_2O_3, especially a 2D Ga_2O_3 monolayer [232], can form a van der Waals heterojunction without constraints on the chemical bonding and interfacial lattice matching, which will be expected to widen the building blocks for novel applications with unprecedented properties and excellent performance. The energy band alignment of such van der Waals heterojunction could be precisely designed by selecting suitable 2D materials, such as transition metal dichalcogenides with different band gaps and working functions so as to optimize the gas selectivity and performance of the Ga_2O_3 gas sensor. Additionally, organic semiconductors offer a viable alternative to conventional inorganic semiconductors in gas sensor applications because of their unusual electrical properties, diversity, large area, and potentially low cost [233]. They show good sensitivities toward many gases or vapors, ranging from organic solvents to inorganic gases. Moreover, it allows for the inexpensive fabrication of novel Ga_2O_3 gas sensors using organic polymers as a platform, owing to their porosity, mechanical flexibility, environmental stability, and solution processability. Therefore, more research should be conducted to fabricate Ga_2O_3-based gas sensors by constructing heterostructures with 2D materials and organic polymers, which may result in some novel features and potential applications.

(ii) Combinations with DFT calculations and machine learning. Hitherto, quite few theoretical investigations have been devoted to understanding the interaction behaviors between gas molecules and Ga_2O_3 in the field of gas sensors. DFT calculations [232,234,235] have been performed to study the adsorption, dissociation, and diffusion characteristics of gas molecules on the surface of Ga_2O_3. It is suggested that the calculated adsorption energy, electron structures of oxygen vacancy, and charge transfer are highly significant in defining the performance of a sensor. Large adsorption energy and significant charge transfer indicate high sensitivity and selectivity towards a specific gas. Alongside DFT, machine learning is considered as an effective data processing approach for developing smart devices with the ability to deal with selectivity and drift problems [236]. The machine learning technique involves data processing of sensor output, dimensionality reduction, and then training a system/network for the predictions. Therefore, DFT and machine learning can be implemented as a powerful tool for studying gas-sensing behavior and provide valuable suggestions and predictions of gas-sensing materials and target gas species.

(iii) Development of optical sensors. Compared to electrical characteristic gas sensors, more attention should be paid on the less-explored sensors using the promising optical properties of Ga_2O_3. In a pioneer study on Ga_2O_3-based optical gas sensors, Reiprich et al. [122] pointed out that the spectral composition of photoluminescence showed a strong intensity difference upon exposure to various gases at room temperature indirectly caused by the negatively charged oxygen ions in Ga_2O_3. Optical gas sensors offer a number of advantages, such as fast responses, minimal drift and high gas specificity [187]. Gas detection can be made in real time and in situ by adopting optical sensors. With the help of the rapid developed technology of Ga_2O_3 optical devices, it is possible to design Ga_2O_3-based optical gas sensors that can measure gas concentrations in the ppm or ppb range with zero cross-response to other gases and high temporal resolution. In this way, optical gas-sensing fills a gap between low-cost electrical/electrochemical sensors with inferior performance and high-end analytic equipment.

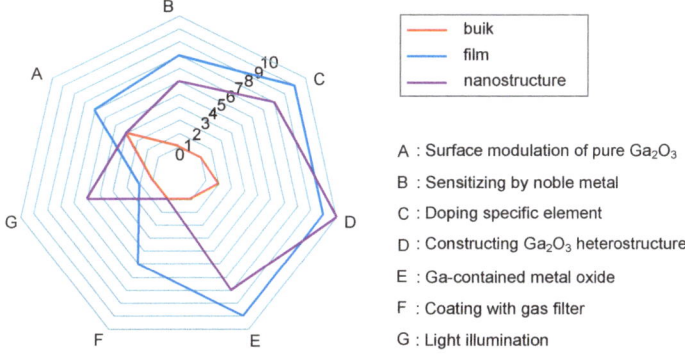

Figure 20. Radar chart of seven performance enhancement strategies of gas sensors made by Ga_2O_3 bulk crystal, thin films, and nanostructures. Original drawing was done by us.

Moreover, the researchers should explore the options for the use of Ga_2O_3 in many other sensor applications such as chemical sensors [237,238] and biosensors [239,240] to detect metal ions, e.g., Ni^{2+} and Fe^{2+}, and DNA sequences in biological fluids, tap water, and so forth. It is believed that Ga_2O_3 will play a significant role in the field of safety, environmental, and medical monitoring systems with the development of device designs and fabrication technologies.

Funding: This work was supported by the National Natural Science Foundation of China (Grant Nos. 12164031, 11864029, 62004078, 12105033, 11774235, and 11933005); National Key R&D Program of China (Grant No. 2022YFE0107400); Natural Science Foundation of Inner Mongolia Autonomous Region of China (Grant Nos. 2020MS01014 and 2020MS06007); the Fundamental Research Funds for the Central Universities (Grant No. DUT20RC(3)042); and the Program for Professor of Special Appointment (Eastern Scholar) at Shanghai Institutions of Higher Learning (GZ2020015).

Acknowledgments: Jun Zhu gratefully acknowledges Xue-Lun Wang for meaning discussions when visiting National Institute of Advanced Industrial Science and Technology (AIST) in Japan.

Conflicts of Interest: The authors declare no conflict of interest.

References

1. Tsao, J.Y.; Chowdhury, S.; Hollis, M.A.; Jena, D.; Johnson, N.M.; Jones, K.A.; Kaplar, R.J.; Rajan, S.; Van de Walle, C.G.; Bellotti, E.; et al. Ultrawide-bandgap semiconductors: Research opportunities and challenges. *Adv. Electron. Mater.* **2018**, *4*, 1600501. [CrossRef]
2. Higashiwaki, M.; Sasaki, K.; Kuramata, A.; Masui, T.; Yamakoshi, S. Gallium oxide (Ga_2O_3) metal-semiconductor field-effect transistors on single-crystal β-Ga_2O_3 (010) substrates. *Appl. Phys. Lett.* **2012**, *100*, 013504. [CrossRef]

3. Pearton, S.; Ren, F.; Mastro, M. *Gallium Oxide: Technology, Devices and Applications*; Elsevier: Amsterdam, The Netherlands, 2019; pp. 439–464. [CrossRef]
4. Higashiwaki, M.; Fujita, S. *Gallium Oxide: Materials Properties, Crystal Growth, and Devices*; Springer Nature: Cham, Switzerland, 2020; pp. 1–11. [CrossRef]
5. Xue, H.; He, Q.; Jian, G.; Long, S.; Pang, T.; Liu, M. An overview of the ultrawide bandgap Ga_2O_3 semiconductor-based Schottky barrier diode for power electronics application. *Nanoscale Res. Lett.* **2018**, *13*, 290. [CrossRef]
6. Liu, Z.; Li, P.; Zhi, Y.; Wang, X.; Chu, X.; Tang, W. Review of gallium oxide based field-effect transistors and Schottky barrier diodes. *Chin. Phys. B* **2019**, *28*, 017105. [CrossRef]
7. Dong, H.; Xue, H.; He, Q.; Qin, Y.; Jian, G.; Long, S.; Liu, M. Progress of power field effect transistor based on ultra-wide bandgap Ga_2O_3 semiconductor material. *J. Semicond.* **2019**, *40*, 011802. [CrossRef]
8. Zhang, H.; Yuan, L.; Tang, X.; Hu, J.; Sun, J.; Zhang, Y.; Zhang, Y. Progress of ultra-wide bandgap Ga_2O_3 semiconductor materials in power MOSFETs. *IEEE Trans. Power Electron.* **2020**, *35*, 5157–5179. [CrossRef]
9. Singh, R.; Lenka, T.R.; Panda, D.K.; Velpula, R.T.; Jain, B.; Bui, H.Q.T.; Nguyen, H.P.T. The dawn of Ga_2O_3 HEMTs for high power electronics—A review. *Mater. Sci. Semicond. Process.* **2020**, *119*, 105216. [CrossRef]
10. Wong, M.H.; Higashiwaki, M. Vertical β-Ga_2O_3 power transistors: A review. *IEEE Trans. Electron Dev.* **2020**, *67*, 3925–3937. [CrossRef]
11. Kumar, S.; Singh, R. Nanofunctional gallium oxide (Ga_2O_3) nanowires/nanostructures and their applications in nanodevices. *Phys. Status Solidi RRL* **2013**, *7*, 781–792. [CrossRef]
12. Higashiwaki, M.; Sasaki, K.; Murakami, H.; Kumagai, Y.; Koukitu, A.; Kuramata, A.; Masui, T.; Yamakoshi, S. Recent progress in Ga_2O_3 power devices. *Semicond. Sci. Technol.* **2016**, *31*, 034001. [CrossRef]
13. Mastro, M.A.; Kuramata, A.; Calkins, J.; Kim, J.; Ren, F.; Pearton, S.J. Opportunities and future directions for Ga_2O_3. *ECS J. Solid State Sci. Technol.* **2017**, *6*, P356–P359. [CrossRef]
14. Pearton, S.J.; Yang, J.; Cary IV, P.H.; Ren, F.; Kim, J.; Tadjer, M.J.; Mastro, M.A. A review of Ga_2O_3 materials, processing, and devices. *Appl. Phys. Rev.* **2018**, *5*, 011301. [CrossRef]
15. Galazka, Z. β-Ga_2O_3 for wide-bandgap electronics and optoelectronics. *Semicond. Sci. Technol.* **2018**, *33*, 113001. [CrossRef]
16. Baldini, M.; Galazka, Z.; Wagner, G. Recent progress in the growth of β-Ga_2O_3 for power electronics applications. *Mater. Sci. Semicond. Process.* **2018**, *78*, 132–146. [CrossRef]
17. Higashiwaki, M.; Jessen, G.H. The dawn of gallium oxide microelectronics. *Appl. Phys. Lett.* **2018**, *112*, 060401. [CrossRef]
18. Zhou, H.; Zhang, J.; Zhang, C.; Feng, Q.; Zhao, S.; Ma, P.; Hao, Y. A review of the most recent progresses of state-of-art gallium oxide power devices. *J. Semicond.* **2019**, *40*, 011803. [CrossRef]
19. Xu, J.; Zheng, W.; Huang, F. Gallium oxide solar-blind ultraviolet photodetectors: A review. *J. Mater. Chem. C* **2019**, *7*, 8753–8770. [CrossRef]
20. Kim, J.; Pearton, S.J.; Fares, C.; Yang, J.; Ren, F.; Kim, S.; Polyakov, A.Y. Radiation damage effects in Ga_2O_3 materials and devices. *J. Mater. Chem. C* **2019**, *7*, 10–24. [CrossRef]
21. Qin, Y.; Long, S.; Dong, H.; He, Q.; Jian, G.; Zhang, Y.; Hou, X.; Tan, P.; Zhang, Z.; Lv, H.; et al. Review of deep ultraviolet photodetector based on gallium oxide. *Chin. Phys. B* **2019**, *28*, 018501. [CrossRef]
22. Guo, D.; Guo, Q.; Chen, Z.; Wu, Z.; Li, P.; Tang, W. Review of Ga_2O_3-based optoelectronic devices. *Mater. Today Phys.* **2019**, *11*, 100157. [CrossRef]
23. Chen, X.; Ren, F.; Gu, S.; Ye, J. Review of gallium-oxide-based solar-blind ultraviolet photodetectors. *Photonics Res.* **2019**, *7*, 381. [CrossRef]
24. Saikumar, A.K.; Nehate, S.D.; Sundaram, K.B. Review-RF sputtered films of Ga_2O_3. *ECS J. Solid State Sci. Technol.* **2019**, *8*, Q3064–Q3078. [CrossRef]
25. Tadjer, M.J.; Lyons, J.L.; Nepal, N.; Freitas, J.A., Jr.; Koehler, A.D.; Foster, G.M. Review-theory and characterization of doping and defects in β-Ga_2O_3. *ECS J. Solid State Sci. Technol.* **2019**, *8*, Q3187–Q3194. [CrossRef]
26. Afzal, A. β-Ga_2O_3 nanowires and thin films for metal oxide semiconductor gas sensors: Sensing mechanisms and performance enhancement strategies. *J. Mater.* **2019**, *5*, 542–557. [CrossRef]
27. Zhai, H.; Wu, Z.; Fang, Z. Recent progress of Ga_2O_3-based gas sensors. *Ceram. Int.* **2022**, *48*, 24213–24233. [CrossRef]
28. McCluskey, M.D. Point defects in Ga_2O_3. *J. Appl. Phys.* **2020**, *127*, 101101. [CrossRef]
29. Zhang, J.; Shi, J.; Qi, D.; Chen, L.; Zhang, K.H.L. Recent progress on the electronic structure, defect, and doping properties of Ga_2O_3. *APL Mater.* **2020**, *8*, 020906. [CrossRef]
30. Bosi, M.; Mazzolini, P.; Seravalli, L.; Fornari, R. Ga_2O_3 polymorphs: Tailoring the epitaxial growth conditions. *J. Mater. Chem. C* **2020**, *8*, 10975–10992. [CrossRef]
31. Hou, X.; Zou, Y.; Ding, M.; Qin, Y.; Zhang, Z.; Ma, X.; Tan, P.; Yu, S.; Zhou, X.; Zhao, X.; et al. Review of polymorphous Ga_2O_3 materials and their solar-blind photodetector applications. *J. Phys. D Appl. Phys.* **2021**, *54*, 043001. [CrossRef]
32. Tak, B.R.; Kumar, S.; Kapoor, A.K.; Wang, D.; Li, X.; Sun, H.; Singh, R. Recent advances in the growth of gallium oxide thin films employing various growth techniques—A review. *J. Phys. D Appl. Phys.* **2021**, *54*, 453002. [CrossRef]
33. Blevins, J.; Yang, G. On optical properties and scintillation performance of emerging Ga_2O_3: Crystal growth, emission mechanisms and doping strategies. *Mater. Res. Bull.* **2021**, *144*, 111494. [CrossRef]

34. Yuan, Y.; Hao, W.; Mu, W.; Wang, Z.; Chen, X.; Liu, Q.; Xu, G.; Wang, C.; Zhou, H.; Zou, Y.; et al. Toward emerging gallium oxide semiconductors: A roadmap. *Fundam. Res.* **2021**, *1*, 697–716. [CrossRef]
35. Huang, H.-C.; Ren, Z.; Chan, C.; Li, X. Wet etch, dry etch, and MacEtch of β-Ga$_2$O$_3$: A review of characteristics and mechanism. *J. Mater. Res.* **2021**, *36*, 4756–4770. [CrossRef]
36. Galazka, Z.; Ganschow, S.; Irmscher, K.; Klimm, D.; Albrecht, M.; Schewski, R.; Pietsch, M.; Schulz, T.; Dittmar, A.; Kwasniewski, A.; et al. Bulk single crystals of β-Ga$_2$O$_3$ and Ga-based spinels as ultra-wide bandgap transparent semiconducting oxides. *Prog. Cryst. Growth Charact. Mater.* **2021**, *67*, 100511. [CrossRef]
37. Sharma, R.; Law, M.E.; Ren, F.; Polyakov, A.Y.; Pearton, S.J. Diffusion of dopants and impurities in β-Ga$_2$O$_3$. *J. Vac. Sci. Technol. A* **2021**, *39*, 060801. [CrossRef]
38. Cooke, J.; Sensale-Rodriguez, B.; Ghadbeigi, L. Methods for synthesizing β-Ga$_2$O$_3$ thin films beyond epitaxy. *J. Phys. Photonics* **2021**, *3*, 032005. [CrossRef]
39. Wang, C.; Zhang, J.; Xu, S.; Zhang, C.; Feng, Q.; Zhang, Y.; Ning, J.; Zhao, S.; Zhou, H.; Hao, Y. Progress in state-of-the-art technologies of Ga$_2$O$_3$ devices. *J. Phys. D Appl. Phys.* **2021**, *54*, 243001. [CrossRef]
40. Kim, H. Control and understanding of metal contacts to β-Ga$_2$O$_3$ single crystals: A review. *SN Appl. Sci.* **2022**, *4*, 27. [CrossRef]
41. Fleischer, M.; Meixner, H. Gallium oxide thin films: A new material for high-temperature oxygen sensors. *Sens. Actuators B Chem.* **1991**, *4*, 437–441. [CrossRef]
42. Fleischer, M.; Meixner, H. Oxygen sensing with long-term stable Ga$_2$O$_3$ thin films. *Sens. Actuators B Chem.* **1991**, *5*, 115–119. [CrossRef]
43. Malik, R.; Tomer, V.K.; Mishra, Y.K.; Lin, L. Functional gas sensing nanomaterials: A panoramic view. *Appl. Phys. Rev.* **2020**, *7*, 021301. [CrossRef]
44. Herbstein, F.H. Diversity amidst similarity: A multidisciplinary approach to phase relationships, solvates, and polymorphs. *Cryst. Growth Des.* **2004**, *4*, 1419–1429. [CrossRef]
45. Roy, R.; Hill, V.; Osborn, E. Polymorphism of Ga$_2$O$_3$ and the system Ga$_2$O$_3$-H$_2$O. *J. Am. Chem. Soc.* **1952**, *74*, 719–722. [CrossRef]
46. Yoshioka, S.; Hayashi, H.; Kuwabara, A.; Oba, F.; Matsunaga, K.; Tanaka, I. Structures and energetics of Ga$_2$O$_3$ polymorphs. *J. Phys. Condens. Matter* **2007**, *19*, 346211. [CrossRef]
47. Zinkevich, M.; Aldinger, F. Thermodynamic assessment of the gallium-oxygen system. *J. Am. Ceram. Soc.* **2004**, *87*, 683–691. [CrossRef]
48. Playford, H.Y.; Hannon, A.C.; Barney, E.R.; Walton, R.I. Structures of uncharacterised polymorphs of gallium oxide from total neutron diffraction. *Chem. Eur. J.* **2013**, *19*, 2803–2813. [CrossRef]
49. Cora, I.; Mezzadri, F.; Boschi, F.; Bosi, M.; Caplovicova, M.; Calestani, G.; Dodony, I.; Pecza, B.; Fornari, R. The real structure of ε-Ga$_2$O$_3$ and its relation to κ-phase. *CrystEngComm* **2017**, *19*, 1509–1516. [CrossRef]
50. Kim, J.; Tahara, D.; Miura, Y.; Kim, B.G. First-principle calculations of electronic structures and polar properties of (κ, ε)-Ga$_2$O$_3$. *Appl. Phys. Express* **2018**, *11*, 061101. [CrossRef]
51. Nishinaka, H.; Ueda, O.; Tahara, D.; Ito, Y.; Ikenaga, N.; Hasuike, N.; Yoshimoto, M. Single-domain and atomically flat surface of κ-Ga$_2$O$_3$ thin films on FZ-grown ε-GaFeO$_3$ substrates via step-flow growth mode. *ACS Omega* **2020**, *5*, 29585–29592. [CrossRef]
52. Jang, S.; Jung, S.; Kim, J.; Ren, F.; Pearton, S.J.; Baik, K.H. Hydrogen sensing characteristics of Pt Schottky diodes on ($\overline{2}$01) and (010) Ga$_2$O$_3$ single crystals. *ECS J. Solid State Sci. Technol.* **2018**, *7*, Q3180–Q3182. [CrossRef]
53. Fleischer, M.; Handrider, W.; Meixner, H. Stability of semiconducting gallium oxide thin films. *Thin Solid Films* **1990**, *190*, 93–102. [CrossRef]
54. Ueda, N.; Hosono, H.; Waseda, R.; Kawazoe, H. Synthesis and control of conductivity of ultraviolet transmitting β-Ga$_2$O$_3$ single crystals. *Appl. Phys. Lett.* **1997**, *70*, 3561–3563. [CrossRef]
55. Varley, J.B.; Weber, J.R.; Janotti, A.; Van de Walle, C.G. Oxygen vacancies and donor impurities in β-Ga$_2$O$_3$. *Appl. Phys. Lett.* **2010**, *97*, 142106. [CrossRef]
56. Feng, Z.; Bhuiyan, A.F.M.A.U.; Karim, M.R.; Zhao, H. MOCVD homoepitaxy of Si-doped (010) β-Ga$_2$O$_3$ thin films with superior transport properties. *Appl. Phys. Lett.* **2019**, *114*, 250601. [CrossRef]
57. Fleischer, M.; Meixner, H. Electron mobility in single and polycrystalline Ga$_2$O$_3$. *J. Appl. Phys.* **1993**, *74*, 300–305. [CrossRef]
58. Irmscher, K.; Galazka, Z.; Pietsch, M.; Uecker, R.; Fornari, R. Electrical properties of β-Ga$_2$O$_3$ single crystals grown by the Czochralski method. *J. Appl. Phys.* **2011**, *110*, 063720. [CrossRef]
59. Ma, N.; Tanen, N.; Verma, A.; Guo, Z.; Luo, T.; Xing, H.; Jena, D. Intrinsic electron mobility limits in β-Ga$_2$O$_3$. *Appl. Phys. Lett.* **2016**, *109*, 212101. [CrossRef]
60. Goto, K.; Konishi, K.; Murakami, H.; Kumagai, Y.; Monemar, B.; Higashiwaki, M.; Kuramata, A.; Yamakoshi, S. Halide vapor phase epitaxy of Si doped β-Ga$_2$O$_3$ and its electrical properties. *Thin Solid Films* **2018**, *666*, 182–184. [CrossRef]
61. Spencer, J.A.; Mock, A.L.; Jacobs, A.G.; Schubert, M.; Zhang, Y.; Tadjer, M.J. A review of band structure and material properties of transparent conducting and semiconducting oxides: Ga$_2$O$_3$, Al$_2$O$_3$, In$_2$O$_3$, ZnO, SnO$_2$, CdO, NiO, CuO, and Sc$_2$O$_3$. *Appl. Phys. Rev.* **2022**, *9*, 011315. [CrossRef]
62. Ogita, M.; Higo, K.; Nakanishi, Y.; Hatanaka, Y. Ga$_2$O$_3$ thin film for oxygen sensor at high temperature. *Appl. Surf. Sci.* **2001**, *175–176*, 721–725. [CrossRef]
63. Ogita, M.; Yuasa, S.; Kobayashi, K.; Yamada, Y.; Nakanishi, Y.; Hatanaka, Y. Presumption and improvement for gallium oxide thin film of high temperature oxygen sensors. *Appl. Surf. Sci.* **2003**, *212–213*, 397–401. [CrossRef]

64. Sui, Y.X.; Liang, H.L.; Chen, Q.S.; Huo, W.X.; Du, X.L.; Mei, Z.X. Room-temperature ozone sensing capability of IGZO-decorated amorphous Ga_2O_3 films. *ACS Appl. Mater. Interfaces* **2020**, *12*, 8929–8934. [CrossRef] [PubMed]
65. Almaev, A.V.; Nikolaev, V.I.; Stepanov, S.I.; Pechnikov, A.I.; Chernikov, E.V.; Davletkildeev, N.A.; Sokolov, D.V.; Yakovlev, N.N.; Kalygina, V.M.; Kopyev, V.V.; et al. Hydrogen influence on electrical properties of Pt-contacted α-Ga_2O_3/ε-Ga_2O_3 structures grown on patterned sapphire substrates. *J. Phys. D Appl. Phys.* **2020**, *53*, 414004. [CrossRef]
66. Almaev, A.; Nikolaev, V.; Butenko, P.; Stepanov, S.; Pechnikov, A.; Yakovlev, N.; Sinyugin, I.; Shapenkov, S.; Scheglov, M. Gas sensors based on pseudohexagonal phase of gallium oxide. *Phys. Status Solidi B* **2022**, *259*, 2100306. [CrossRef]
67. Bartic, M.; Baban, C.; Suzuki, H.; Ogita, M.; Isai, M. β-gallium oxide as oxygen gas sensors at a high temperature. *J. Am. Ceram. Soc.* **2007**, *90*, 2879–2884. [CrossRef]
68. Nakagomi, S.; Kaneko, M.; Kokubun, Y. Hydrogen sensitive Schottky diode based on β-Ga_2O_3 single crystal. *Sens. Lett.* **2011**, *9*, 31–35. [CrossRef]
69. Nakagomi, S.; Ikeda, M.; Kokubun, Y. Comparison of hydrogen sensing properties of Schottky diodes based on SiC and β-Ga_2O_3 single crystal. *Sens. Lett.* **2011**, *9*, 616–620. [CrossRef]
70. Bartic, M. Mechanism of oxygen sensing on β-Ga_2O_3 single-crystal sensors for high temperatures. *Phys. Status Solidi A* **2016**, *213*, 457–462. [CrossRef]
71. Uhlendorf, J.; Galazka, Z.; Schmidt, H. Oxygen diffusion in β-Ga_2O_3 single crystals at high temperatures. *Appl. Phys. Lett.* **2021**, *119*, 242106. [CrossRef]
72. Macri, P.P.; Enzo, S.; Sberveglieri, G.; Groppelli, S.; Perego, C. Unknown Ga_2O_3 structural phase and related characteristics as active layers for O_2 sensors. *Appl. Surf. Sci.* **1993**, *65–66*, 277–282. [CrossRef]
73. Fleischer, M.; Meixner, H. A selective CH_4 sensor using semiconducting Ga_2O_3 thin films based on temperature switching of multigas reactions. *Sens. Actuators B Chem.* **1995**, *25*, 544–547. [CrossRef]
74. Bausewein, A.; Hacker, B.; Fleischer, M.; Meixner, H. Effects of palladium dispersions on gas-sensitive conductivity of semiconducting Ga_2O_3 thin-film ceramics. *J. Am. Ceram. Soc.* **1997**, *80*, 317–323. [CrossRef]
75. Schwebel, T.; Fleischer, M.; Meixner, H.; Kohl, C.-D. CO-sensor for domestic use based on high temperature stable Ga_2O_3 thin films. *Sens. Actuators B Chem.* **1998**, *49*, 46–51. [CrossRef]
76. Josepovits, V.K.; Krafcsik, O.; Kiss, G.; Perczel, I.V. Effect of gas adsorption on the surface structure of β-Ga_2O_3 studied by XPS and conductivity measurements. *Sens. Actuators B Chem.* **1998**, *48*, 373–375. [CrossRef]
77. Flingelli, G.K.; Fleischer, M.; Meixner, H. Selective detection of methane in domestic environments using a catalyst sensor system based on Ga_2O_3. *Sens. Actuators B Chem.* **1998**, *48*, 258–262. [CrossRef]
78. Fleischer, M.; Kornely, S.; Weh, T.; Frank, J.; Meixner, H. Selective gas detection with high-temperature operated metal oxides using catalytic filters. *Sens. Actuators B Chem.* **2000**, *69*, 205–210. [CrossRef]
79. Lang, A.C.; Fleischer, M.; Meixner, H. Surface modifications of Ga_2O_3 thin film sensors with Rh, Ru and Ir clusters. *Sens. Actuators B Chem.* **2000**, *66*, 80–84. [CrossRef]
80. Bene, R.; Pinter, Z.; Perczel, I.V.; Fleischer, M.; Reti, F. High-temperature semiconductor gas sensors. *Vacuum* **2001**, *61*, 275–278. [CrossRef]
81. Frank, J.; Fleischer, M.; Zimmer, M.; Meixner, H. Ozone sensing using In_2O_3 modified Ga_2O_3 thin films. *IEEE Sens. J.* **2001**, *1*, 318–321. [CrossRef]
82. Frank, J.; Meixner, H. Sensor system for indoor air monitoring using semiconducting metal oxides and IR-absorption. *Sens. Actuators B Chem.* **2001**, *78*, 298–302. [CrossRef]
83. Kiss, G.; Pinter, Z.; Perczel, I.V.; Sassi, Z.; Reti, F. Study of oxide semiconductor sensor materials by selected methods. *Thin Solid Films* **2001**, *391*, 216–223. [CrossRef]
84. Lampe, U.; Simon, E.; Pohle, R.; Fleischer, M.; Meixner, H.; Frerichs, H.-P.; Lehmann, M.; Kiss, G. GasFET for the detection of reducing gases. *Sens. Actuators B Chem.* **2005**, *111–112*, 106–110. [CrossRef]
85. Stegmeier, S.; Fleischer, M.; Hauptmann, P. Influence of the morphology of platinum combined with β-Ga_2O_3 on the VOC response of work function type sensors. *Sens. Actuators B Chem.* **2010**, *148*, 439–449. [CrossRef]
86. Stegmeier, S.; Fleischer, M.; Hauptmann, P. Thermally activated platinum as VOC sensing material for work function type gas sensors. *Sens. Actuators B Chem.* **2010**, *144*, 418–424. [CrossRef]
87. Almaev, A.V.; Chernikov, E.V.; Davletkildeev, N.A.; Sokolov, D.V. Oxygen sensors based on gallium oxide thin films with addition of chromium. *Superlattices Microstruct.* **2020**, *139*, 106392. [CrossRef]
88. Almaev, A.V.; Chernikov, E.V.; Novikov, V.V.; Kushnarev, B.O.; Yakovlev, N.N.; Chuprakova, E.V.; Oleinik, V.L.; Lozinskaya, A.D.; Gogova, D.S. Impact of Cr_2O_3 additives on the gas-sensitive properties of β-Ga_2O_3 thin films to oxygen, hydrogen, carbon monoxide, and toluene vapors. *J. Vac. Sci. Technol. A* **2021**, *39*, 023405. [CrossRef]
89. Fleischer, M.; Giber, J.; Meixner, H. H_2-induced changes in electrical conductance of β-Ga_2O_3 thin-film systems. *Appl. Phys. A* **1992**, *54*, 560–566. [CrossRef]
90. Fleischer, M.; Meixner, H. Improvements in Ga_2O_3 sensors for reducing gases. *Sens. Actuators B Chem.* **1993**, *13*, 259–263. [CrossRef]
91. Fleischer, M.; Hollbauer, L.; Meixner, H. Effect of the sensor structure on the stability of Ga_2O_3 sensors for reducing gases. *Sens. Actuators B Chem.* **1994**, *18–19*, 119–124. [CrossRef]

92. Fleischer, M.; Meixner, H. In situ Hall measurements at temperatures up to 1100 degrees C with selectable gas atmospheres. *Meas. Sci. Technol.* **1994**, *5*, 580–583. [CrossRef]
93. Fleischer, M.; Wagner, V.; Hacker, B.; Meixner, H. Comparison of a.c. and d.c. measurement techniques using semiconducting Ga_2O_3 sensors. *Sens. Actuators B Chem.* **1995**, *26–27*, 85–88. [CrossRef]
94. Fleischer, M.; Meixner, H. Sensitive, selective and stable CH_4 detection using semiconducting Ga_2O_3 thin films. *Sens. Actuators B Chem.* **1995**, *26–27*, 81–84. [CrossRef]
95. Fleischer, M.; Seth, M.; Kohl, C.-D.; Meixner, H. A selective H_2 sensor implemented using Ga_2O_3 thin-films which are covered with a gas-filtering SiO_2 layer. *Sens. Actuators B Chem.* **1996**, *35–36*, 297–302. [CrossRef]
96. Fleischer, M.; Seth, M.; Koh, C.-D.; Meixner, H. A study of surface modification at semiconducting Ga_2O_3 thin film sensors for enhancement of the sensitivity and selectivity. *Sens. Actuators B Chem.* **1996**, *35–36*, 290–296. [CrossRef]
97. Reti, F.; Fleischer, M.; Perczel, I.V.; Meixner, H.; Giber, J. Detection of reducing gases in air by β-Ga_2O_3 thin films using self-heated and externally (oven-) heated operation modes. *Sens. Actuators B Chem.* **1996**, *34*, 378–382. [CrossRef]
98. Baban, C.-I.; Toyoda, Y.; Ogita, M. High temperature oxygen sensor using a Pt-Ga_2O_3-Pt sandwich structure. *Jpn. J. Appl. Phys.* **2004**, *43*, 7213–7216. [CrossRef]
99. Baban, C.; Toyoda, Y.; Ogita, M. Oxygen sensing at high temperatures using Ga_2O_3 films. *Thin Solid Films* **2005**, *484*, 369–373. [CrossRef]
100. Bartic, M.; Toyoda, Y.; Baban, C.-I.; Ogita, M. Oxygen sensitivity in gallium oxide thin films and single crystals at high temperatures. *Jpn. J. Appl. Phys.* **2006**, *45*, 5186–5188. [CrossRef]
101. Manandhar, S.; Battu, A.K.; Devaraj, A.; Shutthanandan, V.; Thevuthasan, S.; Ramana, C.V. Rapid response high temperature oxygen sensor based on titanium doped gallium oxide. *Sci. Rep.* **2020**, *10*, 178. [CrossRef]
102. Frank, J.; Fleischer, M.; Meixner, H. Electrical doping of gas-sensitive, semiconducting Ga_2O_3 thin films. *Sens. Actuators B Chem.* **1996**, *34*, 373–377. [CrossRef]
103. Ogita, M.; Saika, N.; Nakanishi, Y.; Hatanaka, Y. Ga_2O_3 thin films for high-temperature gas sensors. *Appl. Surf. Sci.* **1999**, *142*, 188–191. [CrossRef]
104. Juan, Y.M.; Chang, S.-J.; Hsueh, H.T.; Chen, T.C.; Huang, S.W.; Lee, Y.H.; Hsueh, T.J.; Wu, C.L. Self-powered hybrid humidity sensor and dual-band UV photodetector fabricated on back-contact photovoltaic cell. *Sens. Actuators B Chem.* **2015**, *219*, 43–49. [CrossRef]
105. Zhang, Y.; Feng, Z.; Karim, M.R.; Zhao, H. High-temperature low-pressure chemical vapor deposition of β-Ga_2O_3. *J. Vac. Sci. Technol. A* **2020**, *38*, 050806. [CrossRef]
106. Yu, M.-F.; Atashbar, M.Z.; Chen, X. Mechanical and electrical characterization of β-Ga_2O_3 nanostructures for sensing applications. *IEEE Sens. J.* **2005**, *5*, 20–25. [CrossRef]
107. Feng, P.; Xue, X.Y.; Liu, Y.G.; Wan, Q.; Wang, T.H. Achieving fast oxygen response in individual β-Ga_2O_3 nanowires by ultraviolet illumination. *Appl. Phys. Lett.* **2006**, *89*, 112114. [CrossRef]
108. Huang, Y.; Yue, S.; Wang, Z.; Wang, Q.; Shi, C.; Xu, Z.; Bai, X.D.; Tang, C.; Gu, C. Preparation and electrical properties of ultrafine Ga_2O_3 nanowires. *J. Phys. Chem. B* **2006**, *110*, 796–800. [CrossRef] [PubMed]
109. Liu, Z.; Yamazaki, T.; Shen, Y.; Kikuta, T.; Nakatani, N.; Li, Y.X. O_2 and CO sensing of Ga_2O_3 multiple nanowire gas sensors. *Sens. Actuators B Chem.* **2008**, *129*, 666–670. [CrossRef]
110. Arnold, S.P.; Prokes, S.M.; Perkins, F.K.; Zaghloul, M.E. Design and performance of a simple, room-temperature Ga_2O_3 nanowire gas sensor. *Appl. Phys. Lett.* **2009**, *95*, 103102. [CrossRef]
111. Mazeina, L.; Picard, Y.N.; Maximenko, S.I.; Perkins, F.K.; Glaser, E.R.; Twigg, M.E.; Freitas, J.A., Jr.; Prokes, S.M. Growth of Sn-doped β-Ga_2O_3 nanowires and Ga_2O_3-SnO_2 heterostructures for gas sensing applications. *Cryst. Growth Des.* **2009**, *9*, 4471–4479. [CrossRef]
112. Ma, H.L.; Fan, D.W. Influence of oxygen pressure on structural and sensing properties of β-Ga_2O_3 nanomaterial by thermal evaporation. *Chin. Phys. Lett.* **2009**, *26*, 117302. [CrossRef]
113. Cuong, N.D.; Park, Y.W.; Yoon, S.G. Microstructural and electrical properties of Ga_2O_3 nanowires grown at various temperatures by vapor-liquid-solid technique. *Sens. Actuators B Chem.* **2009**, *140*, 240–244. [CrossRef]
114. Mazeina, L.; Perkins, F.K.; Bermudez, V.M.; Arnold, S.P.; Prokes, S.M. Functionalized Ga_2O_3 nanowires as active material in room temperature capacitance-based gas sensors. *Langmuir* **2010**, *26*, 13722–13726. [CrossRef] [PubMed]
115. Mazeina, L.; Bermudez, V.M.; Perkins, F.K.; Arnold, S.P.; Prokes, S.M. Interaction of functionalized Ga_2O_3 NW-based room temperature gas sensors with different hydrocarbons. *Sens. Actuators B Chem.* **2010**, *151*, 114–120. [CrossRef]
116. Ma, H.L.; Fan, D.W.; Niu, X.S. Preparation and NO_2-gas sensing property of individual β-Ga_2O_3 nanobelt. *Chin. Phys. B* **2010**, *19*, 076102. [CrossRef]
117. Jang, Y.-G.; Kim, W.-S.; Kim, D.-H.; Hong, S.-H. Fabrication of Ga_2O_3/SnO_2 core-shell nanowires and their ethanol gas sensing properties. *J. Mater. Res.* **2011**, *26*, 2322–2327. [CrossRef]
118. Nakagomi, S.; Sai, T.; Kokubun, Y. Hydrogen gas sensor with self temperature compensation based on β-Ga_2O_3 thin film. *Sens. Actuators B Chem.* **2013**, *187*, 413–419. [CrossRef]
119. Nakagomi, S.; Yokoyama, K.; Kokubun, Y. Devices based on series-connected Schottky junctions and β-Ga_2O_3/SiC heterojunctions characterized as hydrogen sensors. *J. Sens. Sens. Syst.* **2014**, *3*, 231–239. [CrossRef]

120. Wu, Y.-L.; Luan, Q.; Chang, S.-J.; Jiao, Z.; Weng, W.Y.; Lin, Y.-H.; Hsu, C.L. Highly sensitive β-Ga_2O_3 nanowire isopropyl alcohol sensor. *IEEE Sens. J.* **2014**, *14*, 401–405. [CrossRef]
121. Zhong, W.; Wang, P.W.; Han, X.B.; Yu, D.P. Gas sensor based on Ga_2O_3 nanowires. *J. Chin. Electron Microsc. Soc.* **2014**, *33*, 7–13. [CrossRef]
122. Reiprich, J.; Isaac, N.A.; Schlag, L.; Hopfeld, M.; Ecke, G.; Stauden, T.; Pezoldt, J.; Jacobs, H.O. Corona discharge assisted growth morphology switching of tin-doped gallium oxide for optical gas sensing applications. *Cryst. Growth Des.* **2019**, *19*, 6945–6953. [CrossRef]
123. Krawczyk, M.; Wozniak, P.S.; Szukiewicz, R.; Kuchowicz, M.; Korbutowicz, R.; Teterycz, H. Morphology of Ga_2O_3 nanowires and their sensitivity to volatile organic compounds. *Nanomaterials* **2021**, *11*, 456. [CrossRef] [PubMed]
124. Kim, H.; Jin, C.; An, S.; Lee, C. Fabrication and CO gas-sensing properties of Pt-functionalized Ga_2O_3 nanowires. *Ceram. Int.* **2012**, *38*, 3563–3567. [CrossRef]
125. Park, S.; Kim, H.; Jin, C.; Lee, C. Synthesis, structure, and room-temperature gas sensing of multiple-networked Pd-doped Ga_2O_3 nanowires. *J. Korean Phys. Soc.* **2012**, *60*, 1560–1564. [CrossRef]
126. Jin, C.; Park, S.; Kim, H.; Lee, C. Ultrasensitive multiple networked Ga_2O_3-core/ZnO-shell nanorod gas sensors. *Sens. Actuators B Chem.* **2012**, *161*, 223–228. [CrossRef]
127. Tsai, T.-Y.; Chang, S.-J.; Weng, W.-Y.; Liu, S.; Hsu, C.-L.; Hsueh, H.-T.; Hsueh, T.-J. β-Ga_2O_3 nanowires-based humidity sensors prepared on GaN/sapphire substrate. *IEEE Sens. J.* **2013**, *13*, 4891–4896. [CrossRef]
128. An, S.; Park, S.; Mun, Y.; Lee, C. UV enhanced NO_2 sensing properties of Pt functionalized Ga_2O_3 nanorods. *Bull. Korean Chem. Soc.* **2013**, *34*, 1632–1636. [CrossRef]
129. Park, S.H.; Kim, S.H.; Park, S.Y.; Lee, C. Synthesis and CO gas sensing properties of surface-nitridated Ga_2O_3 nanowires. *RSC Adv.* **2014**, *4*, 63402–63407. [CrossRef]
130. Juan, Y.M.; Chang, S.-J.; Hsueh, H.T.; Wang, S.H.; Weng, W.Y.; Cheng, T.C.; Wu, C.L. Effects of humidity and ultraviolet characteristics on β-Ga_2O_3 nanowire sensor. *RSC Adv.* **2015**, *5*, 84776–84781. [CrossRef]
131. Bui, Q.C.; Largeau, L.; Morassi, M.; Jegenyes, N.; Mauguin, O.; Travers, L.; Lafosse, X.; Dupuis, C.; Harmand, J.-C.; Tchernycheva, M.; et al. GaN/Ga_2O_3 core/shell nanowires growth: Towards high response gas sensors. *Appl. Sci.* **2019**, *9*, 3528. [CrossRef]
132. Abdullah, Q.N.; Ahmed, A.R.; Ali, A.M.; Yam, F.K.; Hassan, Z.; Bououdina, M. Novel SnO_2-coated β-Ga_2O_3 nanostructures for room temperature hydrogen gas sensor. *Int. J. Hydrogen Energy* **2021**, *46*, 7000–7010. [CrossRef]
133. Park, S.; Kim, S.; Sun, G.-J.; Lee, C. Synthesis, structure and ethanol sensing properties of Ga_2O_3-core/WO_3-shell nanostructures. *Thin Solid Films* **2015**, *591*, 341–345. [CrossRef]
134. Sprincean, V.; Caraman, M.; Spataru, T.; Fernandez, F.; Paladi, F. Influence of the air humidity on the electrical conductivity of the β-Ga_2O_3-GaS structure: Air humidity sensor. *Appl. Phys. A* **2022**, *128*, 303. [CrossRef]
135. Girija, K.; Thirumalairajan, S.; Mastelaro, V.R.; Mangalaraj, D. Catalyst free vapor-solid deposition of morphologically different β-Ga_2O_3 nanostructure thin films for selective CO gas sensors at low temperature. *Anal. Methods* **2016**, *8*, 3224–3235. [CrossRef]
136. Almaev, A.V.; Nikolaev, V.I.; Yakovlev, N.N.; Butenko, P.N.; Stepanov, S.I.; Pechnikov, A.I.; Scheglov, M.P.; Chernikov, E.V. Hydrogen sensors based on Pt/α-Ga_2O_3: Sn/Pt structures. *Sens. Actuators B Chem.* **2022**, *364*, 131904. [CrossRef]
137. Yakovlev, N.N.; Nikolaev, V.I.; Stepanov, S.I.; Almaev, A.V.; Pechnikov, A.I.; Chernikov, E.V.; Kushnarev, B.O. Effect of oxygen on the electrical conductivity of Pt-contacted α-Ga_2O_3/ε(κ)-Ga_2O_3 MSM structures on patterned sapphire substrates. *IEEE Sens. J.* **2021**, *21*, 14636–14644. [CrossRef]
138. Krawczyk, M.; Korbutowicz, R.; Szukiewicz, R.; Wozniak, P.S.; Kuchowicz, M.; Teterycz, H. P-type inversion at the surface of β-Ga_2O_3 epitaxial layer modified with Au nanoparticles. *Sensors* **2022**, *22*, 932. [CrossRef]
139. Liu, K.W.; Sakurai, M.; Aono, M. One-step fabrication of β-Ga_2O_3-amorphous-SnO_2 core-shell microribbons and their thermally switchable humidity sensing properties. *J. Mater. Chem.* **2012**, *22*, 12882. [CrossRef]
140. Weng, T.-F.; Ho, M.-S.; Sivakumar, C.; Balraj, B.; Chung, P.-F. VLS growth of pure and Au decorated β-Ga_2O_3 nanowires for room temperature CO gas sensor and resistive memory applications. *Appl. Surf. Sci.* **2020**, *533*, 147476. [CrossRef]
141. Aymerich, E.L.; Gil, G.D.; Moreno, M.; Pellegrino, P.; Rodriguez, A.R. Fabrication, characterization and performance of low power gas sensors based on $(Ga_xIn_{1-x})_2O_3$ nanowires. *Sensors* **2021**, *21*, 3342. [CrossRef]
142. Ge, X.T.; Fang, D.R.; Liu, X.Q. The preparation and gas-sensing properties of Ga_2O_3-NiO complex oxide by sol-gel method. *Acta Phys. Chim. Sin.* **2005**, *21*, 10–15. [CrossRef]
143. Mohammadi, M.R.; Fray, D.J. Semiconductor TiO_2-Ga_2O_3 thin film gas sensors derived from particulate sol-gel route. *Acta Mater.* **2007**, *55*, 4455–4466. [CrossRef]
144. Mohammadi, M.R.; Fray, D.J.; Ghorbani, M. Comparison of single and binary oxide sol-gel gas sensors based on titania. *Solid State Sci.* **2008**, *10*, 884–893. [CrossRef]
145. Hou, Y.; Jayatissa, A.H. Low resistive gallium doped nanocrystalline zinc oxide for gas sensor application via sol-gel process. *Sens. Actuators B Chem.* **2014**, *204*, 310–318. [CrossRef]
146. Trinchi, A.; Galatsis, K.; Wlodarski, W.; Li, Y.X. A Pt/Ga_2O_3-ZnO/SiC Schottky diode-based hydrocarbon gas sensor. *IEEE Sens. J.* **2003**, *3*, 548–553. [CrossRef]
147. Trinchi, A.; Li, Y.X.; Wlodarski, W.; Kaciulis, S.; Pandolfi, L.; Russo, S.P.; Duplessis, J.; Viticoli, S. Investigation of sol-gel prepared Ga-Zn oxide thin films for oxygen gas sensing. *Sens. Actuators A Phys.* **2003**, *108*, 263–270. [CrossRef]

148. Li, Y.X.; Trinchi, A.; Wlodarski, W.; Galatsis, K.; Kalantar-zadeh, K. Investigation of the oxygen gas sensing performance of Ga_2O_3 thin films with different dopants. *Sens. Actuators B Chem.* **2003**, *93*, 431–434. [CrossRef]
149. Trinchi, A.; Kaciulis, S.; Pandolfi, L.; Ghantasala, M.K.; Li, Y.X.; Wlodarski, W.; Viticoli, S.; Comini, E.; Sberveglieri, G. Characterization of Ga_2O_3 based MRISiC hydrogen gas sensors. *Sens. Actuators B Chem.* **2004**, *103*, 129–135. [CrossRef]
150. Trinchi, A.; Wlodarski, W.; Li, Y.X. Hydrogen sensitive Ga_2O_3 Schottky diode sensor based on SiC. *Sens. Actuators B Chem.* **2004**, *100*, 94–98. [CrossRef]
151. Trinchi, A.; Wlodarski, W.; Faglia, G.; Ponzoni, A.; Comini, E.; Sberveglieri, G. High temperature hydrocarbon sensing with Pt-thin Ga_2O_3-SiC diodes. *Mater. Sci. Forum* **2005**, *483–485*, 1033–1036. [CrossRef]
152. Trinchi, A.; Wlodarski, W.; Li, Y.X.; Faglia, G.; Sberveglieri, G. Pt/Ga_2O_3/SiC MRISiC devices: A study of the hydrogen response. *J. Phys. D Appl. Phys.* **2005**, *38*, 754–763. [CrossRef]
153. Bagheri, M.; Khodadadi, A.A.; Mahjoub, A.R.; Mortazavi, Y. Strong effects of gallia on structure and selective responses of Ga_2O_3-In_2O_3 nanocomposite sensors to either ethanol, CO or CH_4. *Sens. Actuators B Chem.* **2015**, *220*, 590–599. [CrossRef]
154. Lin, H.-J.; Baltrus, J.P.; Gao, H.; Ding, Y.; Nam, C.-Y.; Ohodnicki, P.; Gao, P.-X. Perovskite nanoparticle-sensitized Ga_2O_3 nanorod arrays for CO detection at high temperature. *ACS Appl. Mater. Interfaces* **2016**, *8*, 8880–8887. [CrossRef] [PubMed]
155. Lin, H.-J.; Gao, H.; Gao, P.-X. UV-enhanced CO sensing using Ga_2O_3-based nanorod arrays at elevated temperature. *Appl. Phys. Lett.* **2017**, *110*, 043101. [CrossRef]
156. Chen, H.; Hu, J.B.; Li, G.-D.; Gao, Q.; Wei, C.D.; Zou, X.X. Porous Ga-In bimetallic oxide nanofibers with controllable structures for ultrasensitive and selective detection of formaldehyde. *ACS Appl. Mater. Interfaces* **2017**, *9*, 4692–4700. [CrossRef]
157. Zhang, Y.L.; Jia, C.W.; Kong, Q.; Fan, N.Y.; Chen, G.; Guan, H.T.; Dong, C.J. ZnO-decorated In/Ga oxide nanotubes derived from bimetallic In/Ga MOFs for fast acetone detection with high sensitivity and selectivity. *ACS Appl. Mater. Interfaces* **2020**, *12*, 26161. [CrossRef]
158. Pilliadugula, R.; Krishnan, N. Gas sensing performance of GaOOH and β-Ga_2O_3 synthesized by hydrothermal method: A comparison. *Mater. Res. Express* **2018**, *6*, 025027. [CrossRef]
159. Pilliadugula, R.; Krishnan, N.G. Effect of pH dependent morphology on room temperature NH_3 sensing performances of β-Ga_2O_3. *Mater. Sci. Semicond. Process.* **2020**, *112*, 105007. [CrossRef]
160. Wang, J.; Jiang, S.S.; Liu, H.L.; Wang, S.H.; Pan, Q.J.; Yin, Y.D.; Zhang, G. P-type gas-sensing behavior of Ga_2O_3/Al_2O_3 nanocomposite with high sensitivity to NO_x at room temperature. *J. Alloys Compd.* **2020**, *814*, 152284. [CrossRef]
161. Zhang, B.; Lin, H.-J.; Gao, H.Y.; Lu, X.X.; Nam, C.-Y.; Gao, P.-X. Perovskite-sensitized β-Ga_2O_3 nanorod arrays for highly selective and sensitive NO_2 detection at high temperature. *J. Mater. Chem. A* **2020**, *8*, 10845–10854. [CrossRef]
162. Pilliadugula, R.; Gopalakrishnan, N. Room temperature ammonia sensing performances of pure and Sn doped β-Ga_2O_3. *Mater. Sci. Semicond. Process.* **2021**, *135*, 106086. [CrossRef]
163. Demin, I.E.; Kozlov, A.G. In_2O_3-Ga_2O_3 thin films for ammonia sensors of petrochemical industry safety systems. *AIP Conf. Proc.* **2018**, *2007*, 050004. [CrossRef]
164. Demin, I.E.; Kozlov, A.G. Selectivity of the gas sensor based on the 50% In_2O_3-50% Ga_2O_3 thin film in dynamic mode of operation. *J. Phys. Conf. Ser.* **2018**, *944*, 012027. [CrossRef]
165. Demin, I.E. Increasing the selectivity of semiconductor gas sensors working at sinusoidal-varying temperature for machine industry safety systems. *J. Phys. Conf. Ser.* **2019**, *1260*, 032008. [CrossRef]
166. Demin, I.E. Selection of methods for increasing gas selectivity on the example of a sensor system based on In_2O_3-Ga_2O_3 semiconductor films. *J. Phys. Conf. Ser.* **2019**, *1210*, 012032. [CrossRef]
167. Demin, I.E. Reducing the energy consumption of semiconductor methane sensors for gas alarm systems. *AIP Conf. Proc.* **2019**, *2141*, 050020. [CrossRef]
168. Wei, Z.H.; Akbari, M.K.; Hai, Z.Y.; Ramachandran, R.K.; Detavernier, C.; Verpoort, F.; Kats, E.; Xu, H.Y.; Hu, J.; Zhuiykov, S. Ultra-thin sub-10 nm Ga_2O_3-WO_3 heterostructures developed by atomic layer deposition for sensitive and selective C_2H_5OH detection on ppm level. *Sens. Actuators B Chem.* **2019**, *287*, 147–156. [CrossRef]
169. Jochum, W.; Penner, S.; Fottinger, K.; Kramer, R.; Rupprechter, G.; Klotzer, B. Hydrogen on polycrystalline β-Ga_2O_3: Surface chemisorption, defect formation, and reactivity. *J. Catal.* **2008**, *256*, 268–277. [CrossRef]
170. Vorobyeva, N.; Rumyantseva, M.; Platonov, V.; Filatova, D.; Chizhov, A.; Marikutsa, A.; Bozhev, I.; Gaskov, A. Ga_2O_3 (Sn) oxides for high-temperature gas sensors. *Nanomaterials* **2021**, *11*, 2938. [CrossRef]
171. Pandeeswari, R.; Jeyaprakash, B.G. High sensing response of β-Ga_2O_3 thin film towards ammonia vapours: Influencing factors at room temperature. *Sens. Actuators B Chem.* **2014**, *195*, 206–214. [CrossRef]
172. Yan, J.-T.; Lee, C.-T. Improved detection sensitivity of Pt/β-Ga_2O_3/GaN hydrogen sensor diode. *Sens. Actuators B Chem.* **2009**, *143*, 192–197. [CrossRef]
173. Lee, C.-T.; Yan, J.-T. Investigation of a metal-insulator-semiconductor Pt/mixed Al_2O_3 and Ga_2O_3 insulator/AlGaN hydrogen sensor. *J. Electrochem. Soc.* **2010**, *157*, J281. [CrossRef]
174. Lee, C.-T.; Yan, J.-T. Sensing mechanisms of Pt/β-Ga_2O_3/GaN hydrogen sensor diodes. *Sens. Actuators B Chem.* **2010**, *147*, 723–729. [CrossRef]
175. Shafiei, M.; Hoshyargar, F.; Motta, N.; O'Mullane, A.P. Utilizing p-type native oxide on liquid metal microdroplets for low temperature gas sensing. *Mater. Des.* **2017**, *122*, 288–295. [CrossRef]

176. Saidi, S.A.; Tang, J.; Yang, J.; Han, J.; Daeneke, T.; O'Mullane, A.P.; Kalantar-Zadeh, K. Liquid metal-based route for synthesizing and tuning gas-sensing elements. *ACS Sens.* **2020**, *5*, 1177–1189. [CrossRef]
177. Sivasankaran, B.R.; Balaji, M. Novel gallium oxide/reduced graphene oxide nanocomposite for ammonia gas sensing application. *Mater. Lett.* **2021**, *288*, 129386. [CrossRef]
178. Ji, H.; Zeng, W.; Li, Y. Gas sensing mechanisms of metal oxide semiconductors: A focus review. *Nanoscale* **2019**, *11*, 22664–22684. [CrossRef]
179. Fleischer, M.; Meixner, H. Fast gas sensors based on metal oxides which are stable at high temperatures. *Sens. Actuators B Chem.* **1997**, *43*, 1–10. [CrossRef]
180. Fleischer, M.; Meixner, H. Thin-film gas sensors based on high-temperature-operated metal oxides. *J. Vac. Sci. Technol. A* **1999**, *17*, 1866–1872. [CrossRef]
181. Hoefer, U.; Frank, J.; Fleischer, M. High temperature Ga_2O_3-gas sensors and SnO_2-gas sensors: A comparison. *Sens. Actuators B Chem.* **2001**, *78*, 6–11. [CrossRef]
182. Fleischer, M. Advances in application potential of adsorptive-type solid state gas sensors: High-temperature semiconducting oxides and ambient temperature GasFET devices. *Meas. Sci. Technol.* **2008**, *19*, 042001. [CrossRef]
183. Pohle, R.; Weisbrod, E.; Hedler, H. Enhancement of MEMS-based Ga_2O_3 gas sensors by surface modifications. *Proc. Eng.* **2016**, *168*, 211–215. [CrossRef]
184. Weisz, P.B.; Prater, C.D. Interpretation of measurements in experimental catalysis. *Adv. Catal.* **1954**, *6*, 143. [CrossRef]
185. Jaaniso, R.; Tan, O.K. *Semiconductor Gas Sensors*; Woodhead Publishing: Duxford, UK, 2019; pp. 27–34. [CrossRef]
186. Yamazoe, N. New approaches for improving semiconductor gas sensors. *Sens. Actuators B Chem.* **1991**, *5*, 7–19. [CrossRef]
187. Korotcenkov, G. *Handbook of Gas Sensor Materials*; Springer: New York, NY, USA, 2013; pp. 49–116. [CrossRef]
188. Frank, J.; Fleischer, M.; Meixner, H. Gas-sensitive electrical properties of pure and doped semiconducting Ga_2O_3 thick films. *Sens. Actuators B Chem.* **1998**, *48*, 318–321. [CrossRef]
189. Frank, J.; Fleischer, M.; Meixner, H.; Feltz, A. Enhancement of sensitivity and conductivity of semiconducting Ga_2O_3 gas sensors by doping with SnO_2. *Sens. Actuators B Chem.* **1998**, *49*, 110–114. [CrossRef]
190. Schwebel, T.; Fleischer, M.; Meixner, H. A selective, temperature compensated O_2 sensor based on Ga_2O_3 thin films. *Sens. Actuators B Chem.* **2000**, *65*, 176–180. [CrossRef]
191. Fleischer, M.; Meixner, H. Sensing reducing gases at high temperatures using long-term stable Ga_2O_3 thin films. *Sens. Actuators B Chem.* **1992**, *6*, 257–261. [CrossRef]
192. Hacker, B.; Fleischer, M.; Meixner, H. Topography and performance of gas-sensing devices: An AFM study. *Scanning* **1993**, *15*, 291–294. [CrossRef]
193. Weh, T.; Fleischer, M.; Meixner, H. Optimization of physical filtering for selective high temperature H_2 sensors. *Sens. Actuators B Chem.* **2000**, *68*, 146–150. [CrossRef]
194. Weh, T.; Frank, J.; Fleischer, M.; Meixner, H. On the mechanism of hydrogen sensing with SiO_2 modified high temperature Ga_2O_3 sensors. *Sens. Actuators B Chem.* **2001**, *78*, 202–207. [CrossRef]
195. Wiesner, K.; Knozinger, H.; Fleischer, M.; Meixner, H. Working mechanism of an ethanol filter for selective high-temperature methane gas sensors. *IEEE Sens. J.* **2002**, *2*, 354–359. [CrossRef]
196. Biskupski, D.; Geupel, A.; Wiesner, K.; Fleischer, M.; Moos, R. Platform for a hydrocarbon exhaust gas sensor utilizing a pumping cell and a conductometric sensor. *Sensors* **2009**, *9*, 7498–7508. [CrossRef] [PubMed]
197. Giber, J.; Perczel, I.V.; Gerblinger, J.; Lampe, U.; Fleischer, M. Coadsorption and cross sensitivity on high temperature semiconducting metal oxides: Water effect on the coadsorption process. *Sens. Actuators B Chem.* **1994**, *18–19*, 113–118. [CrossRef]
198. Reti, F.; Fleischer, M.; Meixner, H.; Giber, J. Effect of coadsorption of reducing gases on the conductivity of β-Ga_2O_3 thin films in the presence of O_2. *Sens. Actuators B Chem.* **1995**, *18–19*, 573–577. [CrossRef]
199. Reti, F.; Fleischer, M.; Meixner, H.; Giber, J. Influence of water on the coadsorption of oxidizing and reducing gases on the β-Ga_2O_3 surface. *Sens. Actuators B Chem.* **1994**, *18*, 138–142. [CrossRef]
200. Reti, F.; Fleischer, M.; Gerblinger, J.; Lampe, U.; Varhegyi, E.B.; Perczel, V.; Meixner, H.; Giber, J. Comparison of the water effect on the resistance of different semiconducting metal oxides. *Sens. Actuators B Chem.* **1995**, *26–27*, 103–107. [CrossRef]
201. Pohle, R.; Fleischer, M.; Meixner, H. In situ infrared emission spectroscopic study of the adsorption of H_2O and hydrogen-containing gases on Ga_2O_3 gas sensors. *Sens. Actuators B Chem.* **2000**, *68*, 151–156. [CrossRef]
202. Varhegyi, E.B.; Perczel, I.V.; Gerblinger, J.; Fleischer, M.; Meixner, H.; Giber, J. Auger and SIMS study of segregation and corrosion some semiconducting oxide gas-sensor materials. *Sens. Actuators B Chem.* **1994**, *18–19*, 569–572. [CrossRef]
203. Hovhannisyan, R.V.; Khondkaryan, H.D.; Aleksanyan, M.S.; Arakelyan, V.M.; Semerjyan, B.O.; Gasparyan, F.V.; Aroutiounian, V.M. Static and noise characteristics of nanocomposite gas sensors. *J. Contemp. Phys.* **2014**, *49*, 151–157. [CrossRef]
204. Dyndal, K.; Zarzycki, A.; Andrysiewicz, W.; Grochala, D.; Marszalek, K.; Rydosz, A. CuO-Ga_2O_3 thin films as a gas-sensitive material for acetone detection. *Sensors* **2020**, *20*, 3142. [CrossRef]
205. Lundstrom, I.; Shivaraman, S.; Svensson, C.; Lundkvist, L. A hydrogen-sensitive MOS field-effect transistor. *Appl. Phys. Lett.* **1975**, *26*, 55–57. [CrossRef]
206. Leu, M.; Doll, T.; Flietner, B.; Lechner, J.; Eisele, I. Evaluation of gas mixtures with different sensitive layers incorporated in hybrid FET structures. *Sens. Actuators B Chem.* **1994**, *18–19*, 678–681. [CrossRef]

207. Geistlinger, H. Accumulation layer model for Ga_2O_3 thin-film gas sensors based on the Volkenstein theory of catalysis. *Sens. Actuators B Chem.* **1994**, *18*, 125–131. [CrossRef]
208. Geistlinger, H.; Eisele, I.; Flietner, B.; Winter, R. Dipole- and charge transfer contributions to the work function change of semiconducting thin films: Experiment and theory. *Sens. Actuators B Chem.* **1996**, *34*, 499–505. [CrossRef]
209. Shin, W.; Hong, S.; Jeong, Y.; Jung, G.; Park, J.; Kim, D.; Park, B.-G.; Lee, J.-H. Effects of channel length scaling on the signal-to-noise ratio in FET-type gas sensor with horizontal floating-gate. *IEEE Electron Device Lett.* **2022**, *43*, 442–445. [CrossRef]
210. Imanaka, N.; Tamura, S.; Adachi, G. Ammonia sensor based on ionically exchanged NH_4^+-gallate solid electrolytes. *Electrochem. Solid State Lett.* **1999**, *1*, 282. [CrossRef]
211. Westphal, D.; Jakobs, S.; Guth, U. Gold-composite electrodes for hydrocarbon sensors based on YSZ solid electrolyte. *Ionics* **2001**, *7*, 182–186. [CrossRef]
212. Zosel, J.; Westphal, D.; Jakobs, S.; Muller, R.; Guth, U. Au-oxide composites as HC-sensitive electrode material for mixed potential gas sensors. *Solid State Ion.* **2002**, *152–153*, 525. [CrossRef]
213. Zosel, J.; Ahlborn, K.; Muller, R.; Westphal, D.; Vashook, V.; Guth, U. Selectivity of HC-sensitive electrode materials for mixed potential gas sensors. *Solid State Ion.* **2002**, *169*, 115–119. [CrossRef]
214. Zhang, W.-F.; Schmidt-Zhang, P.; Guth, U. Electrochemical studies on cells M/YSZ/Pt (M = Pt, Pt-Ga_2O_3) in NO, O_2, N_2 gas mixtures. *Solid State Ion.* **2004**, *169*, 121–128. [CrossRef]
215. Shuk, P.; Bailey, E.; Zosel, J.; Guth, U. New advanced in situ carbon monoxide sensor for the process application. *Ionics* **2009**, *15*, 131–138. [CrossRef]
216. Wu, N.Q.; Chen, Z.; Xu, J.H.; Chyu, M.; Mao, S.X. Impedance-metric Pt/YSZ/Au-Ga_2O_3 sensor for CO detection at high temperature. *Sens. Actuators B Chem.* **2005**, *110*, 49–53. [CrossRef]
217. Yan, S.C.; Wan, L.J.; Li, Z.S.; Zhou, Y.; Zou, Z.G. Synthesis of a mesoporous single crystal Ga_2O_3 nanoplate with improved photoluminescence and high sensitivity in detecting CO. *Chem. Commun.* **2010**, *46*, 6388. [CrossRef]
218. Korotcenkov, G.; Han, S.H.; Cho, B.K. Material design for metal oxide chemiresistive gas sensors. *J. Sens. Sci. Technol.* **2013**, *22*, 1–17. [CrossRef]
219. Bartic, M.; Ogita, M.; Isai, M.; Baban, C.; Suzuki, H. Oxygen sensing properties at high temperatures of β-Ga_2O_3 thin films deposited by the chemical solution deposition method. *J. Appl. Phys.* **2007**, *102*, 023709. [CrossRef]
220. Wang, D.; Lou, Y.L.; Wang, R.; Wang, P.P.; Zheng, X.J.; Zhang, Y.; Jiang, N. Humidity sensor based on Ga_2O_3 nanorods doped with Na^+ and K^+ from GaN powder. *Ceram. Int.* **2015**, *41*, 14790–14797. [CrossRef]
221. Potje-Kamloth, K. Semiconductor junction gas sensors. *Chem. Rev.* **2008**, *108*, 367–399. [CrossRef]
222. Ratko, A.; Babushkin, O.; Baran, A.; Baran, S. Sorption and gas sensitive properties of In_2O_3 based ceramics doped with Ga_2O_3. *J. Eur. Ceram. Soc.* **1998**, *18*, 2227–2232. [CrossRef]
223. Silver, A.T.; Juarez, A.S. SnO_2: Ga thin films as oxygen gas sensor. *Mater. Sci. Eng. B* **2004**, *110*, 268–271. [CrossRef]
224. Bagheri, M.; Khodadadi, A.A.; Mahjoub, A.R.; Mortazavi, Y. Highly sensitive gallia-SnO_2 nanocomposite sensors to CO and ethanol in presence of methane. *Sens. Actuators B Chem.* **2013**, *188*, 45–52. [CrossRef]
225. Du, L.T.; Li, H.Y.; Li, S.; Liu, L.; Li, Y.; Xu, S.Y.; Gong, Y.M.; Cheng, Y.L.; Zeng, X.G.; Liang, Q.C. A gas sensor based on Ga-doped SnO_2 porous microflowers for detecting formaldehyde at low temperature. *Chem. Phys. Lett.* **2018**, *713*, 235–241. [CrossRef]
226. Kevin, M.; Tho, W.H.; Ho, G.W. Transferability of solution processed epitaxial Ga:ZnO films; tailored for gas sensor and transparent conducting oxide applications. *J. Mater. Chem.* **2012**, *22*, 16442. [CrossRef]
227. Vorobyeva, N.; Rumyantseva, M.; Filatova, D.; Konstantinova, E.; Grishina, D.; Abakumov, A.; Turner, S.; Gaskov, A. Nanocrystalline ZnO(Ga): Paramagnetic centers, surface acidity and gas sensor properties. *Sens. Actuators B Chem.* **2013**, *182*, 555–564. [CrossRef]
228. Rashid, T.-R.; Phan, D.-T.; Chung, G.-S. Effect of Ga-modified layer on flexible hydrogen sensor using ZnO nanorods decorated by Pd catalysts. *Sens. Actuators B Chem.* **2014**, *193*, 869–876. [CrossRef]
229. Girija, K.G.; Somasundaram, K.; Debnath, A.K.; Topkar, A.; Vatsa, R.K. Enhanced H_2S sensing properties of gallium doped ZnO nanocrystalline films as investigated by DC conductivity and impedance spectroscopy. *Mater. Chem. Phys.* **2018**, *214*, 297–305. [CrossRef]
230. Fleischer, M.; Meixner, H. Selectivity in high-temperature operated semiconductor gas-sensors. *Sens. Actuators B Chem.* **1998**, *52*, 179–187. [CrossRef]
231. Anichini, C.; Czepa, W.; Pakulski, D.; Aliprandi, A.; Ciesielski, A.; Samori, P. Chemical sensing with 2D materials. *Chem. Soc. Rev.* **2018**, *47*, 4860–4908. [CrossRef]
232. Zhao, J.L.; Huang, X.R.; Yin, Y.H.; Liao, Y.K.; Mo, H.W.; Qian, Q.K.; Guo, Y.Z.; Chen, X.L.; Zhang, Z.F.; Hua, M.Y. Two-dimensional gallium oxide monolayer for gas-sensing application. *J. Phys. Chem. Lett.* **2021**, *12*, 5813–5820. [CrossRef]
233. Wang, S.; Li, H.; Huang, H.; Cao, X.; Chen, X.; Cao, D. Porous organic polymers as a platform for sensing applications. *Chem. Soc. Rev.* **2022**, *51*, 2031–2080. [CrossRef]
234. Yang, Y.; Zhang, P. Dissociation of H_2 molecule on the β-Ga_2O_3 (100)B surface: The critical role of oxygen vacancy. *Phys. Lett. A* **2010**, *374*, 4169–4173. [CrossRef]
235. Nagarajan, V.; Chandiramouli, R. Methane adsorption characteristics on β-Ga_2O_3 nanostructures: DFT investigation. *Appl. Surf. Sci.* **2015**, *344*, 65–78. [CrossRef]

236. Yaqoob, U.; Younis, M.I. Chemical gas Sensors: Recent developments, challenges, and the potential of machine Learning—A review. *Sensors* **2021**, *21*, 2877. [CrossRef] [PubMed]
237. Mohmed, M.; Attia, N.; Elashery, S. Greener and facile synthesis of hybrid nanocomposite for ultrasensitive iron (II) detection using carbon sensor. *Microporous Mesoporous Mater.* **2021**, *313*, 110832. [CrossRef]
238. Elashery, S.; Attia, N.; Oh, H. Design and fabrication of novel flexible sensor based on 2D Ni-MOF nanosheets as a preliminary step toward wearable sensor for onsite Ni (II) ions detection in biological and environmental samples. *Anal. Chim. Acta* **2022**, *1197*, 339518. [CrossRef] [PubMed]
239. Rahman, T.; Masui, T.; Ichiki, T. Single-crystal gallium oxide-based biomolecular modified diode for nucleic acid sensing. *Jpn. J. Appl. Phys.* **2015**, *54*, 04DL08. [CrossRef]
240. Das, M.; Chakraborty, T.; Lin, C.H.; Lin, R.; Kao, C.H. Screen-printed Ga_2O_3 thin film derived from liquid metal employed in highly sensitive pH and non-enzymatic glucose recognition. *Mater. Chem. Phys.* **2022**, *278*, 125652. [CrossRef]

Communication

Femtosecond Laser Modification of Silica Optical Waveguides for Potential Bragg Gratings Sensing

Jian Chen [1], Ji-Jun Feng [1,*], Hai-Peng Liu [1], Wen-Bin Chen [1], Jia-Hao Guo [1], Yang Liao [2], Jie Shen [3], Xue-Feng Li [3], Hui-Liang Huang [3] and Da-Wei Zhang [1]

[1] Shanghai Key Laboratory of Modern Optical System, Engineering Research Center of Optical Instrument and System, Ministry of Education, School of Optical-Electrical and Computer Engineering, University of Shanghai for Science and Technology, Shanghai 200093, China
[2] Key Laboratory of High Field Laser Physics and CAS Center for Excellence in Ultra-Intense Laser Science, Shanghai Institute of Optics and Fine Mechanics, Chinese Academy of Sciences, Shanghai 201800, China
[3] Shanghai Honghui Optics Communication Tech. Corp., Shanghai 201822, China
* Correspondence: fjijun@usst.edu.cn

Abstract: The optimum femtosecond laser direct writing of Bragg gratings on silica optical waveguides has been investigated. The silica waveguide has a 6.5×6.5 μm^2 cross-sectional profile with a 20-μm-thick silicon dioxide cladding layer. Compared with conventional grating inscribed on fiber platforms, the silica planar waveguide circuit can realize a stable performance as well as a high-efficiency coupling with the fiber. A thin waveguide cladding layer also facilitates laser focusing with an improved spherical aberration. Different from the circular fiber core matching with the Gaussian beam profile, a 1030-nm, 400-fs, and 190-nJ laser is optimized to focus on the top surface of the square silica waveguide, and the 3rd-order Bragg gratings are inscribed successfully. A 1.5-mm long uniform Bragg gratings structure with a reflectivity of 90% at a 1548.36-nm wavelength can be obtained. Cascaded Bragg gratings with different periods are also inscribed in the planar waveguide. Different reflection wavelengths can be realized, which shows great potential for wavelength multiplexing-related applications such as optical communications or sensing.

Keywords: gratings; ultrafast; integrated optics devices; waveguides; microstructure fabrication

1. Introduction

Wavelength-division multiplexing (WDM) devices can effectively improve data transmission bandwidth and have been extensively utilized in optics communications and multiparameter sensings [1–3]. Several approaches for WDM have been investigated, including Mach–Zehnder interferometers (MZIs) [4,5], microring resonators [6], arrayed waveguide gratings (AWGs) [7,8], thin film filters (TFFs) [9], and Bragg gratings [10]. MZIs and microring-based devices are usually limited by free spectral range (FSR) and channel bandwidth. For tunable AWGs, different approaches have been proposed in order to relax the limitations imposed by wavelength tolerances [11]. Multimode waveguides, double-peaked electric field distribution, and thermo-optic effect can be used for wavelength tuning, but they will increase the insertion loss and fabrication complexity [8]. TFFs have the advantages of a narrow passband, low loss, and good temperature stability. However, the fabrication process is a little complex and sometimes more than 100 layers may need to be coated [12,13]. The Bragg grating can reflect the designed wavelength and is considered a promising candidate for WDM systems. With the development of integrated photonic circuit technology, silicon waveguide-based Bragg grating has received more attention. However, its high coupling and polarization-dependent loss as well as the temperature sensitivity are not so favorable [14]. For the silica waveguide, the cross-sectional profile is usually square, which can work polarization independently. The waveguide mode matches the pattern of single-mode fiber, resulting in a low coupling loss [15]. The silica waveguide-based AWGs have been developed by the typical semiconductor lithographic process, which requires a high precision alignment. Furthermore, it is not so convenient to adjust the grating parameter for manufacturing error compensation. On the other hand, femtosecond laser direct writing technology can realize three-dimensional, low-cost, and flexible waveguide fabrication, which can induce permanent refractive index changes in transparent materials [16,17].

Till now, a large number of femtosecond laser-inscribed Bragg gratings have been reported in fibers [18,19]. The stacking inscription technique has been adopted to prepare the highest reflectivity Bragg grating for mid-infrared applications [20]. However, the cylindrical geometry of the fiber would cause an aberration when the laser is focused inside with air-based lenses, and adaptive optics aberration compensation should be adopted [21]. The aberrations can also be improved by placing fibers in refractive index matching fluids or ferrules, which are not so convenient and are unsuitable for mass manufacturing [22,23]. Some Bragg gratings are also realized by irradiating the fiber core using the laser interference pattern generated by the phase mask without eliminating spherical aberration. However, a phase mask has a fixed structure parameter and can only work for a specific resonance wavelength, which is slightly expensive and time-consuming [24]. Nevertheless, maintaining the roundness, symmetry, and mode–field profile of the fiber grating by the femtosecond laser direct writing is still challenging. A silica-based planar lightwave circuit (PLC) can naturally avoid aberration during laser writing, which has stable performance and can be mass-produced [25,26]. However, the femtosecond laser direct writing of the silica-based PLC waveguide for Bragg gratings has not yet been reported, to the best of our knowledge.

In the following, the optimum preparation of Bragg gratings on silica waveguides by femtosecond laser direct writing is presented. Compared with the fiber Bragg grating, the waveguide case can be easily prepared with no need for aberration correction. The fabrication reproducibility is competitive and the device performance is also stable. Most importantly, the waveguide gratings can be monolithically integrated with other photonic integrated devices, which will benefit some system-on-chip applications. The 3rd-order Bragg gratings with a length of 1.5 mm are fabricated with an extinction ratio of 9.5 dB and reflectivity of about 90% at a 1548.36-nm resonance wavelength. Moreover, cascaded Bragg gratings are also inscribed to achieve multi-wavelength reflection. Detailed experimental results and interaction mechanisms are analyzed and discussed.

2. Bragg Grating Design and Fabrication

The silica chip was prepared by photolithography and reactive ion etching on a 4-inch wafer. A 20-μm-thick SiO_2 cladding layer was then deposited by flame hydrolysis deposition (FHD), which could provide a convenient refractive index modification range and smooth film profile, a low propagation loss (0.01 dB/cm), low coupling and reflection loss (less than 0.1 dB) due to almost the same refractive index and mode field diameter as conventional single-mode fiber, excellent physical and chemical stability, as well as an inexpensive large-scale fabrication [27]. Then, the waveguides were cut to a size of $30 \times 2.5 \times 1$ mm^3, with facets polished and packaged with input/output fiber as shown in Figure 1a. The waveguide has a germanium-doped silica core with a 6.5-μm-wide square profile, as shown in Figure 1b. The refractive indices of the core and cladding layer are 1.463 and 1.444, respectively, at a wavelength of 1550 nm. Such a refractive index also benefits from femtosecond laser inscription [28]. The waveguide can work on polarization independently from the calculated effective refractive index of about 1.447 for both transverse electric (TE) and transverse magnetic (TM) modes, whose profiles can match well with that of the fiber, resulting in a low mode overlap loss and Fresnel reflection between chip and fiber. Tight packaging can also improve the Fresnel reflection. In the experiment, the actual connection loss of the silica-based packaged device was measured to be 0.2 dB, which includes coupling loss as well as propagation loss. Some commercial

PLC devices can realize a coupling loss of less than 0.1 dB, whereas the current waveguide can be further optimized to realize a better fiber-to-chip coupling.

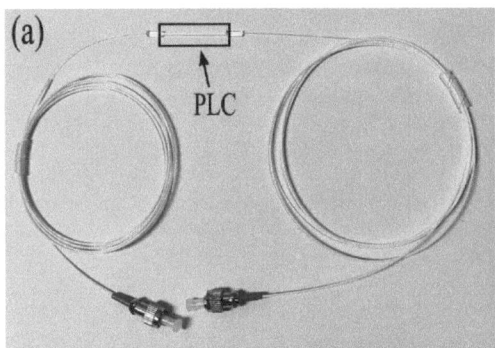 

Figure 1. (a) Photo of fiber packaged silica chip. (b) Cross-sectional SEM image of the silica waveguide.

For the waveguide Bragg grating, it is a structure that periodically modulates the effective refractive index of a waveguide. When a wave enters the grating region, changes in refractive index can cause periodic reflections. If all the reflections are in phase, many reflections combine with constructive interference and a strong reflected signal will appear. That is Bragg resonance wavelength λ_B, which can be expressed as [29]

$$\lambda_B = 2n_{eff}\Lambda / m, \tag{1}$$

where n_{eff} is the effective refractive index of the propagating mode, Λ is the grating period, and $m = 1, 2, 3, \ldots$, is the grating order. For uniform Bragg gratings, the maximum reflectivity R_{max} at the Bragg wavelength can be calculated by the transmission spectrum [23]

$$R_{max} = 1 - 10^{-(T/10)}, \tag{2}$$

where T is the transmittance in decibels (dB) at λ_B. The coupling coefficient κ depends on the refractive index modulation (Δn), mode overlap factor (η), and Bragg wavelength, which can be expressed as [30]

$$\kappa = \pi \Delta n \eta / \lambda_B. \tag{3}$$

The spectrum bandwidth of Bragg grating with length L can be obtained by [31]

$$\Delta \lambda = \lambda_B^2 / n_{eff} \sqrt{\frac{\kappa^2}{\pi^2} + \frac{1}{L^2}}. \tag{4}$$

The bandwidth of weakly modulated Bragg gratings ($\kappa L < 1$) is length-limited, and a longer grating has a narrower bandwidth. In contrast, for strongly modulated Bragg gratings ($\kappa L > 3$), the light does not penetrate the full length of the grating. Thus, the bandwidth is directly proportional to the coupling coefficient and almost independent of length. Therefore, by reducing the coupling coefficient, it is possible to narrow the linewidth of a strong modulation Bragg grating.

A schematic illustration of the femtosecond laser inscription process is shown in Figure 2a. The system consists of a 1030-nm femtosecond laser (YL-20, Anyang, China) with a pulse width of 400 fs and a repetition rate of 25–5000 kHz. After the collimated beam passes through the electric shutter, it is introduced to the upper gantry bracket by a mirror, then goes through a half-waver-plate, a dichroic mirror, and an objective lens (Mitutoyo (Kawasaki, Japan), 20×, NA 0.40). The lens has a working distance of 20.35 mm, which can focus the laser to a spot size of about 1 µm. The electric shutter is used to control the opening and blocking of the pulse train of the laser pulse. The dichroic mirror can transmit visible light and reflect light at a 1030-nm wavelength. A CCD camera with a coaxial light source is installed for real-time monitoring of the inscription process. Due to the transparency of the PLC chip, a transmitted illumination system is adopted. A three-dimensional air flotation platform (Aerotech Inc., Pittsburgh, PA, USA) is used for the motion control, where the objective lens is mounted on the Z-axis platform to facilitate the adjustment of the laser focal length, and the XY-axis air flotation platform is used for the motion of the PLC chip. Although the resolution of the air flotation platform can provide a fabrication resolution of 0.5 nm, the preparation resolution is about ±3 nm due to the influence of repeatability. Figure 2b shows the image of the PLC chip placed on the moving stage, and the waveguide can be displayed on the monitoring screen. The laser inscription system is integrated on a custom granite base, which is highly resistant to disturbance with excellent positioning stability.

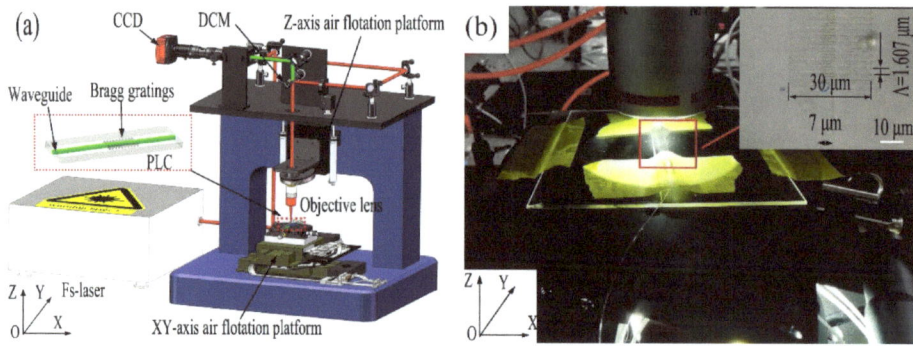

Figure 2. (**a**) Schematic illustration of Bragg gratings fabrication. (**b**) Photo for the PLC chip on the moving stage. Inset: the microscope image of the fabricated gratings.

After precisely adjusting the position of the silica chip in vertical and horizontal directions, the waveguides were inscribed with different laser energy, engraving speed, grating length, and focusing position. The transmission and reflection spectra of Bragg gratings were monitored in real-time during the inscription process by using an amplified spontaneous emission (ASE) laser, a circulator, and an optical spectrum analyzer (OSA) (Yokogawa (Tokyo, Japan), AQ6370C). It should be noted that when characterizing the polarization characteristic, the ASE light source needs to connect a polarizer and a polarization controller. TE light is presented for the performance characterization, though the device is polarization-independent. According to Equation (1), the grating periods should be 0.536, 1.071, and 1.607 µm for the 1st, 2nd, and 3rd grating orders, respectively, with consideration of the effective refractive index of 1.447 at a Bragg wavelength of 1550 nm. Though the coupling coefficient is higher for the low grating order [31], the short grating period usually needs an oil-immersion objective lens with a high numerical aperture such as 1.25, which greatly increases the system complexity. On the other hand, the grating can be inscribed more conveniently for the 3rd-order grating, whose coupling coefficient is also comparable to the lower-order gratings when the duty cycle reaches more than 75% [31]. The grating duty cycle is defined as the ratio of the width of the laser writing line to the grating period. For the inscribing of the waveguide grating, the laser energy, scanning speed, and focusing position should be optimized. The optimum 3rd-order Bragg gratings could be obtained with high reflectance and low loss at a laser power of 190 nJ, engraving speed of 0.15 mm/s, and focusing position just above the waveguide. Since the refractive index change is not so easy to be characterized by the scanning microscope, we measure the top-surface grating profile by a microscope as in Figure 2b. The silica-based PLC was placed in a marked position on the moving stage. There is a cut mark at the edge of the waveguide, which is used to ensure that the adjustment is started from the same position every time since we can clearly observe the mark through the imaging system. The pitch, rotation, and focusing positions were constantly adjusted so that the waveguide remained clearly imaged over the whole processing range. Then the platform moved to the initial mark position for laser writing. Good repeatability can be guaranteed and we can obtain almost the same performance when we repeated the process 10 times with the same laser writing parameters. It can also be confirmed by the consistency of the resonance wavelengths for the single and multi-grating structures.

3. Characterization of Uniform and Cascaded Bragg Gratings

The 3rd-order Bragg gratings were fabricated successfully as in the inset of Figure 2b, with a grating duty cycle of about 75% and a writing line length of 30 μm. When preparing a 3rd-order Bragg grating, the coupling coefficient increases with the duty cycle (50–75%), and the highest coefficient can be achieved with a duty cycle of 75% [31]. For the duty cycle, it should be optimized to obtain a maximum extinction ratio for the resonance peak. Here, it was adjusted with varying laser energy. A maximum extinction ratio of about 9.5 dB can be obtained with a laser energy of 190 nJ, corresponding to a duty cycle of about 75% [31], roughly matching the shape in the microscope image as in Figure 2b. The coupling efficiency affects the extinction ratio and reflectivity of the Bragg grating. The reflectivity of the Bragg resonance can be inferred from the observed transmission dip according to Equation (2). When the coupling coefficient is optimal, the extinction ratio and reflectance will increase with the length. During the grating writing, the optimal reflectivity can be obtained by observing the extinction ratio with the change of grating length. With the increase in the total grating length, the reflectivity and extinction ratio of the transmission peak will increase and then saturate at a certain length. More gratings will not improve the reflectance much at the Bragg wavelength but cause higher transmission loss. During the fabrication, the total length of the grating is optimized to be around 1.5 mm.

The transmission loss is slightly high here. Actually, the loss may be mainly caused by the laser fabrication condition such as the pulse width, laser energy, focusing position, etc., which may influence the grating profile and the relative position to the waveguide. The laser with too large energy will cause large waveguide loss or even damage the waveguide, whereas a too small energy laser is difficult to form refractive index modulation. The focusing position will influence the loss and the pulse width will affect the interaction mechanism. When the focusing depth is 20 to 25 μm, the laser energy has a wider adjustment range and is easy to form refractive index modulation [32]. We further optimized the focusing position and engraving speed to improve the bandwidth and extinction ratio of the Bragg gratings. The optimum 3rd-order Bragg gratings could be obtained with high reflectance and low loss at a focusing position of 20 μm and engraving speed of 0.15 mm/s, as shown in Figure 3. Further optimization is still needed and rapid thermal annealing may also help to improve the transmission loss [33]. The reflectance can reach 90% for the total grating length of 1.5 mm at a Bragg wavelength of 1548.36 nm. For the fabricated Bragg grating, the polarization-dependent loss is also very low and the device performs almost the same for both polarizations. A slightly different resonance wavelength from 1550 nm means that the refractive index change is about 1.449, which is also similar to the reported value [32].

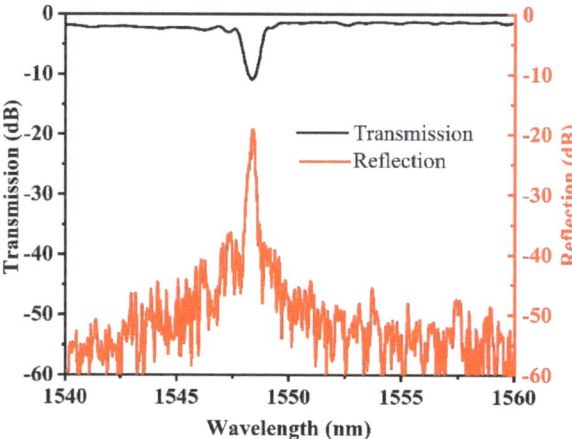

Figure 3. Measured transmission and reflection spectra of the Bragg grating.

Since good repeatability can be obtained, cascaded Bragg gratings can also be prepared. Actually, multi-wavelength multiplexing also has important applications in multiparameter sensing and optical communication [34]. The multiplexing of fiber Bragg gratings can be used for the multi-wavelength ring laser [35]. To demonstrate the potential of multi-parameter sensing, a series of cascaded Bragg waveguide gratings were inscribed on a silica waveguide by the femtosecond laser direct writing technology. The cascaded grating structure consists of many uniform Bragg gratings with total length L and is separated by equal intervals d, as shown in Figure 4. Here, both the grating length and interval between two Bragg gratings are about 1.5 mm. Different gratings can be cascaded such as two gratings with periods of 1.60 and 1.61 μm, whose transmission spectra are shown in Figure 5a. The corresponding resonance wavelengths are 1543.8 and 1553.7 nm, respectively, with an insertion loss of about 2 dB. More gratings can also be prepared, such as 5 gratings with periods of 1.60, 1.61, 1.62, 1.63, and 1.64 μm. The obtained transmission spectrum is shown in Figure 5b. The corresponding resonance wavelengths are 1543.8, 1553.7, 1563.4, 1572.3, and 1581.4 nm, respectively, with a free spectral range of about 9.7 nm. The insertion loss increases to about 5 dB. The resonance wavelength of the grating with the same period is identical for the two and five cascaded cases, which confirms the good repeatability of the system. For cascaded gratings, the flatness of the chip needs to be carefully adjusted before the grating writing.

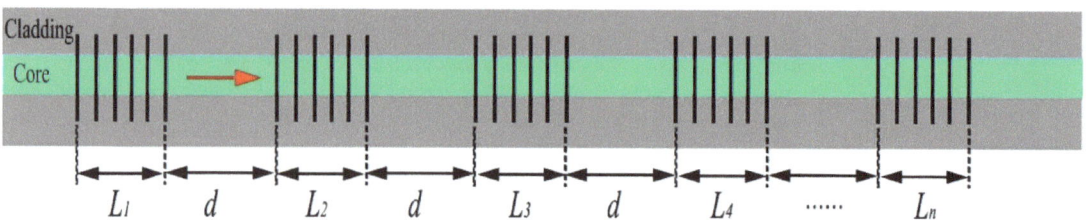

Figure 4. Schematic diagram of the cascaded Bragg gratings.

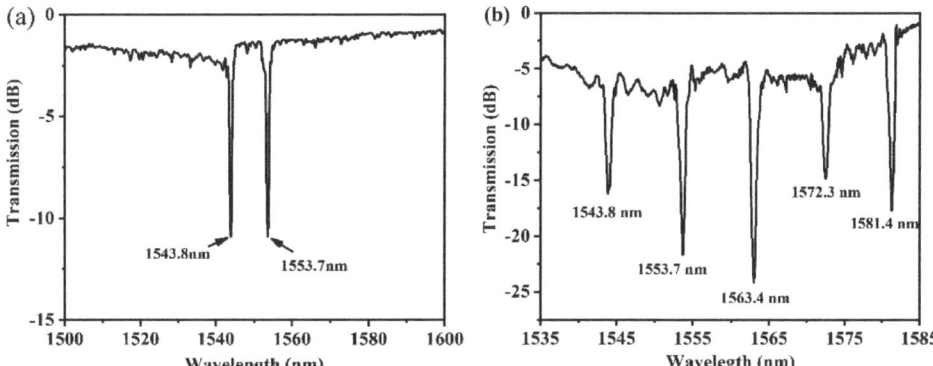

Figure 5. Transmission spectra of (**a**) two and (**b**) five cascaded Bragg gratings, respectively.

Multiphoton nonlinear absorption effect plays an important role in the femtosecond laser direct writing on a silica-based PLC waveguide by FHD processing [32]. Besides, there is still much space for the improvement of grating performance such as the loss by optimizing the femtosecond laser writing process. The successful preparation of the Bragg grating on the silica waveguide shows the potential of the post-trimming or even three-dimensional structure fabrication on a conventional two-dimensional planar waveguide. Compared with the fiber Bragg grating, the Bragg gratings on photonic integrated circuits are more stable and attractive [36]. Based on the waveguide structure, the femtosecond laser processing can easily achieve multi-wavelength multiplexing [37] or multiparameter sensing-related [38] applications. Bragg gratings are widely used as a sensor to measure refractive index, temperature, strain, and other parameters; any bending and strain of the grating section must be avoided when measuring refractive index and temperature, which is usually achieved by being placed in a capillary or straightened for a fiber case [39,40]. The presented PLC Bragg grating can avoid the cross-sensitivity caused by stress and bending, thus greatly simplifying the measurement and improving the detection sensitivity. For example, Zhang et al. proposed a highly sensitive waveguide magnetic field sensor device via laser direct writing, which is an order of magnitude higher sensitivity than similar sensors [41]. Cascaded grating structures with different periods on a planar waveguide platform can also achieve different information encoding, such as gray code [42,43]. Microchannels can be prepared on the top surface of each grating, which can be used for multichannel biosensors or multiparameter sensing, as illustrated in Figure 6. The sensor chip is connected to an ASE laser and OSA for characterization, which can be used to determine microbial function, sample concentration, detection, and cell screening [44,45]. However, for the real fabrication, the processing condition such as for optofluidic channel needs more time to optimize. Further work still needs to be conducted for practical applications. Nonetheless, the presented waveguide platform fabricated by semiconductor process and laser inscription can be applied in the fields such as lab-on-chip sensing [46], optical computing [47], and so forth.

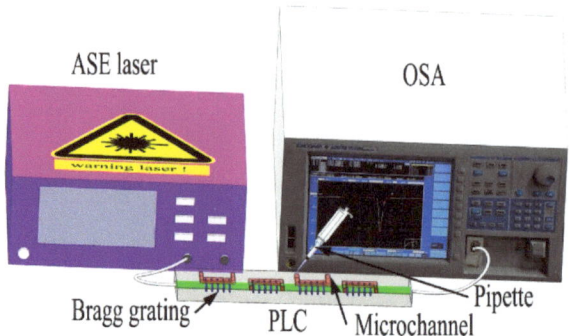

Figure 6. Schematic illustration of multiparameter sensing.

4. Conclusions

To summarize, Bragg gratings were inscribed at the desired resonance wavelength in the silica PLC chips by femtosecond laser direct writing. A reflectivity of approximately 90% can be realized for the 3rd-order Bragg gratings at a wavelength of 1548.36 nm. Good repeatability can be obtained and the presented fabrication process is more convenient and suitable for mass production. In addition, cascaded Bragg gratings were also prepared on the silica waveguide with slight variations in the period. Multi-wavelength reflection can be obtained, which is beneficial for multiplexing or multiparameter sensing. The presented femtosecond laser post-trimming method will facilitate more complex structure fabrication for conventional semiconductor planar waveguide platforms.

Author Contributions: Conceptualization, J.-J.F. and D.-W.Z.; experiment, J.C., H.-P.L., W.-B.C., J.-H.G., Y.L., J.S., X.-F.L. and H.-L.H.; writing, J.C., J.-J.F., J.-H.G. and D.-W.Z. All authors have read and agreed to the published version of the manuscript.

Funding: This work was supported by the National Natural Science Foundation of China (11774235, 61705130, 11727812, and 11933005), Shanghai Rising-Star Program (19QA1406100), and Program for Professor of Special Appointment (Eastern Scholar) at Shanghai Institutions of Higher Learning.

Institutional Review Board Statement: Not applicable.

Informed Consent Statement: Not applicable.

Data Availability Statement: Data supporting the results reported in this paper may be obtained from the authors upon reasonable request.

Conflicts of Interest: The authors declare no conflict of interest.

References

1. Yu, X.-J.; Dong, X.; Chen, X.-F.; Tian, C.; Liu, S.-C. Large-scale multilongitudinal mode fiber laser sensor array with wavelength/frequency division multiplexing. *J. Lightw. Technol.* **2017**, *35*, 2299–2305. [CrossRef]
2. Liu, Z.; Zhang, J.; Li, X.; Wang, L.; Li, J.; Xue, C.; An, J.; Cheng, B. 25 × 50 Gbps wavelength division multiplexing silicon photonics receiver chip based on a silicon nanowire-arrayed waveguide grating. *Photonics Res.* **2019**, *7*, 659–663. [CrossRef]
3. Yeh, C.-H.; Lin, W.-P.; Jiang, S.-Y.; Hsieh, S.-E.; Hsu, C.-H.; Chow, C.-W. Integrated fiber-FSO WDM access system with fiber fault protection. *Electronics* **2022**, *11*, 2101. [CrossRef]
4. Mikkelsen, J.C.; Bois, A.; Lordello, T.; Mahgerefteh, D.; Menezo, S.; Poon, J.K.S. Polarization-insensitive silicon nitride Mach-Zehnder lattice wavelength demultiplexers for CWDM in the O-band. *Opt. Express* **2018**, *26*, 30076–30084. [CrossRef]
5. Munk, D.; Katzman, M.; Kaganovskii, Y.; Inbar, N.; Misra, A.; Hen, M.; Priel, M.; Feldberg, M.; Tkachev, M.; Bergman, A.; et al. Eight-channel silicon-photonic wavelength division multiplexer with 17 GHz spacing. *IEEE J. Sel. Top. Quantum Electron.* **2019**, *25*, 1–10. [CrossRef]
6. Fujiwara, M.; Wakabayashi, R.; Sasaki, M.; Takeoka, M. Wavelength division multiplexed and double-port pumped time-bin entangled photon pair generation using Si ring resonator. *Opt. Express* **2017**, *25*, 3445–3453. [CrossRef]

7. Pitris, S.; Dabos, G.; Mitsolidou, C.; Alexoudi, T.; Heyn, P.D.; Campenhout, J.V.; Broeke, R.; Kanellos, G.T.; Pleros, N. Silicon photonic 8 × 8 cyclic Arrayed Waveguide Grating Router for O-band on-chip communication. *Opt. Express* **2018**, *26*, 6276–6284. [CrossRef] [PubMed]
8. Chung, K.-F.; Fu, P.-H.; Shih, Y.-T.; Chiu, H.-K.; Shih, T.-T.; Huang, D.-W. Demonstration of a bi-directionally tunable arrayed waveguide grating with ultra-low thermal power using S-shaped architecture and parallel-circuit configuration. *Opt. Express* **2022**, *30*, 25842–25854. [CrossRef]
9. Kuo, W.-K.; Lin, W.-S.; Yang, S.-W. Thin-film tunable bandpass filter for spectral shift detection in surface plasmon resonance sensors. *Opt. Lett.* **2020**, *45*, 65–68. [CrossRef]
10. Saber, M.G.; Xing, Z.; Patel, D.; Fiky, E.E.; Abadia, N.; Wang, Y.; Jacques, M.; Osman, M.M.; Plant, D.V. A CMOS compatible ultracompact silicon photonic optical add-drop multiplexer with misaligned sidewall Bragg gratings. *IEEE Photonics J.* **2017**, *9*, 6601010. [CrossRef]
11. Tan, D.T.H.; Grieco, A.; Fainman, Y. Towards 100 channel dense wavelength division multiplexing with 100GHz spacing on silicon. *Opt. Express* **2014**, *22*, 10408–10415. [CrossRef] [PubMed]
12. Domash, L.; Wu, M.; Nemchuk, N.; Ma, E. Tunable and switchable multiple-cavity thin film filters. *J. Lightw. Technol.* **2004**, *22*, 126–135. [CrossRef]
13. Zhu, X.-Y.; Yu, K.-X.; Zhu, X.-F.; Wu, C. Wavelength-tunable deep-ultraviolet thin-film filter: Design and experimental demonstration. *Appl. Opt.* **2021**, *60*, 10199–10206. [CrossRef] [PubMed]
14. Mulugeta, T.; Rasras, M. Silicon hybrid (de)multiplexer enabling simultaneous mode and wavelength-division multiplexing. *Opt. Express* **2015**, *23*, 943–949. [CrossRef] [PubMed]
15. Han, S.-J.; Park, J.; Yoo, S.; Yu, K. Lateral silicon photonic grating-to-fiber coupling with angle-polished silica waveguide blocks. *Opt. Express* **2020**, *28*, 8811–8818. [CrossRef]
16. Wei, D.-Z.; Wang, C.-W.; Wang, H.-J.; Hu, X.-P.; Wei, D.; Fang, X.-Y.; Zhang, Y.; Wu, D.; Hu, Y.-L.; Li, J.-W.; et al. Experimental demonstration of a three-dimensional lithium niobate nonlinear photonic crystal. *Nat. Photonics* **2018**, *12*, 596–600. [CrossRef]
17. Hu, Y.-L.; Yuan, H.-W.; Liu, S.-L.; Ni, J.-C.; Lao, Z.-X.; Xin, C.; Pan, D.; Zhang, Y.-Y.; Zhu, W.-L.; Li, J.-W.; et al. Chiral assemblies of laser-printed micropillars directed by asymmetrical capillary force. *Adv. Mater.* **2020**, *32*, 2002356. [CrossRef]
18. Wang, Z.; Tan, D.-Z.; Qiu, J.-R. Single-shot photon recording for three-dimensional memory with prospects of high capacity. *Opt. Lett.* **2020**, *45*, 6274–6277. [CrossRef]
19. Bharathan, G.; Fernandez, T.T.; Ams, M.; Carrée, J.Y.; Poulain, S.; Poulain, M.; Fuerbach, A. Femtosecond laser direct-written fiber Bragg gratings with high reflectivity and low loss at wavelengths beyond 4 μm. *Opt. Lett.* **2020**, *45*, 4316–4319. [CrossRef]
20. Yu, Y.-J.; Shi, J.-D.; Han, F.; Sun, W.-J.; Feng, X. High-precision fiber Bragg gratings inscription by infrared femtosecond laser direct-writing method assisted with image recognition. *Opt. Express* **2020**, *28*, 8937–8948. [CrossRef]
21. Bharathan, G.; Fernandez, T.T.; Ams, M.; Woodward, R.I.; Hudson, D.D.; Fuerbach, A. Optimized laser-written ZBLAN fiber Bragg gratings with high reflectivity and low loss. *Opt. Lett.* **2019**, *44*, 423–426. [CrossRef] [PubMed]
22. Salter, P.S.; Woolley, M.J.; Morris, S.M.; Booth, M.J.; Fells, J.A.J. Femtosecond fiber Bragg grating fabrication with adaptive optics aberration compensation. *Opt. Lett.* **2018**, *43*, 5993–5996. [CrossRef] [PubMed]
23. Ertorer, E.; Haque, M.; Li, J.-Z.; Herman, P.R. Femtosecond laser filaments for rapid and flexible writing of fiber Bragg grating. *Opt. Express* **2018**, *26*, 9323–9331. [CrossRef] [PubMed]
24. Huang, B.; Xu, Z.-W.; Shu, X.-W. Dual interference effects in a line-by-line inscribed fiber Bragg grating. *Opt. Lett.* **2020**, *45*, 2950–2953. [CrossRef] [PubMed]
25. Shirata, M.; Fujisawa, T.; Sakamoto, T.; Matsui, T.; Nakajima, K.; Saitoh, K. Design of small mode-dependent-loss scrambling-type mode (de)multiplexer based on PLC. *Opt. Express* **2020**, *28*, 9653–9665. [CrossRef] [PubMed]
26. Geng, Y.; Cui, W.-W.; Sun, J.-W.; Chen, X.-X.; Yin, X.-J.; Deng, G.-W.; Zhou, Q.; Zhou, H. Enhancing the long-term stability of dissipative Kerr soliton microcomb. *Opt. Lett.* **2020**, *45*, 5073–5076. [CrossRef]
27. Takahashi, H. Planar lightwave circuit devices for optical communication: Present and future. *Proc. SPIE Act. Passiv. Opt. Compon WDM Commun.* **2003**, *3*, 520–531. [CrossRef]
28. Nasu, Y.; Kohtoku, M.; Hibino, Y. Low-loss waveguides written with a femtosecond laser for flexible interconnection in a planar lightwave circuit. *Opt. Lett.* **2005**, *30*, 723–725. [CrossRef]
29. Mihailov, S.J.; Smelser, C.W.; Grobnic, D.; Walker, R.B.; Lu, P.; Ding, H.; Unruh, J. Bragg gratings written in all-SiO$_2$ and Ge-doped core fibers with 800-nm femtosecond radiation and a phase mask. *J. Lightwave Technol.* **2004**, *22*, 94–100. [CrossRef]
30. Marshall, G.D.; Williams, R.J.; Jovanovic, N.; Steel, M.J.; Withford, M.J. Point-by-point written fiber-Bragg gratings and their application in complex grating designs. *Opt. Express* **2010**, *18*, 19844–19859. [CrossRef]
31. Ams, M.; Dekker, P.; Gross, S.; Withford, M.J. Fabricating waveguide Bragg gratings (WBGs) in bulk materials using ultrashort laser pulses. *Nanophotonics* **2017**, *6*, 743–763. [CrossRef]
32. Nasu, Y.; Kohtoku, M.; Hibino, Y.; Inoue, Y. Waveguide interconnection in silica-based plana lightwave circuit using femtosecond laser. *J. Lightw. Technol.* **2009**, *27*, 4033–4039. [CrossRef]
33. Wu, X.Y.; Feng, J.-J.; Liu, X.-T.; Zeng, H.-P. Effects of rapid thermal annealing on aluminum nitride waveguides. *Opt. Mater. Express* **2020**, *10*, 3073–3080. [CrossRef]
34. Davis, J.A.; Li, A.; Alshamrani, N.; Fainman, Y. Silicon photonic chip for 16-channel wavelength division (de-)multiplexing in the O-band. *Opt. Express* **2020**, *28*, 23620–23627. [CrossRef] [PubMed]

35. Kim, C.S.; Han, Y.-G.; Lee, S.B.; Jung, E.J.; Lee, T.H.; Park, J.S.; Jeong, M.Y. Individual switching of multi-wavelength lasing outputs based on switchable FBG filters. *Opt. Express* **2007**, *15*, 3702–3707. [CrossRef]
36. Chen, J.; Feng, J.-J.; Yan, J.-C.; Yao, Q.; Zhang, D.-W. Highly sensitive detection of water salinity and surface height using a double fiber grating system fabricated by femtosecond laser. *Opt. Fiber Technol.* **2021**, *66*, 102658. [CrossRef]
37. Xia, X.; Lang, T.-T.; Zhang, L.-B.; Yu, Z.-H. Reduction of non-uniformity for a 16 × 16 arrayed waveguide grating router based on silica waveguides. *Appl. Opt.* **2019**, *58*, 1139–1145. [CrossRef]
38. Yang, D.; Wu, T.-S.; Wang, Y.-P.; Cao, W.-P.; Zhang, H.-X.; Liu, Z.-H.; Yang, Z.-N. A multi-parameter integrated sensor based on selectively filled D-Shaped photonic crystal fiber. *Materials* **2022**, *15*, 2811. [CrossRef]
39. Hope, B.; Dutz, F.F.; Bosselmann, T.; Willsch, M.; Koch, A.W.; Roths, J. Iterative matrix algorithm for high precision temperature and force decoupling in multi-parameter FBG sensing. *Opt. Express* **2018**, *26*, 12092–12105. [CrossRef]
40. Zhao, J.; Xu, J.-S.; Wang, C.-X.; Liu, Y.-P.; Yang, Z.-Q. Experimental demonstration of multi-parameter sensing based on polarized interference of polarization-maintaining few-mode fibers. *Opt. Express* **2020**, *28*, 20372–20378. [CrossRef]
41. Zhang, D.-W.; Zhang, Z.-H.; Wei, H.-M.; Qiu, J.-R.; Krishnaswamy, S. Direct laser writing spiral Sagnac waveguide for ultrahigh magnetic field sensing. *Photon. Res.* **2021**, *9*, 1984–1991. [CrossRef]
42. Abbasi, M.; Sadeghi, M.; Adelpour, Z. All-optical graphene-based plasmonic binary to gray code converter. *Opt. Quant. Electron.* **2022**, *54*, 142. [CrossRef]
43. Monteiro, J.; Pedro, A.; Silva, A.J. A gray code model for the encoding of grid cells in the entorhinal cortex. *Neural Comput. Appl.* **2022**, *34*, 2287–2306. [CrossRef]
44. Sugioka, K.; Cheng, Y. Femtosecond laser processing for optofluidic fabrication. *Lab Chip* **2012**, *12*, 3576–3589. [CrossRef] [PubMed]
45. Aghakhani, A.; Cetin, H.; Erkoc, P.; Tombak, G.-I.; Sitti, M. Flexural wave-based soft attractor walls for trapping microparticles and cells. *Lab Chip* **2021**, *21*, 582–596. [CrossRef]
46. Liu, Z.-M.; Xu, J.; Lin, Z.-J.; Qi, J.; Li, X.-L.; Zhang, A.-D.; Lin, J.-T.; Chen, J.-F.; Fang, Z.-W.; Song, Y.-P.; et al. Fabrication of single-mode circular optofluidic waveguides in fused silica using femtosecond laser microfabrication. *Opt. Laser Technol.* **2021**, *141*, 107118. [CrossRef]
47. Ahmadpour, A.; Sharif, A.H.; Chenaghlou, F.B. Design and comprehensive analysis of an ultra-fast fractional-order temporal differentiator based on a plasmonic Bragg grating microring resonator. *Opt. Express* **2021**, *29*, 36257–36272. [CrossRef]

Article

Probing Non-Equilibrium Pair-Breaking and Quasiparticle Dynamics in Nb Superconducting Resonators Under Magnetic Fields

Joong-Mok Park [1,2], Zhi Xiang Chong [2], Richard H. J. Kim [1,2], Samuel Haeuser [2], Randy Chan [2], Akshay A. Murthy [3], Cameron J. Kopas [4], Jayss Marshall [4], Daniel Setiawan [4], Ella Lachman [4], Joshua Y. Mutus [4], Kameshwar Yadavalli [4], Anna Grassellino [3], Alex Romanenko [3] and Jigang Wang [1,2,*]

1. Ames National Laboratory, U.S. Department of Energy, Ames, IA 50011, USA; joongmok@iastate.edu (J.-M.P.); rkim@iastate.edu (R.H.J.K.)
2. Department of Physics and Astronomy, Iowa State University, Ames, IA 50011, USA; ianchong@iastate.edu (Z.X.C.); shaeuser@iastate.edu (S.H.); rkchan@iastate.edu (R.C.)
3. Fermi National Accelerator Laboratory, Batavia, IL 60510, USA; amurthy@fnal.gov (A.A.M.); annag@fnal.gov (A.G.); aroman@fnal.gov (A.R.)
4. Rigetti Computing, Berkeley, CA 94710, USA; ckopas@rigetti.com (C.J.K.); jayss@rigetti.com (J.M.); danielosetiawan@gmail.com (D.S.); elachman@rigetti.com (E.L.); jmutus@rigetti.com (J.Y.M.); kyadavalli@rigetti.com (K.Y.)
* Correspondence: jgwang@ameslab.gov

Abstract: We conducted a comprehensive study of the non-equilibrium dynamics of Cooper pair breaking, quasiparticle (QP) generation, and relaxation in niobium (Nb) cut from superconducting radio-frequency (SRF) cavities, as well as various Nb resonator films from transmon qubits. Using ultrafast pump–probe spectroscopy, we were able to isolate the superconducting coherence and pair-breaking responses. Our results reveal both similarities and notable differences in the temperature- and magnetic-field-dependent dynamics of the SRF cavity and thin-film resonator samples. Moreover, femtosecond-resolved QP generation and relaxation under an applied magnetic field reveals a clear correlation between non-equilibrium QPs and the quality factor of resonators fabricated by using different deposition methods, such as DC sputtering and high-power impulse magnetron sputtering. These findings highlight the pivotal influence of fabrication techniques on the coherence and performance of Nb-based quantum devices, which are vital for applications in superconducting qubits and high-energy superconducting radio-frequency applications.

Keywords: ultrafast pump–probe; superconductivity; superconducting quantum computer; non-equilibrium quasiparticles; superconducting radio frequency (SRF)

1. Introduction

Most modern quantum computers are built on superconducting (SC) transmon qubits incorporating Josephson junctions (JJs) made of aluminum oxide between two aluminum (Al) electrodes [1–3]. Niobium (Nb) is widely used for RF resonators due to its high critical temperature and well-established lithographic patterning, which enables the control, readout, and coupling of multiple qubits in superconducting quantum circuits [4–9]. To achieve scalable multi-qubit systems, qubit states require long relaxation times (T_1) and dephasing times (T_2), typically greater than 100 µs [10–12]. Significant efforts have been made to study pair-breaking and loss mechanisms to extend T_1 and T_2 [1,13,14]. Microwave dielectric losses at metal–dielectric interfaces, such as at Nb–substrate boundaries, have been one of the limiting factors for qubit lifetimes. Intrinsic losses in bulk materials should theoretically permit lifetimes exceeding 1 ms. Despite being a type II superconductor with a high critical temperature T_c = 9.3 K, Nb used in microwave circuits for transmon qubits has been identified as a potential source of decoherence. Therefore, understanding the decoherence and loss mechanisms in Nb superconducting states is crucial for improving qubit performance. Shorter-than-expected T_1 and T_2 times are often attributed to two-level system (TLS) losses below 1 K and non-equilibrium quasiparticle (QP) generation above 1 K, stemming from factors such as nonuniform surface morphology, defects, native Nb oxide, and ionizing radiation [15–18].

The ultrafast pump–probe technique using femtosecond (fs) laser pulses is a powerful and versatile method for investigating non-equilibrium Cooper pair breaking and quasi-particle dynamics in superconductors. Transient signals measured after ultrafast excitation provide exclusive insight into the out-of-equilibrium processes arising from the conversion between the superconducting condensate and quasiparticles (QPs). In our fs-resolved transient reflectivity scheme, weak 1.2 eV laser pulses break a small fraction of Cooper pairs in the Nb samples. This non-thermal depletion of the superconducting condensate induces a noticeable change in reflectivity at the probe frequency, as illustrated schematically in Figure 1a. This scheme allows us to measure the transient changes in reflectivity as a function of the time delay, temperature, and magnetic field in Nb superconducting states. According to the well-established Rothwarf–Taylor (R-T) model [19], $\Delta R_{T\to 0}/\Delta R(T) - 1$ directly corresponds to the thermal quasiparticle density n_T in the weak perturbation regime, where the depletion of superfluid density $\Delta n/n_s \ll 1$ [20].

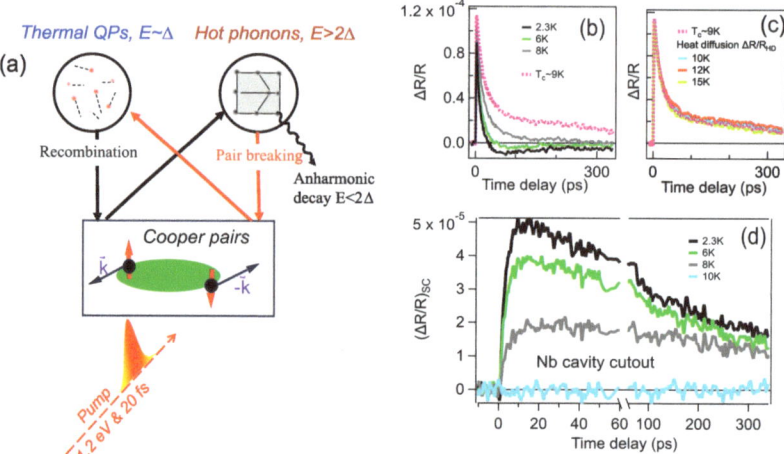

Figure 1. Temperature-dependent photoinduced reflectivity change $\Delta R/R$ of Nb SRF cavity cutout sample. (**a**) Schematic diagram of pair-breaking mechanism in superconducting Nb with ultrafast optical pump having photon energy $\hbar\omega \gg 2\Delta$. Thermal QPs are generated by high-frequency phonon via pair breaking. (**b**,**c**) Measured pump–probe $\Delta R/R$ dynamics for 2 mm thick Nb cavity cutout at 2.3 K, 6 K, and 8 K SC states and at 10 K, 12 K, and 15 K normal states above Tc. (**d**) Superconducting $\Delta R/R_{SC}$ signals are obtained with subtraction from average normal-state data.

To differentiate the quasiparticle (QP) recombination process from electron and phonon temperature changes driven by laser heating, experiments are conducted at two temperature ranges: above and below Tc. The superconducting response is obtained by isolating the low-temperature signal from the high-temperature one [21]. For instance, optical pump–probe experiments on Pb [22] show that femtosecond laser pulses rapidly induce electronic transition, followed by a fast decay (~1 ps) as energy transfers to QPs and high-frequency phonons (HFPs). HFPs, which have energies exceeding the superconducting gap, break Cooper pairs until majority of energy dissipates through electron–phonon interactions or thermal diffusion. In this weak perturbation regime, the photoinduced QP density is much smaller than the thermal-equilibrium QP density. To maintain a weak perturbation, the laser fluence should be kept low (<10 $\mu J/cm^2$) [21–24] to avoid breaking a large number of Cooper pairs. Initially, photo-excited hot electrons equilibrate rapidly, followed by a slower decay of HFPs. Once the photoinduced QPs reach a quasi-equilibrium state with HFPs within a few picoseconds, a slower decay process ensues, driven by HFP anharmonic decay and bi-molecular QP recombination within the phonon bottleneck regime. The dynamic pair-breaking and recombination process induced by the ultrafast laser must be modeled as a non-equilibrium state, in which pair recombination is coupled with high-frequency phonons and Cooper pairs [19,20,25,26].

Moreover, a magnetic field is anticipated to significantly influence the non-equilibrium QP generation and decay processes through magnetic vortex formation and field penetration, offering a valuable approach to assessing the robustness of superconducting coherence. However, ultrafast pump–probe experiments exploring the magnetic field dependence of non-equilibrium superconducting dynamics have not yet been conducted on Nb superconductors. Despite several pump–probe studies on QP dynamics in unconventional superconductors such as cuprates [20,27,28] and pnictides [29–31], fs studies of equilibrium QP dynamics on conventional BCS-type SCs are relatively rare.

In this article, we conducted an in-depth investigation into the non-equilibrium QP generation and relaxation dynamics in superconducting Nb sourced from radio-frequency cavities, as well as various Nb resonator films fabricated using deposition techniques such as DC sputtering and high-power impulse magnetron sputtering (HiPIMS). We obtained QP densities from pump–probe signals under different temperature and magnetic field conditions. Our results reveal both shared characteristics and distinct differences in the temperature- and magnetic-field-dependent behavior of Nb samples designed for SRF and qubit applications. Significantly, fs-resolved QP relaxation under an applied magnetic field demonstrated a clear correlation between the QP density and the quality factor of Nb thin-film resonators. These findings provide valuable insights into non-equilibrium QP dynamics in Nb materials and lay the groundwork for systematic measurements that could guide strategies to develop highly coherent quantum devices.

2. Materials and Methods

Nb thin films fabricated with different growth methods and bulk Nb from a superconducting radio-frequency (SRF) cavity used in the experiments are summarized in Table 1. Sputter-coated Nb thin-film samples of thickness t = 175 nm were grown by HiPIMS, DC high-power sputtering, and DC low-to-high (DC LH)-power sputtering on intrinsic Si (001) wafers at Rigetti Computing. The deposition rate for HiPIMS was ~5.1 nm/min, and the DC high-power deposition rate was ~25 nm/min. For the DC LH sample, the low-power deposition rate was 5.3 nm/min to 30 nm thickness, followed by 145 nm deposition with high power. The average grain sizes of sputtered samples were 44 nm for HiPIMS, 69 nm for DC high, and 65 nm for DC LH. A thick t = 2 mm SRF sample was obtained as an SRF cavity cutout from Fermi National Accelerator Laboratory. The polycrystalline SRF sample has a large average grain size of about 50 μm, compared to 44–69 nm for sputter-coated samples. Detailed sample information and growth methods are described elsewhere [32,33]. The SRF cavity cutout sample was mechanically polished for optical measurement. To avoid oxidation, the sample was polished with oil-based diamond powders, rinsed with isopropyl alcohol, and then dried with nitrogen gas. Sputter-coated thin-film samples were cut to 10 × 10 mm² with a diamond pen from a 3-inch wafer. All samples were kept in a dry box before mounting in a vacuum cryostat to avoid contamination and moisture absorption. The thermal conductivity of Nb itself is ~100 kW/m·K, much higher than Si ~ 80 W/m·K near T_c.

Table 1. Summary of Nb samples used in this paper. HiPIMS, DC high, and DC LH samples are sputter-coated on intrinsic Si wafers. Bulk SRF sample is cut out from SRF cavity.

Sample	Fabrication Method	Thickness	Average Grain Size	Substrate
HiPIMS	HiPIMS sputter	175 nm	44 nm	Si (001)
DC high	DC sputter (high power)	175 nm	69 nm	Si (001)
DC LH	DC sputter (low to high power)	175 nm (30 nm low, then 145 nm high)	65 nm	Si (001)
SRF	cavity cutout	2 mm	50 μm	bulk

The quasiparticle density and decay dynamics induced by ultrafast excitation are influenced by intrinsic processes, such as hot phonons and the photon bottleneck effect, as well as extrinsic processes, including pinning and pair breaking caused by defects and grain boundaries. In the studied multi-crystallized samples, extrinsic processes dominate. As shown in Table 1, grain size, defective boundaries, and other impurities are strongly correlated with the growth methods, significantly impacting the superconducting and transport properties of the samples, such as T_c, RRR values, and grain topology. Consequently, the thermalized QP density and decay dynamics are closely linked to the growth methods, which produce distinctly different grain boundaries and pinning centers. Additionally, in the mixed state of type II superconductors, vortex cores formed under a magnetic field can trap QPs and influence their relaxation dynamics. The formation and pinning of these vortices are strongly influenced by grain boundaries and impurities, underscoring the pivotal role of growth methods in shaping QP density and dynamics. These effects will be investigated in detail below using ultrafast spectroscopy.

Our ultrafast pump–probe measurements were performed using a femtosecond pulsed laser with a 7 Tesla dry cryostat. The experimental setup employed a normal reflection geometry for pump–probe spectroscopy [34]. A 1250 nm laser output was split into two amplified arms via optical fibers, generating 20 fs light pulses at 1500 nm. One arm served as the pump beam, while the other functioned as the probe beam at 750 nm, generated through second-harmonic generation. The pump and probe beams were overlapped using a dichroic beam combiner and focused with a 100 mm focal length lens in a collinear geometry. Samples were mounted facing the top window in the cryostat, with the magnetic field applied perpendicular to the sample surface.

A silicon balanced detector paired with a lock-in amplifier was used to detect reflectivity changes, with a 40 kHz mechanical chopper placed in the pump beam path. The temporal overlap of the pump and probe beams was controlled via a motorized linear delay stage in the pump path. The reflected probe signal was directed to the balanced detector, with a reference beam of equal optical path length focused onto the detector's reference channel to cancel out noise. A 900 nm short-pass filter was used to block the 1500 nm pump beam entirely. The laser's focal point diameter was approximately 100 μm, with pump fluences ranging from 1 to 7 μJ/cm² for fluence dependence and the probe fluence kept at around ~0.3 μJ/cm² to ensure minimal disturbance. A common pump fluence of 3 μJ/cm² was used for most measurements to achieve both a high signal-to-noise ratio and operation within the weak perturbation regime. The optical skin depth of Nb, calculated as $\delta = \lambda/4\pi k$, was found to be 19 nm at 750 nm and 15 nm at 1.5 µm based on optical constants [35], as shown in Appendix A Figure A1. The analysis of QP dynamics under varying magnetic fields, along with a comprehensive comparison of the ultrafast superconducting behavior across different Nb samples, will be discussed in detail in the following sections. Additionally, the thin-film sample's residual resistivity ratio (RRR = R(290 K)/R(10 K)) and the power-dependent quality factor Q_i were measured using fabricated Nb resonators at the qubit operating frequency of 5 GHz.

3. Results and Discussion

Figure 1b,c present the representative pump–probe responses of the photoinduced reflectivity change, $\Delta R/R$, at 2.3 K, 6 K, and 8 K in the superconducting (SC) state, as well as in the normal state above the T_c of the Nb cavity cutout sample. Both $\Delta R/R$ signals exhibit a sharp rise time of approximately 3 ps, followed by a slower decay lasting over 300 ps. The rapid increase is attributed to QP generation from initial pair breaking by the laser pulse, followed by electron–phonon scattering, while the slower decay is mainly due to high-frequency phonon relaxation and thermal diffusion. Thermal diffusion primarily occurs vertically rather than laterally, as the laser focal spot, measuring tens of micrometers, is significantly larger than the optical skin depth (~15 nm). In Figure 1b, the $\Delta R/R$ signal includes both the photoinduced QP and HFP generation and the thermalization process, as well as the vertical thermal diffusion. The latter components of $\Delta R/R$ shown in Figure 1c exhibit minimal temperature dependence for small temperature changes ($\leq$10 K). Subtracting $\Delta R/R$ from the average normal-state $\Delta R/R$ isolates the photoinduced QP dynamics in the SC state $\Delta R/R_{SC}$, as shown in Figure 1d. This $\Delta R/R_{SC}$ component, absent above the critical temperature, shows both a slower rise time of ~5 ps and a slower decay compared to the dynamics observed prior to subtraction. Our data clearly demonstrate the capability to independently measure the dynamics of quasiparticle generation/relaxation from the SC condensate and hot-phonon thermal diffusion processes. This methodology was subsequently extended to investigate additional Nb samples fabricated through various techniques.

A theoretical QP dynamic model was developed for the non-equilibrium states measured in Figure 1d after laser excitation [19,20,25,26]. Rothwarf and Taylor (R-T) proposed two coupled equations to relate the QP recombination and non-negligible high-frequency phonons created by the initial energetic QP during laser excitation. The coupled photoinduced HFP pair breaking and bi-molecular QP recombination process governs the long QP decay dynamics in the phonon bottleneck regime [19], as shown in Figure 1d. In the R-T model, phonons generated from quasiparticle recombination have a high likelihood of being absorbed in a subsequent pair-breaking process. This high-frequency phonon can then break a Cooper pair, creating two additional QPs, as illustrated in Figure 1a. The recovery of superconductivity results from the decay of photoinduced non-equilibrium QPs. This hot-phonon-mediated pair-breaking and recombination process has a longer lifetime than the intrinsic hot-electron decay. QP dynamics also require a detailed balance between phonon generation, decay, and the phonon-driven pair-breaking process [25]. Since the laser is focused on a small area of the sample, hot-phonon diffusion should also be considered to analyze the $\Delta R/R_{SC}$ component for non-equilibrium QP dynamics in SC states.

Next, we present the ultrafast dynamics of quasiparticles and phonons in thin-film Nb resonators. Figure 2a,c,e show temperature-dependent $\Delta R/R$, and Figure 2b,d,f show subtracted superconducting components $\Delta R/R_{SC}$ from normal-state average traces of DC high-power, DC LH-power, and HiPIMS Nb samples. We emphasize two key points. First, the SRF cavity sample exhibits significantly better thermal diffusion compared to the thin-film samples, which accounts for the faster QP decay times (Figure 1d) observed in the Nb cavity cutout sample compared to Nb thin films. The SRF cavity sample is polycrystalline with an average grain size of 50 μm, whereas the HiPIMS Nb thin film has a grain size of ~10 s of nm. Smaller grains lead to increased grain boundaries and defects, which act as QP pinning centers that slow recombination. Moreover, HFP diffusion within the optical depth is significantly more efficient in the SRF cavity sample due to its larger grain size. Second, the photoinduced superconducting $\Delta R/R_{SC}$ signals in Figure 2b,d,f approach zero as the temperature nears T_C. The signal amplitude directly correlates with QP generation, while the time dependence reflects the QP relaxation dynamics. The temperature-dependent thermal-equilibrium QP density n_T is estimated from $n_T(T) \propto Q(T \to 0 \text{ K})/Q(T) - 1$, where $Q(T)$ is the peak intensity in $\Delta R/R_{SC}$ in Figure 2b,d,f. $Q(T \to 0 \text{ K})$ is obtained from the lowest temperature value at 2.2 K. The thermal-equilibrium QP density n_T measured in Figure 2g agrees well with Equation (1) [20],

$$n_T(T) = n(0)\sqrt{2n\Delta_{SC} k_B T}\, e^{-\Delta_{SC}/k_B T}, \tag{1}$$

where $n(0)$ is the electronic density of states in the unit cell and $2\Delta_{SC}$ is the superconducting energy gap. And, the phonon density N_T at thermal equilibrium is shown in Equation (2) [20].

$$N_T(T) = (36\nu \Delta_{SC}^2 T)/\omega_D^3 \, e^{-2\Delta_{SC}/k_B T}, \tag{2}$$

where ω_D is the Debye energy, ν is the number of atoms in the unit cell. The initial photoinduced QP density Δn and excited phonon density ΔN are small compared to n_T and N_T in the weak perturbation limit, i.e., $\Delta n/n_s \ll 1$.

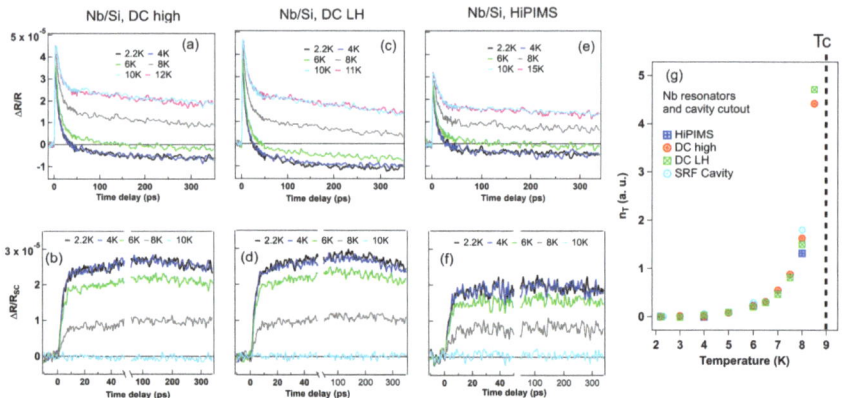

Figure 2. Temperature-dependent photoinduced reflectivity change $\Delta R/R$ of Nb thin films and QP density of Nb samples. (**a,c,e**) Temperature-dependent $\Delta R/R$ of Nb thin-film samples from T = 2.2 K to T = 15 K for top figures. (**b,d,f**) Superconducting state contributions in $\Delta R/R_{SC}$ components subtracted from average normal stage values for bottom figures. (**g**) Temperature-dependent equilibrium QP densities of HiPIMS, DC high power, DC LH power, SRF cavity samples. Pump fluence is set to be 3.0 μJ/cm².

Because the incident photon energy ~1 eV is much larger than the SC gap 2Δ ~3 meV, the initial energetic QP after laser excitation on the sample generates an HFP. After energetic QP dissipation in a few ps, QP generation and recombination are driven by balanced states between HFP pair breaking and QP recombination, as shown in Figure 1a's schematic, i.e., the strong phonon bottleneck region. Thermal-equilibrium QP densities, n_T, derived from measurements, exhibit an exponential increase with temperature, as shown in Figure 2g. Additionally, in the SC state, the high-frequency phonon density dissipates heat through electron–phonon interactions and diffusion. The pump–probe signal, $\Delta R/R$, in SC states includes temperature-independent contributions below T_c, such as rapid thermal and carrier diffusion [30]. The SC contribution, $\Delta R/R_{SC}$, is again isolated by subtracting the normal-state signal measured above T_c, which is proportional to $Q(T)$. As shown in Appendix A Figure A2, this quantity diminishes as the temperature approaches T_c.

Pump–probe experiments reveal that thin-film samples with smaller grain sizes exhibit higher quasiparticle densities, longer QP relaxation times back to the condensate, and lower Q factors. These effects can be attributed to quasiparticles becoming trapped in grain boundaries and defect centers, which inhibit the reformation of the condensate state. Samples with smaller grains inherently have larger grain boundary areas and potentially a higher density of defects. These trapped quasiparticles undergo a delayed recombination process back into the condensate, thereby extending the relaxation time.

Figure 3a,b show the magnetic field dependence of $\Delta R/R$ on the Nb SRF cavity at 2.2 K and 10 K. The results demonstrate a strong magnetic field dependence of $\Delta R/R$ at 2.2 K, while no clear field dependence is observed at 10 K. The extracted SC component $\Delta R/R_{SC}$ of the SRF cavity reveals a distinct magnetic field dependence in QP generation, with the majority of Cooper pairs breaking at B > 700 mT, as shown in Figure 3c. The $\Delta R/R$ behavior of the HiPIMS sample, as shown in Figure 3d,e, resembles that of the SRF cavity samples. The subtracted superconducting component, $\Delta R/R_{SC}$, of the HiPIMS sample also displays magnetic-field-dependent QP dynamics, though a majority of Cooper pairs break at a lower magnetic field of B∼600 mT, as shown in Figure 3f. Intriguingly, there is a clear difference in the $\Delta R/R_{SC}$ components under the external magnetic field between the SRF cavity (Figure 3e) and Fe resonator (Figure 3f) samples. First, HiPIMS has a much longer relaxation time and also requires a lower magnetic field to break the majority of Cooper pairs, as shown in Figure 3f, compared to the cavity cutout sample in Figure 3c. Other DC sputter samples show similar trends to HiPIMS. Second, the thermalized QP densities $n_T(B)$s extracted from photoinduced $\Delta R/R_{SC}$ signals are plotted together for HIPIMS, DC high, DC LH, and the SRF cavity, as shown in Figure 3g with guided lines. The SRF cavity sample exhibits the lowest quasiparticle population and the most robust superconducting properties, showing strong resistance to external magnetic fields. Notably, a distinct threshold magnetic field is required to induce a significant quasiparticle population. The quasiparticle density generation behavior in the SRF cavity sample is succeeded in resilience by the DC high, DC LH, and HiPIMS samples, exhibiting progressively higher densities in that order. Moreover, the QP signals of these thin-film samples decrease steadily from 0 to 400 mT without a threshold, whereas the SRF cavity sample shows an uneven reduction in QP signals, generating QPs only at sufficiently high fields. Among the thin-film samples, the magnetic-field-induced quasiparticle density exhibits an increasingly steep slope in the order of DC high, DC LH, and HiPIMS. These findings underscore the SRF cavity sample's superior superconducting properties and high tolerance to pair-breaking perturbations, followed by the DC high, DC LH, and HiPIMS samples, making them progressively suitable for applications requiring highly coherent quantum devices.

We can attribute these differences to the grain heterogeneity and magnetic field penetration into the sample. The SRF cavity requires a stronger magnetic field to break SC states due to the large grains that are more robust in external magnetic field penetration. The low QP density in $n_T(B)$ indicates that a strong magnetic field is needed to penetrate the sample. Additionally, DC samples tend to produce films with larger grain sizes compared to HiPIMS ones. This difference in our observations arises due to the distinct deposition processes and energy distributions in the two techniques [32]. In general, large-grained samples inherently have fewer grain boundaries as flux pinning centers. Thus, it is expected that a magnetic field can easily penetrate the HiPIMS sample. Therefore, among sputter-coated samples, the DC high sample is robust in magnetic field penetration. Assuming that thermal diffusion in sputter samples is not much different due to the similar overall thickness of the Nb and interface layers, one expects that the magnetic field can penetrate more easily and break more QPs in small-grain thin films.

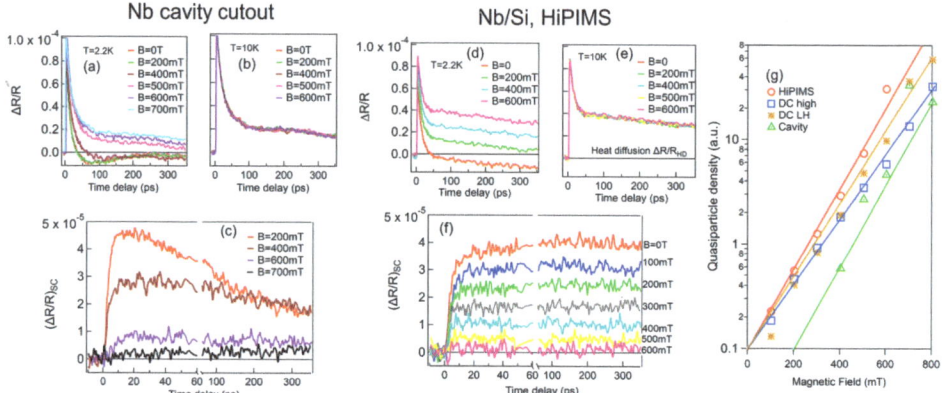

Figure 3. Magnetic-field-dependent $\Delta R/R$ of Nb samples. (**left**) Nb SRF cavity cutout $\Delta R/R$: (**a**) low temperature at T = 2.2 K, (**b**) normal state at T = 10 K, (**c**) SC component $\Delta R/R_{SC}$ subtracted from average 10 K value. (**middle**) Thin-film Nb HiPIMS $\Delta R/R$: (**d**) low temperature at T = 2.2 K, (**e**) normal state at T = 10 K, (**f**) $\Delta R/R_{SC}$ subtraction from average 10 K value. (**g**) Magnetic-field-dependent thermal-equilibrium QP densities $n_T(B)$ of HiPIMS, DC high, DC LH, and SRF cavity samples. Pump fluence is set to be 3.0 μJ/cm^2.

Finally, the fs $\Delta R/R_{SC}$ dynamics under an applied magnetic field shown in Figure 3g reveals a clear correlation between non-equilibrium QPs and the quality factor of resonators fabricated by using different deposition methods. The thin-film Nb RRR = R(290 K)/R(10 K) and Nb resonator quality factor Q_i are measured and presented in Figure 4. The Nb thin-film RRR values are 4.55 for HiPIMS, 6.66 for DC LH, and 6.53 for DC high. The low RRR value of HiPIMS is consistent with a smaller grain size compared to DC samples. Loss tangents $tan(\delta)$ are also obtained with $tan(\delta) = 1/Q_i$ from measured Q_i. Figure 4 shows the loss tangent $tan(\delta)$ and Q_i of RF resonators with different deposition methods. In Figure 4a–c, different colors represent different device power scans on the same samples. The power-dependent scans show high loss tangents at low power in general. The box plots in Figure 4d,e show median values for different scans. From the loss tangent box plot, HiPIMS has the highest loss, followed by DC LH and DC high. The median values of Q_i are 0.1712 × 10^6 for HiPIMS, 0.3064 × 10^6 for DC LH, and 1.141 × 10^6 for DC high. Because the Q_i values were measured with the resonator only, the estimated upper limits T_1s for the qubit device are 13.86 μs for HiPIMS, 26.15 μs for DC LH, and 55.80 μs for DC high from $T_1 = Q_i/(2\pi f)$ with an f = 5 GHz qubit operating frequency. These values align with the $\Delta R/R_{SC}$ results in Figure 3g, indicating that a smaller QP population and greater resistance to magnetic fields correlate with a higher Q_i for Nb film samples. The Nb cavity cutout sample shows a significantly higher SRF cavity Q factor, consistent with this conclusion.

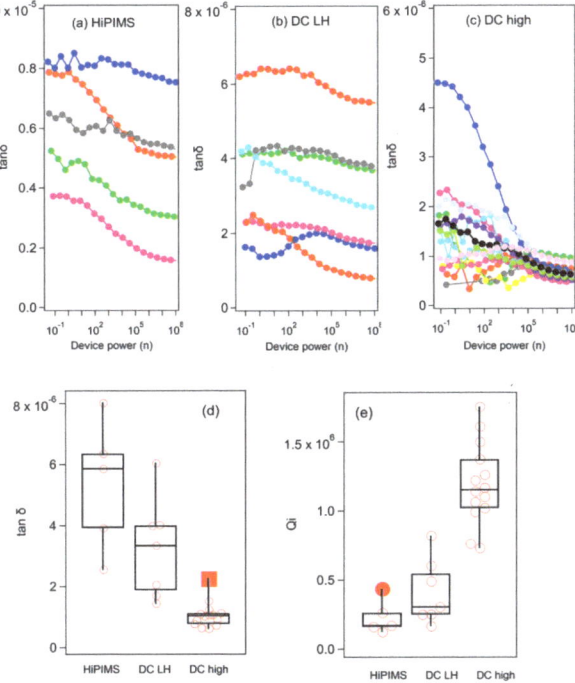

Figure 4. Power-dependent microwave characterization of Nb thin-film resonators. Loss tangent $tan(\delta)$ power spectra for three selected resonators made with (**a**) HiPIMS, (**b**) DC LH, (**c**) DC high samples. The device power (n) is converted to no. of photons operating at 5 GHz frequency. Different color traces are from different measurement scans. (**d**) Box plot of averaged loss tangent and (**e**) averaged internal Q_i of Nb thin-film resonators. Each point in box plots is median value for one measurement, showing variation between different measurement sweeps. Filled points are outlier values far from median values.

4. Conclusions

We measured femtosecond-resolved transient reflectivity in Nb thin-film resonators and SRF cavity cutout samples, examining non-equilibrium QP dynamics in the superconducting states across various temperatures and magnetic fields. The thermalized QP density and QP decay dynamics below the critical temperature display notably different behaviors based on the sample growth methods. Distinct differences in QP lifetime, density, and thermal diffusion among Nb samples are attributed to grain boundaries and defects.

The thermal diffusion contrasts significantly between thin films and bulk polycrystalline SRF cavities, with the SRF cavity exhibiting more efficient heat conduction. Magnetic-field-dependent measurements reveal clear behavioral distinctions between the two sample types, with the SRF sample showing stronger superconductivity against defects and faster QP relaxation characteristics that are highly favorable for applications requiring coherent quantum devices.

Within the thin-film samples, the DC high sample shows the lowest loss tangent in RF resonators, followed by DC LH, while HiPIMS has the highest loss tangent, which we attribute to grain boundary effects and grain size variations from different deposition methods. Magnetic-field-dependent thermal-equilibrium QP density measurements are consistent with this trend.

Our study further suggests that, in addition to a larger grain size and fewer defects to reduce TLS losses and QP pinning, Nb thin-film resonators fabricated on high-conductivity substrates such as sapphire could enhance transmon qubit performance. These findings emphasize the critical impact of fabrication techniques on the coherence and performance of Nb-based quantum devices, essential for applications in superconducting qubits and high-energy superconducting radio-frequency systems.

Finally, our results warrant a quantitative investigation of non-thermal pair breaking under a magnetic field, which could provide deeper insights into the quantum dynamics of superconducting vortex states.

Author Contributions: Investigation, J.-M.P., Z.X.C., R.H.J.K., S.H., R.C., A.A.M., C.J.K., J.M., D.S., E.L., J.Y.M., K.Y., A.G. and A.R.; Resources, A.A.M., C.J.K., J.M., D.S., E.L., J.Y.M., K.Y., A.G. and A.R.; Data curation, J.-M.P., Z.X.C., R.H.J.K., S.H. and R.C.; Writing—original draft, J.-M.P.; Writing—review & editing, J.W.; Supervision, J.Y.M. and J.W. All authors have read and agreed to the published version of the manuscript.

Funding: This work was supported by the U.S. Department of Energy, Office of Science, National Quantum Information Science Research Centers, Superconducting Quantum Materials and Systems Center (SQMS), under Contract No. DE-AC02-07CH11359. The Ames Laboratory is operated for the U.S. Department of Energy by Iowa State University under Contract No. DE-AC02-07CH11358.

Institutional Review Board Statement: Not applicable.

Informed Consent Statement: Not applicable.

Data Availability Statement: The original contributions presented in this study are included in the article. Further inquiries can be directed to the corresponding author.

Acknowledgments: We thank the Rigetti fab team members for the processing development and fabrication of studied specimens.

Conflicts of Interest: Authors Cameron J. Kopas, Jayss Marshall, Daniel Setiawan, Ella Lachman, Joshua Y. Mutus and Kameshwar Yadavalli were employed by Rigetti Computing. The remaining authors declare that the research was conducted in the absence of any commercial or financial relationships that could be construed as a potential conflict of interest.

Abbreviations

The following abbreviations are used in this manuscript:

Al	aluminum
fs	femtosecond
HiPIMS	high-power impulse magnetron sputtering
HFP	high-frequency phonon
JJ	Josephson junction
Nb	niobium
QP	quasiparticle
R-T	Rothwarf and Taylor
SC	superconducting
SRF	superconducting radio frequency
TLS	two-level system

Appendix A

Appendix A.1

Skin depth $\delta = \lambda/(4\pi k)$ of Nb is plotted in Figure A1. Complex optical constants $n + ik$ are from Palik's handbook.

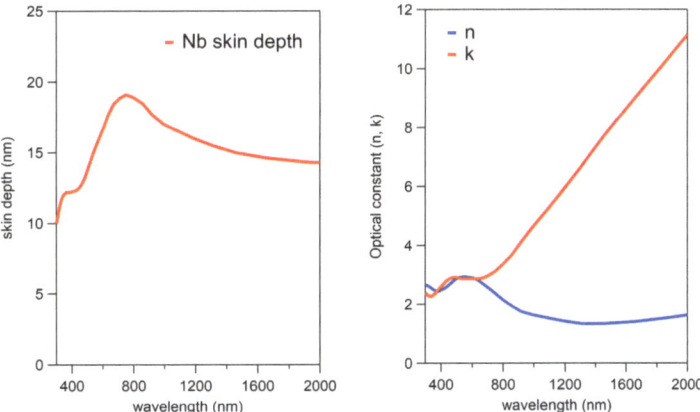

Figure A1. (**left**) Optical skin depth of Nb with wavelength. (**right**) Real and imaginary parts of optical constant $n + ik$ of Nb.

Appendix A.2

The photoinduced QP density of the superconducting component, Q(T), of the Nb DC low–high sample is plotted in Figure A2 with a pump fluence of 3 µJ/cm^2. Q(T) is the peak value obtained from $\Delta R/R$ in the superconducting state by subtracting from the average normal-state values measured at temperatures (T > T_c). Q(T) approaches 0 as the temperature gets closer to T_c. The overall shape of Q(T) indicates that the pump fluence is a weak perturbation region and the photoinduced QP density is smaller than the thermal-equilibrium QP density.

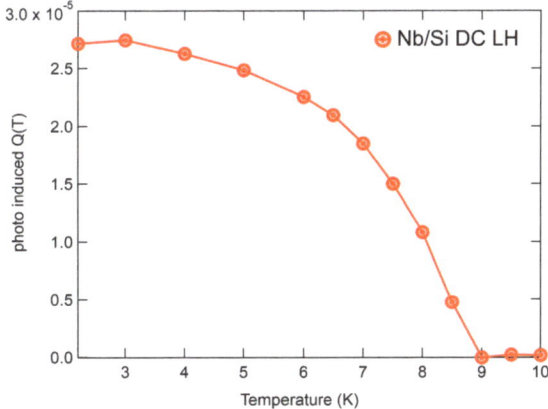

Figure A2. Peak value of photoinduced QP density Q(T) of Nb DC low–high sample.

References

1. Siddiqi, I. Engineering high-coherence superconducting qubits. *Nat. Rev. Mater.* **2021**, *6*, 875–891. [CrossRef]
2. Krantz, P.; Kjaergaard, M.; Yan, F.; Orlando, T.P.; Gustavsson, S.; Oliver, W.D. A quantum engineer's guide to superconducting qubits. *Appl. Phys. Rev.* **2019**, *6*, 021318. [CrossRef]
3. Kjaergaard, M.; Schwartz, M.E.; Braumüller, J.; Krantz, P.; Wang, J.I.; Gustavsson, S.; Oliver, W.D. Superconducting Qubits: Current State of Play. *Annu. Rev. Condens. Matter. Phys.* **2020**, *11*, 369–95. [CrossRef]
4. Premkumar, A.; Weil, C.; Hwang, S.; Jäck, B.; Place, A.P.; Waluyo, I.; Hunt, A.; Bisogni, V.; Pelliciari, J.; Barbour, A.; et al. Microscopic relaxation channels in materials for superconducting qubits. *Commun. Mater.* **2021**, *2*, 72. [CrossRef]

5. Blais, A.; Huang, R.S.; Wallraff, A.; Girvin, S.M.; Schoelkopf, R.J. Cavity quantum electrodynamics for superconducting electrical circuits: An architecture for quantum computation. *Phys. Rev. A* **2004**, *69*, 062320. [CrossRef]
6. Oliver, W.D.; Welander, P.B. Materials in superconducting quantum bits. *MRS Bull.* **2013**, *38*, 816–825. [CrossRef]
7. Blais, A.; Grimsmo, A.L.; Girvin, S.M.; Wallraff, A. Circuit quantum electrodynamics. *Rev. Mod. Phys.* **2021**, *93*, 025005. [CrossRef]
8. Wallraff, A.; Schuster, D.I.; Blais, A.; Frunzio, L.; Huang, R.S.; Majer, J.; Kumar, S.; Girvin, S.M.; Schoelkopf, R.J. Strong coupling of a single photon to a superconducting qubit using circuit quantum electrodynamics. *Nature* **2004**, *431*, 162–167. [CrossRef] [PubMed]
9. Harrelson, T.F.; Sheridan, E.; Kennedy, E.; Vinson, J.; N'Diaye, A.T.; Altoé, M.V.P.; Schwartzberg, A.; Siddiqi, I.; Ogletree, D.F.; Scott, M.C.; et al. Elucidating the local atomic and electronic structure of amorphous oxidized superconducting niobium films. *Appl. Phys. Lett.* **2021**, *119*, 244004. [CrossRef]
10. Carroll, M.; Rosenblatt, S.; Jurcevic, P.; Lauer, I.; Kandala, A. Dynamics of superconducting qubit relaxation times. *npj Quantum Inf.* **2022**, *8*, 132. [CrossRef]
11. Zhang, E.J.; Srinivasan, S.; Sundaresan, N.; Bogorin, D.F.; Martin, Y.; Hertzberg, J.B.; Timmerwilke, J.; Pritchett, E.J.; Yau, J.B.; Wang, C.; et al. High-performance superconducting quantum processors via laser annealing of transmon qubit. *Sci. Adv.* **2022**, *8*, eabi6690. [CrossRef] [PubMed]
12. Bravyi, S.; Dial, O.; Gambetta, J.M.; Gil, D.; Nazario, Z. The future of quantum computing with superconducting qubits. *J. Appl. Phys.* **2022**, *132*, 160902. [CrossRef]
13. Smirnov, N.S.; Krivko, E.A.; Solovyova, A.A.; Ivanov, A.I.; Rodionov, I.A. Wiring surface loss of a superconducting transmon qubit. *Sci. Rep.* **2024**, *14*, 7326. [CrossRef] [PubMed]
14. Verjauw, J.; Acharya, R.; Van Damme, J.; Ivanov, T.; Lozano, D.P.; Mohiyaddin, F.A.; Wan, D.; Jussot, J.; Vadiraj, A.M.; Mongillo, M.; et al. Path toward manufacturable superconducting qubits with relaxation times exceeding 0.1 ms. *npj Quantum Inf.* **2022**, *8*, 93. [CrossRef]
15. De Graaf, S.E.; Faoro, L.; Ioffe, L.B.; Mahashabde, S.; Burnett, J.J.; Lindström, T.; Kubatkin, S.E.; Danilov, A.V.; Tzalenchuk, A.Y. Two-level systems in superconducting quantum devices due to trapped quasiparticles. *Sci. Adv.* **2020**, *6*, eabc5055. [CrossRef]
16. Müller, C.; Cole, J.H.; Lisenfeld, J. Towards understanding two-level-systems in amorphous solids: Insights from quantum circuits. *Rep. Prog. Phys.* **2019**, *82*, 124501. [CrossRef] [PubMed]
17. Ristè, D.; Bultink, C.C.; Tiggelman, M.J.; Schouten, R.N.; Lehnert, K.W.; DiCarlo, L. Millisecond charge-parity fluctuations and induced decoherence in a superconducting transmon qubit. *Nat. Commun.* **2013**, *4*, 1913. [CrossRef]
18. Vepsäläinen, A.P.; Karamlou, A.H.; Orrell, J.L.; Dogra, A.S.; Loer, B.; Vasconcelos, F.; Kim, D.K.; Melville, A.J.; Niedzielski, B.M.; Yoder, J.L.; et al. Impact of ionizing radiation on superconducting qubit coherence. *Nature* **2020**, *584*, 551–556. [CrossRef] [PubMed]
19. Rothwarf, A.; Taylor, B.N. Measurement of Recombination Lifetimes in Superconductors. *Phys. Rev. Lett.* **1967**, *19*, 27. [CrossRef]
20. Kabanov, V.V.; Demsar, J.; Mihailovic, D. Kinetics of a Superconductor Excited with a Femtosecond Optical Pulse. *Phys. Rev. Lett.* **2005**, *95*, 147002. [CrossRef]
21. Stojchevska, L.; Kusar, P.; Mertelj, T.; Kabanov, V.V.; Toda, Y.; Yao, X.; Mihailovic, D. Mechanisms of nonthermal destruction of the superconducting state and melting of the charge-density-wave state by femtosecond laser pulses. *Phys. Rev. B* **2011**, *84*, 180507(R). [CrossRef]
22. Federici, J.F.; Greene, B.I.; Saeta, P.N.; Dykaar, D.R.; Sharifi, F.; Dynes, R.C. Direct picosecond measurement of photoinduced Cooper-pair breaking in lead. *Phys. Rev. B* **1992**, *46*, 11153. [CrossRef] [PubMed]
23. Yang, X.; Zhao, X.; Vaswani, C.; Sundahl, C.; Song, B.; Yao, Y.; Cheng, D.; Liu, Z.; Orth, P.P.; Mootz, M.; et al. Ultrafast nonthermal terahertz electrodynamics and possible quantum energy transfer in the Nb3Sn superconductor. *Phys. Rev. B* **2019**, *99*, 094504. [CrossRef]
24. Cheng, B.; Cheng, D.; Lee, K.; Luo, L.; Chen, Z.; Lee, Y.; Wang, B.Y.; Mootz, M.; Perakis, I.E.; Shen, Z.X.; et al. Evidence for d–wave superconductivity of infinite-layer nickelates from low-energy electrodynamics. *Nat. Mater.* **2024**, *23*, 775. [CrossRef] [PubMed]
25. Rothwarf, A.; Sai-Halasz, G.A.; Langenberg, D.N. Quasiparticle Lifetimes and Microwave Response in Nonequilibrium Superconductors. *Phys. Rev. Lett.* **1974**, *23*, 212. [CrossRef]
26. Chang, J.-J.; Scalapino, D.J. Kinetic-equation approach to nonequilibrium superconductivity. *Phys. Rev. B* **1977**, *15*, 2651. [CrossRef]
27. Gedik, N.; Blake, P.; Spitzer, R.C.; Orenstein, J.; Liang, R.; Bonn, D.A.; Hardy, W.N. Single-quasiparticle stability and quasiparticle-pair decay in YBa2Cu3O6.5. *Phys. Rev. B* **2004**, *70*, 014504. [CrossRef]
28. Hinton, J.P.; Thewalt, E.; Alpichshev, Z.; Mahmood, F.; Koralek, J.D.; Chan, M.K.; Veit, M.J.; Dorow, C.J.; Barišić, N.; Kemper, A.F.; et al. The rate of quasiparticle recombination probes the onset of coherence in cuprate superconductors. *Sci. Rep.* **2016**, *6*, 25962. [CrossRef]
29. Chia, E.E.M.; Talbayev, D.; Zhu, Ji.; Yuan, H.Q.; Park, T.; Thompson, J.D.; Panagopoulos, C.; Chen, G.F.; Luo, J.L.; Wang, N.L.; et al. Ultrafast Pump-Probe Study of Phase Separation and Competing Orders in the Underdoped (Ba,K)Fe2As2 Superconductor. *Phys. Rev. Lett.* **2010**, *104*, 027003. [CrossRef]
30. Lin, K.H.; Wang, K.J.; Chang, C.C.; Wen, Y.C.; Lv, B.; Chu, C.W.; Wu, M.K. Ultrafast dynamics of quasiparticles and coherent acoustic phonons in slightly underdoped (BaK)Fe2As2. *Sci. Rep.* **2016**, *6*, 25962. [CrossRef]

31. Yang, X.; Luo, L.; Mootz, M.; Patz, A.; Bud'ko, S.L.; Canfield, P.C.; Perakis, I.E.; Wang, J. Nonequilibrium Pair Breaking in Ba(Fe1-xCox)2As2 Superconductors: Evidence for Formation of a Excitonic State. *Phys. Rev. Lett.* **2018**, *121*, 267001. [CrossRef] [PubMed]
32. Oh, J.S.; Kopas, C.J.; Marshall, J.; Fang, X.; Joshi, K.R.; Datta, A.; Ghimire, S.; Park, J.M.; Kim, R.; Setiawan, D.; et al. Correlating Deposition Method, Microstructure, and Performance of Nb/Si-based Superconducting Coplanar Waveguide Resonators. *Acta Mater.* **2024**, *276*, 120153. [CrossRef]
33. Checchin, M.; Martinello, M.; Grassellino, A.; Aderhold, S.; Chandrasekaran, S.K.; Melnychuk, O.S.; Posen, S.; Romanenko, A.; Sergatskov, D.A. Frequency dependence of trapped flux sensitivity in SRF cavities. *Appl. Phys. Lett.* **2018**, *112*, 072601. [CrossRef]
34. Cheng, D.; Song, B.; Kang, J.-H.; Sundahl, C.; Edgeton, A.L.; Luo, L.; Park, J.-M.; Collantes, Y.G.; Hellstrom, E.E.; Mootz, M.; et al. Study of Elastic and Structural Properties of BaFe2As2 Ultrathin Film Using Picosecond Ultrasonics. *Materials* **2023**, *16*, 7031. [CrossRef]
35. Palik, E.D. (Ed.) *Handbook of Optical Constants of Solids*; Academic Press: Cambridge, MA, USA, 1991.

Disclaimer/Publisher's Note: The statements, opinions and data contained in all publications are solely those of the individual author(s) and contributor(s) and not of MDPI and/or the editor(s). MDPI and/or the editor(s) disclaim responsibility for any injury to people or property resulting from any ideas, methods, instructions or products referred to in the content.

Article

Dye-Sensitized Solar Cells with Modified TiO$_2$ Scattering Layer Produced by Hydrothermal Method

Yu-Shyan Lin [1,2,*] and Wei-Hung Chen [1]

[1] Department of Materials Science and Engineering, National Dong Hwa University, Hualien 974301, Taiwan
[2] Department of Opto-Electronic Engineering, National Dong Hwa University, Hualien 974301, Taiwan
* Correspondence: yslindh@gms.ndhu.edu.tw; Tel.: +886-3-8903218

Abstract: This work proposes dye-sensitized solar cells (DSSCs) with various photoanode designs. A hydrothermal method is used to synthesize hydrangea-shaped TiO$_2$ (H-TiO$_2$) aggregates. The X-ray diffraction (XRD) pattern of H-TiO$_2$ reveals only an anatase phase. No peaks of any other phases are detected, indicating that the hydrangea-shaped TiO$_2$ is phase-pure. The size of the synthesized H-TiO$_2$ is approximately 300 nm to 2 μm, and its particle size is suitable for use in the scattering layer of a DSSC. Mixing the P25 TiO$_2$ into the H-TiO$_2$ aggregate with the best mixing ratio can significantly improve the conversion efficiency of DSSCs. When the ratio of H-TiO$_2$:P25 TiO$_2$ = 3:7, the scattering layer has the optimal parameters, as determined experimentally. The optimal structure is a double layer that is formed of five layers of P25 TiO$_2$ plus a single scattering layer. An open circuit voltage (V_{oc}) of 0.77 V, short-circuit current (J_{sc}) of 15.26 mA/cm^2, fill factor (FF) of 0.71, conversion efficiency (η) of 8.33%, and charge-collection efficiency (η_{cc}) of 0.96 are obtained from the optimally designed photoelectrode. To the best of the authors' knowledge, this work is the first in which large particles of hydrangea are mixed with small particles of P25 TiO$_2$ in various proportions to form a scattering layer.

Keywords: DSSC; hydrothermal; scattering; hydrangea; aggregate

Academic Editors: Jijun Feng and Shengli Pu

Received: 7 December 2024
Revised: 26 December 2024
Accepted: 28 December 2024
Published: 9 January 2025

Citation: Lin, Y.-S.; Chen, W.-H. Dye-Sensitized Solar Cells with Modified TiO$_2$ Scattering Layer Produced by Hydrothermal Method. *Materials* **2025**, *18*, 278. https://doi.org/10.3390/ma18020278

Copyright: © 2025 by the authors. Licensee MDPI, Basel, Switzerland. This article is an open access article distributed under the terms and conditions of the Creative Commons Attribution (CC BY) license (https://creativecommons.org/licenses/by/4.0/).

1. Introduction

Global warming causes the greenhouse effect and global climate anomalies. In recent years, these issues have attracted substantial attention. The underlying causes are excessive carbon dioxide emissions and pollution. Accordingly, renewable energy is valued and is gradually being developed. The advantage of renewable energy is that it is inexhaustible. It includes solar energy, which is indispensable. Solar energy is abundant. Such advantages make solar energy a green energy that attracts much attention and has led to the establishment of an industry that is highly valued around the world [1,2].

Silicon-based solar cells dominate the commercial market at the present time. However, their manufacturing needs high vacuum equipment, which results in high manufacturing costs.

Compound solar cells can be mainly divided into II–VI and III–V series. Many material systems have been extensively and intensively studied [3–7]. The main materials of the II–VI compound solar cells are CdTe, CuIn$_{1-x}$Ga$_x$Se$_2$ (CIGS), etc. [3–5]. However, due to the high toxicity of CdTe and CIGS, and the limited raw materials required, there are still concerns about environmental protection and sustainability. The main materials of the III–V compound solar cells are GaAs, GaInP, etc. [6,7]. Its conversion efficiency is very high and has been used on space satellites. However, it is difficult to manufacture on a large scale, which is also a shortcoming of this type of solar cell.

A dye-sensitized solar cell (DSSC) is a third-generation solar cell and has the following characteristics: (1) its cost is not high; (2) its manufacturing process is easy, and the cost of the required equipment is not high; (3) it can be produced with a large area; (4) it is less affected by the angle of sunlight and high-temperature environment [8]. These properties make DSSC a competent contender not only in replacing other solar cell technologies but also in building integrated photovoltaic applications.

Titanium dioxide (TiO_2) has been successfully used in metal-oxide-semiconductor high-electron mobility transistors (MOS-HEMTs) owing to its high dielectric constant [9]. Generally, most working electrodes in DSSC are made of TiO_2, mainly because it has the following characteristics: (1) a high specific surface area and roughness factor, (2) porosity, (3) high conductivity, (4) transparency, and (5) high chemical stability [8].

The proper use of readily available solar energy can alleviate the current problems of environmental pollution and global warming. Solar energy is not only a renewable energy but also a clean energy. The several ways to improve the efficiency of dye-sensitized solar cells target include the (1) working electrode, (2) counter electrode, (3) dye, and (4) electrolyte. Light-scattering is a well-known method for boosting the optical absorption of the photoelectrode in a DSSC [10–29]. This research is concerned with improving the photoelectric conversion efficiency of solar cells by focusing on the scattering layer in the photoanode. The manufacturing process is simple, and experimental results reveal success.

Hydrangeas MoS_2 has been successfully synthesized and applied to the counter electrode of a DSSC [30]. Although hydrangea-shaped TiO_2 (H-TiO_2) has been used as a scattering layer [15], the relevant research is quite limited. In this investigation, the scattering layer in the photoanode is modified. The effects of the mixing ratio of H-TiO_2 to P25 TiO_2 on the characteristics of DSSCs are studied comprehensively.

Table 1 compares the performance of our studied DSSC and the previously reported DSSCs with various scattering layers [15–29]. The short-circuit photocurrent density (J_{sc}), open-circuit voltage (V_{oc}), fill factor (FF), and conversion efficiency (η) are listed. The conversion efficiency of our studied DSSC with the 3H7C scattering layer compares favorably to the results of other authors [15–29]. Our experimental results clearly demonstrate that the studied DSSC with an optimally designed photoanode is a candidate for high-performance solar cells.

Table 1. Comparison of the photovoltaic characteristics of our proposed DSSC with those of the previously reported DSSCs.

Light Scattering Layer	J_{sc} (mA/cm^2)	V_{oc} (V)	FF	η (%)	References
H-TiO$_2$:P25 TiO$_2$ = 3:7 (5-layer P25 TiO$_2$)	15.26	0.77	0.71	8.33	This work
Coral-like TiO$_2$	13.28	0.71	0.71	6.7	[15]
Hydrangea-likeTiO$_2$	14.03	0.74	0.72	7.5	[15]
Hollow TiO$_2$ nanoparticles (HTNPs)	16.26	0.68	0.72	8.08	[16]
TiO$_2$ hierarchical micro-spheres and nanobelts	17.86	0.72	0.63	8.08	[17]
TiO$_2$ hollow microspheres (THS) (1 wt%)	12.02	0.69	0.59	5.01	[18]
Anatase TiO$_2$ nanowires with nanoscale whiskers	12.72	0.74	0.63	5.98	[19]
popcorn-like TiO$_2$	13.95	0.77	0.7	7.56	[20]
worms-like TiO$_2$ nanostructures	14.77	0.77	0.62	7.05	[21]
TiO$_2$ microspheres	13.32	0.78	0.63	6.49	[22]
TiO$_2$ nanobelts	16.1	0.69	0.63	7.85	[23]
TiO$_2$ nanoleaf	14.00	0.69	0.53	5.12	[24]
Nanofiber-structured TiO$_2$	12.6	0.7	0.69	6.00	[25]
Flower-like TiO$_2$	16.07	0.65	0.62	6.48	[26]
TiO$_2$/graphene quantum dot (GQD)	14.22	0.69	0.51	5.01	[27]
50-nm ZnO/30-nm ZnO	18.99	0.67	0.46	5.87	[28]
TiO$_2$/7.5% graphene	15.64	0.71	0.63	7.08	[29]

2. Experimental

2.1. Preparation of Working Electrode Paste

In this experiment, commercial Degussa P25 TiO$_2$ ready-made powder (80% anatase, 20% rutile) (UniRegion Bio-Tech, Hsinchu, Taiwan) was used to make the working electrode. To increase the dye adsorption capacity, high-viscosity (30–50 mPs) and low-viscosity (5–15 mPas) ethyl cellulose (Sigma Aldrich, St. Louis, MA, USA) were used to control the viscosity of the paste and increase the porosity after sintering [31]. The boiling point of ethanol is low. In order to avoid cracking during sintering due to the poor heat resistance of the film, α-terpineol (95%) (Showa Chemical Industry Co., Ltd., Tokyo, Japan) was first added as a mixing agent of the paste [31]. Then pure ethanol (>99.8%) (ECHO Chemical Co., Ltd., Miaoli, Taiwan) was added dropwise to dissolve the ethyl cellulose. Finally, a stirring hot plate (CORNING PC-420D, Corning, NY, USA) was used to stir at 650 rpm/85 °C until the ethanol had evaporated.

2.2. Preparation of Scattering Layer Paste

First, hydrangea-shaped TiO$_2$ nano aggregates were prepared. A 40 mL volume of de-ionized (DI) water was added to the beaker. A 0.3958 g mass of ammonium hexafluorotitanate (IV) (99.99%) (Acros Organics, Geel, Belgium) was added to the DI water and stirred with a magnet for 30 min until dissolved. Then, 4.8048 g of urea (CON$_2$H$_4$) (99%) (Showa Chemical Industry Co., Ltd., Tokyo, Japan) was added to the above solution, with continued stirring with a magnet until it had completely dissolved. Finally, 1 mL of polyethylene glycol (PEG-600) (Shimakyu Co., Ltd., Osaka, Japan) was dropped into the solution, which was stirred for 30 min. The PEG could modify the TiO$_2$ surface [32] or it prevented the TiO$_2$ film from cracking when the film was dried [33].

The above mixture was poured into a Teflon cup and put into a hydrothermal autoclave. The accessories for this hydrothermal process were purchased from an agent (Shin, Hualien, Taiwan). The hydrothermal temperature was 180 °C, and this temperature was maintained for 12 h. After the hydrothermal reaction was completed, the mixture was cooled to room temperature. DI water and ethanol were used in the centrifugation step. This latter step was repeated twice. Finally, the mixture was put into an oven (DOS30, DENGYNG INSTRUMENTS CO., LTD, New Taipei City, Taiwan) to dry at 80 °C for 24 h. Thus, H-TiO$_2$ powder was obtained. Figure 1 shows the key process steps in making H-TiO$_2$ powder.

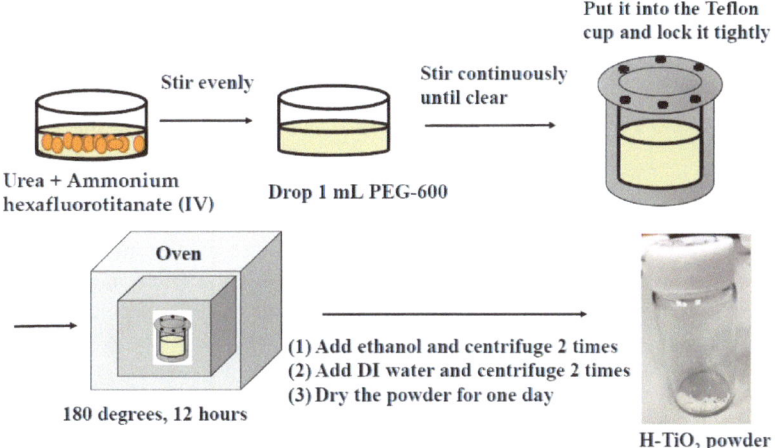

Figure 1. Key processing steps to obtain H-TiO$_2$ powder.

Finally, the H-TiO$_2$ powder was mixed with commercial P25 TiO$_2$ (C-TiO$_2$) powder, as indicated in Table 2, which provides the nomenclature for scattering paste. For example, paste 3H7C contained a mixture of H-TiO$_2$ and commercial P25 TiO$_2$ in a ratio of 3:7. The production process of the scattering layer paste was similar to the process for preparing the working electrode paste. Ethyl cellulose was added to adjust the viscosity and increase the porosity after sintering. α-terpineol was added dropwise as a mixture, and a stirring hot plate was used to stir until the ethanol had evaporated.

Table 2. Ratios of constituents of scattering paste with relevant notation.

Symbols of Scattering Paste	(H-TiO$_2$):Commercial P25 TiO$_2$ Ratio
9H1C	9:1
7H3C	7:3
5H5C	5:5
3H7C	3:7
1H9C	1:9

2.3. DSSC Fabrication

To prevent corrosion by direct contact between the electrolyte and the FTO glass substrate (Ruilong Optoelectronics Co., Ltd., Miaoli, Taiwan), a dense layer of TiO$_2$ nanoparticles had to be formed on the glass substrate [34]. Accordingly, for pretreatment, the substrate was soaked in a solution of 40 mM TiCl$_4$ (99.9%) (Showa Chemical Industry Co., Ltd., Tokyo, Japan) in de-ionized water for 30 min at 70 °C. It was then annealed at 450 °C for 30 min. The dense layer was thus produced by hydrolysis. The hydrolysis reaction is as follows.

$$TiCl_4 + 2H_2O \rightarrow TiO_2 + 4 HCl \tag{1}$$

Then, TiO$_2$ paste was coated on the pretreated FTO glass by screen printing to complete the working electrode. In order to increase the adsorption of dye on the TiO$_2$ thin film, a step similar to TiCl$_4$ pretreatment, called TiCl$_4$ post-treatment, was carried out [35,36].

N719 dye (ECHO Chemical Co., Ltd., Miaoli, Taiwan) was mixed with ethanol to form a 5×10^{-4} M dye solution. The solution was oscillated for 1 h using an ultrasonic cleaner (DC300, DELTA, ULTRASONIC CO., LTD., New Taipei City, Taiwan) to disperse the dye in the ethanol. Finally, the working electrode was soaked in N719 dye and put in a dark environment at 30 °C for 24 h. Acetonitrile (ACN) solution was used as the solvent in the commercial electrolyte solution (Ruilong Optoelectronics Co., Ltd., Miaoli, Taiwan). To inject the electrolyte into the DSSCs, two tiny holes had to be drilled in the FTO glass of the counter electrode. After the FTO glass had been cleaned as described in the previous steps, 3M tape was used to define the working area before it was coated with commercial platinum (Pt) paste (Ruilong Optoelectronics Co., Ltd., Miaoli, Taiwan).

DSSC assembly is the most important part of the process because of the possibility of the leakage of toxic electrolytes. The steps, over which much care must be taken, are as follows: The photoanode and Pt-coated counter-electrodes were assembled into a sealed sandwich-type cell. Efficient sealing was obtained by heating the two electrodes with hot-melt Surlyn (Ruilong Optoelectronics Co., Ltd., Miaoli, Taiwan), which served as a spacer between the electrodes. The electrolyte solution was injected through the pre-drilled holes on the counter-electrode, and the openings were sealed with a piece of glass.

3. Results and Discussion

3.1. Material and Device Analyses

The X-ray diffractometer that was used in this experiment was a Rigaku D/MAX-2500 (Rigaku, Tokyo, Japan). It was a low-angle X-ray diffractometer with a power of 18 kW. A field-emission scanning electron microscope (FE-SEM) (JEOL-7000F, Tokyo, Japan) was used to observe the morphology and surface properties of materials. A ultraviolet-visible (UV-Vis) spectrometer (Jasco V-650, Tokyo, Japan) was also used.

Photocurrent density-voltage (J-V) characteristics were measured under illumination by a simulated AM1.5G solar light from a Class AAA 550-W Xenon lamp solar simulator (ABET Technologies Sun 3000, Abet Technologies, Inc., Milford, CT, USA). Intensity-modulated photovoltage spectroscopy (IMVS) and intensity-modulated photocurrent spectroscopy (IMPS) were carried out by an electrochemical workstation (Zennium, Zahner, Germany).

3.1.1. X-Ray Diffraction Analysis of the Scattering Layer

Figure 2a displays the X-ray diffraction (XRD) patterns of the H-TiO$_2$. The * symbol represents a signal from the FTO glass substrate. Jade 5.0 software is used to fit the X-ray diffraction pattern. The XRD diffraction pattern proves that the hydrangea-shaped H-TiO$_2$ is an anatase phase. 2θ = 25.3°, 37.8°, 48.1°, 53.9°, 55.1°, and 62.7° correspond to the lattice planes (101), (004), (200), (105), (211), and (204) respectively. No characteristic peak of the rutile phase is obtained.

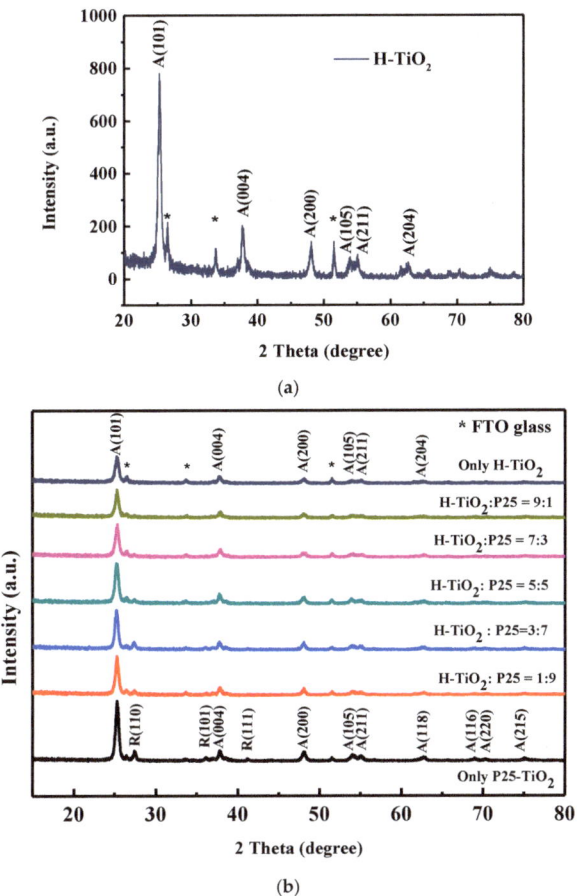

Figure 2. X-ray diffraction patterns of (**a**) H-TiO$_2$ and (**b**) comparison of our studied films. (The * symbol represents a signal from the FTO glass substrate.)

Figure 2b compares the XRD spectra for our studied films. The bottom line in Figure 2b shows the fitting result for the XRD diffraction pattern of P25 TiO$_2$. P25 TiO$_2$ has an obvious peak at 2θ = 25.35°, which is characteristic of the anatase phase plane (1,0,1) (JCPDS card number 21-1272). Characteristic peaks of the planes (1,1,0), (1,0,1), and (1,1,1) of the P25 TiO$_2$ rutile phase are also observed, revealing that P25 TiO$_2$ has anatase and rutile phases. An obvious characteristic peak near 2θ = 25.3° is the anatase phase. A larger amount of H-TiO$_2$ yields a smaller (1,0,1) peak because larger hydrangea-type H-TiO$_2$ particles yield a smaller peak.

3.1.2. Surface Morphology of Scattering Layer

The surface morphology of the scattering layer is studied using a FE-SEM at an accelerating voltage of 15 kV and working distances of 8.7–8.9 mm. The scattering layer has large particles of hydrangea powder and small particles of P25-TiO$_2$ powder.

Figure 3a displays the surface morphology of the P25 TiO$_2$ film. The magnification is 10,000. The P25-TiO$_2$ particles have an average size of around 20 nm. These particles are small and, therefore, have a high specific surface area. Figure 3b shows the surface morphology of the hydrothermally synthesized TiO$_2$ film. The magnification is 30,000. The degree of aggregation is high, favoring the scattering of light.

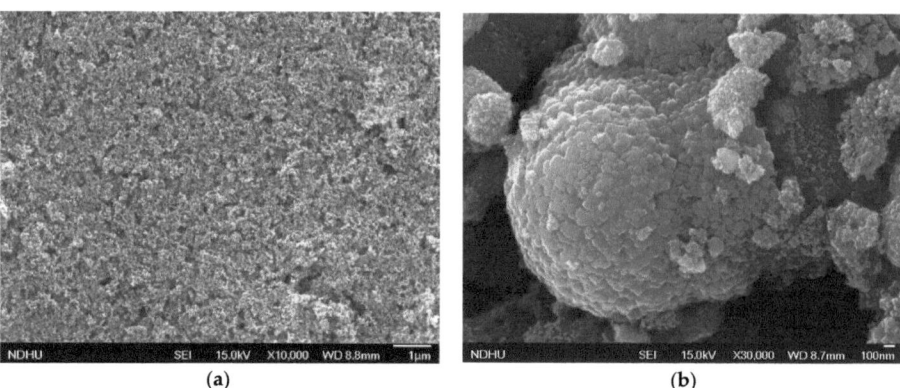

(a)   (b)

Figure 3. FE-SEM images showing the surface morphology of (**a**) P25 TiO$_2$ and (**b**) H-TiO$_2$ films.

Figure 4 displays FE-SEM images of the scattering layer with different ratios of constituent materials, including H-TiO$_2$ only. The size of H-TiO$_2$ is about 300 nm to 2 μm. The magnification is 10,000. The figure clearly reveals that after adding P25 TiO$_2$, the characteristics of the film are relatively dense. The small particles of P25 TiO$_2$ increase surface area and dye adsorption capacity.

Figure 4. *Cont.*

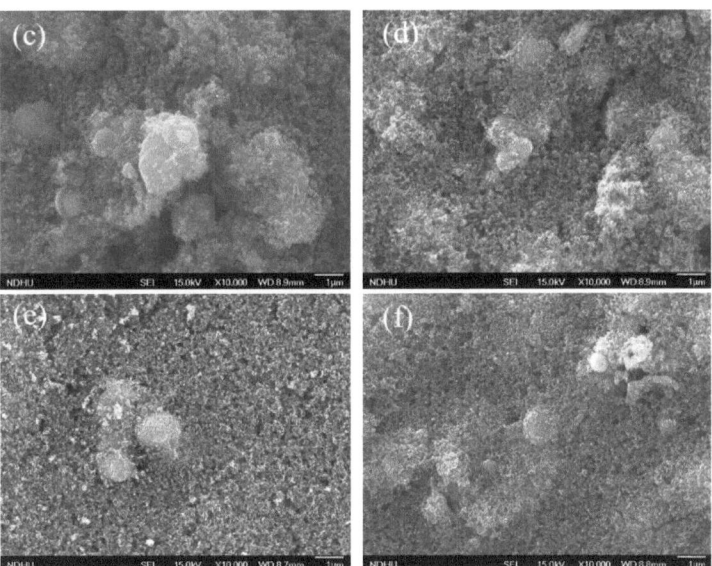

Figure 4. FE-SEM images of TiO$_2$ with different proportions between H-TiO$_2$ and P25 TiO$_2$. (**a**) H-TiO$_2$ only; (**b**) H-TiO$_2$:P25 TiO$_2$ = 9:1; (**c**) H-TiO$_2$:P25 TiO$_2$ = 7:3; (**d**) H-TiO$_2$:P25 TiO$_2$ = 5:5; (**e**) H-TiO$_2$:P25 TiO$_2$ = 3:7; (**f**) H-TiO$_2$:P25 TiO$_2$ = 1:9.

3.1.3. Absorption Spectrum

The amount of dye that is absorbed by the TiO$_2$ film is quantified by UV-vis spectroscopy desorption [11,18]. The absorption value at 515 nm was used to calculate the number of adsorbed N719 dye molecules, according to the Beer–Lambert law [12,35,37],

$$A_{dye} = \varepsilon_{dye} \, l \, C_{dye}, \tag{2}$$

where A_{dye} is the absorbance of UV–visible light at a wavelength of 515 nm; ε_{dye} is the molar extinction coefficient of the dye [10,12]; l is the path length of an optical cuvette (1 cm); and C_{dye} is the molar dye concentration in the NaOH solution [10,12].

Figure 5 shows the absorbance of dye desorbed from the different scattering layers by NaOH. Table 3 lists the dye loading amount. Experiments indicate that the addition of small-particle P25 TiO$_2$ increases the dye adsorption capacity. When the ratio is H-TiO$_2$:P25 TiO$_2$ = 3:7, the N719 dye loading capacity reaches its maximum value, which is 184.3 nmol/cm^2. Accordingly, mixing H-TiO$_2$ and P25 TiO$_2$ in an appropriate ratio favors the adsorption of dye on the film, improving photovoltaic capacity.

Table 3. Dye-loading of scattering layer film.

Sample	Dye Loading Amount (nmol/cm^2)
H-TiO$_2$	162.0
9H1C	162.3
7H3C	168.3
5H5C	169.5
3H7C	184.3
1H9C	182.5

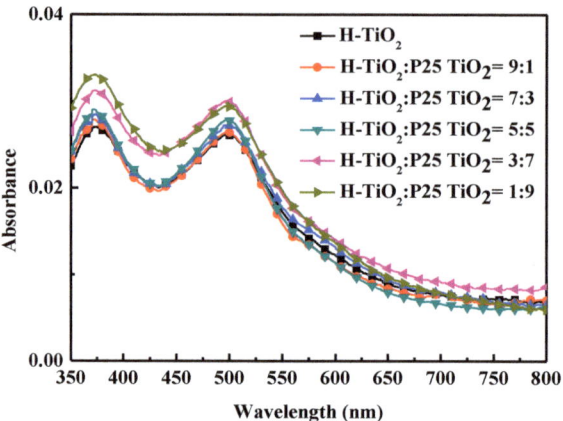

Figure 5. Absorption spectra of desorbed dye from the different scattering layers.

3.2. Photovoltaic Characterization

3.2.1. Analysis of DSSCs with 1-Layer P25 TiO$_2$ and Single Scattering Layer

Screen printing is used to form single-layer P25 TiO$_2$, and single-layer scattering layers are coated with different proportions of P25 TiO$_2$ and H-TiO$_2$. The 1-layer P25 TiO$_2$ combined with a 1-layer H-TiO$_2$ scattering layer [(H-TiO$_2$): P25-TiO$_2$ ratio = 10:0] is first studied to examine its potential effectiveness in DSSCs. Then, P25 TiO$_2$ is mixed into the H-TiO$_2$ scattering layer in various ratios to maximize the conversion efficiency.

Photocurrent density-voltage (*J-V*) curves are plotted to characterize DSSCs directly under illumination. The photoelectrode has an area of 0.16 cm^2. Figure 6 plots the *J-V* curves of DSSCs with 1-layer P25 TiO$_2$ and different scattering layers. Table 4 provides the characteristic photovoltaic parameters of interest, which are obtained from Figure 6. Measurements are made of all solar cells under the same illumination conditions at an illumination level of 100 mW/cm^2. Table 4 reveals that after the outermost P25 TiO$_2$ layer is replaced with a H-TiO$_2$ scattering layer, the short-circuit photocurrent and conversion efficiency are improved.

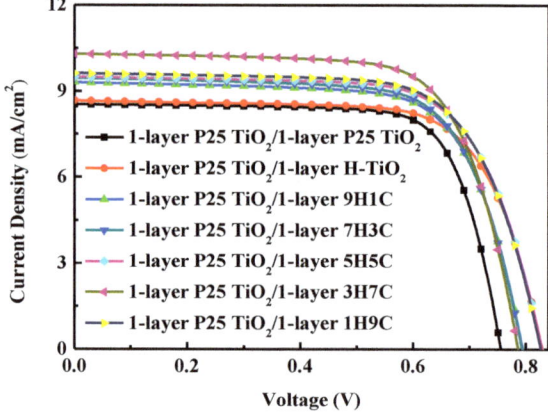

Figure 6. *J-V* characteristics of illuminated DSSCs with 1-layer P25 TiO$_2$ and single scattering layer.

Table 4. Photovoltaic properties of DSSCs with 1-layer working electrode and single scattering layer.

Photoanode	J_{sc} (mA/cm^2)	V_{oc} (V)	FF	η (%)
1-layer P25/1-layer P25	8.54	0.76	0.74	4.82
1-layer P25/1-layer H-TiO$_2$	8.68	0.83	0.71	5.09
1-layer P25/1-layer 9H1C	9.31	0.80	0.7	5.22
1-layer P25/1-layer 7H3C	9.47	0.79	0.71	5.29
1-layer P25/1-layer 5H5C	9.49	0.83	0.69	5.46
1-layer P25/1-layer 3H7C	9.93	0.77	0.73	5.58
1-layer P25/1-layer 1H9C	9.63	0.83	0.69	5.51

The solar energy-to-electricity conversion efficiency (η) of DSSC is expressed as [1,12,38].

$$\eta = \frac{J_{sc} V_{oc} FF}{P_{in}} \times 100\% \qquad (3)$$

where J_{sc} is the short-circuit photocurrent density (mA/cm^2); V_{oc} is the open-circuit voltage (V); P_{in} is the incident light power per unit area. The fill factor (FF) is expressed as [12,38]

$$FF = \frac{P_{max}}{J_{sc} V_{oc}} = \frac{J_{max} V_{max}}{J_{sc} V_{oc}} \qquad (4)$$

where J_{max} and V_{max} are the current and voltage, respectively, at the maximum power point in the J-V curves of the solar cells.

To confirm the effects of the H-TiO$_2$ scattering layer, reflectance was measured. Figure 7 displays the measured reflectance of P25 TiO$_2$ and H-TiO$_2$ films without adsorbed dye. The reflectance of P25 TiO$_2$ gradually decreases in the band above 500 nm, indicating that the reflectance of P25 TiO$_2$ becomes worse and the scattering becomes weaker. H-TiO$_2$ retains a high reflectance at wavelengths above 500 nm, so it exhibits stronger overall scattering.

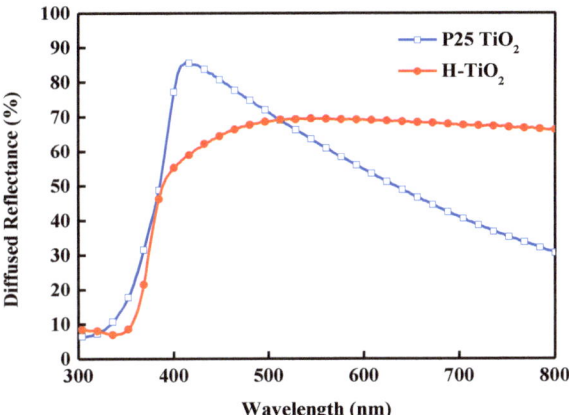

Figure 7. Diffused reflectance spectra of P25 TiO$_2$ and H-TiO$_2$ films without adsorbed dye.

The above results prove that this scattering layer can indeed improve photovoltaic performance. In this experiment, when the (H-TiO$_2$): P25 TiO$_2$ ratio = 3:7, the DSSC generates the best short-circuit photocurrent and highest conversion efficiency. Increasing the ratio of P25 TiO$_2$ not only increases the surface area but also increases the adsorption of dyes, improves photon scattering ability, and, thereby, increases the short-circuit current. However, if too much P25 TiO$_2$ is added, then the scattering becomes weak because the proportion of hydrangea TiO$_2$ is too small, so the short-circuit current is reduced. These results can be compared with those in Table 3. When the scattering layer ratio (H-TiO$_2$):P25 TiO$_2$ = 3:7, the device has the highest dye loading and J$_{SC}$ of the studied structures.

3.2.2. Analysis of DSSCs with 5-Layer P25 TiO$_2$ and Single Scattering Layer

In Section 3.2.1, the 1-layer P25 TiO$_2$ DSSCs with different scattering layers are investigated. Experimental results demonstrate that the DSSC with the 3H7C scattering layer has the highest photoelectric conversion efficiency. Consequently, DSSCs with 5-layer P25 TiO$_2$ and a single scattering layer with various parameters are also investigated to find the best (H-TiO$_2$):P25 TiO$_2$ ratio. Figure 8 plots the illuminated J-V curves, and Table 5 summarizes the photovoltaic characteristics. The thickness of the photoanode is about 25 mm. Coating the five-layer P25 TiO$_2$ and single-layer 3H7C scattering layer yielded the highest current density of 15.26 mA/cm^2, the largest open-circuit voltage of 0.77 V, the largest filling factor of 0.71, and the highest conversion efficiency of 8.33% of any of the studied devices. Measurements of more than three DSSCs of the same type were made. The variations in all DSSC characteristics were less than 3%. The ratio of P25 TiO$_2$ and H-TiO$_2$ is 3:7 is posited to maximize connectivity, resulting in no excessive fracture surfaces on the electrode surface and improved dye adsorption capacity.

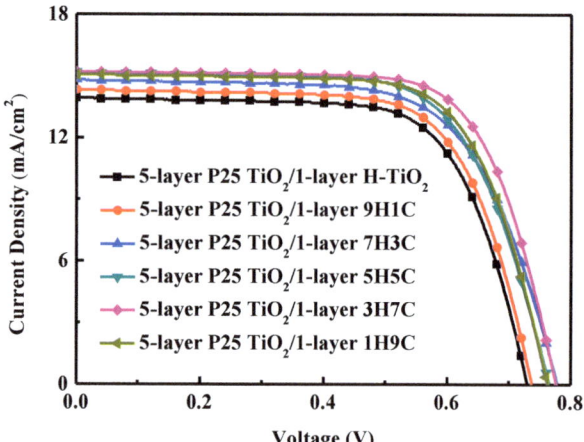

Figure 8. J-V characteristics of illuminated DSSCs with a 5-layer P25 TiO$_2$ and single scattering layer.

Table 5. Photovoltaic characteristics of DSSCs with a 5-layer working electrode and single scattering layer.

Photoanode	J_{sc} (mA/cm^2)	V_{oc} (V)	FF	η (%)
5-layer P25/1-layer H-TiO$_2$	13.94	0.74	0.68	7.03
5-layer P25/1-layer 9H1C	14.31	0.74	0.69	7.29
5-layer P25/1-layer 7H3C	14.81	0.76	0.68	7.62
5-layer P25/1-layer 5H5C	15.13	0.76	0.68	7.84
5-layer P25/1-layer 3H7C	15.26	0.77	0.71	8.33
5-layer P25/1-layer 1H9C	15.06	0.76	0.7	7.99

Furthermore, IMVS and IMPS are used to evaluate the charge-collection efficiency (η_{cc}) of the studied DSSCs. The recombination time (τ_r) can be calculated from the equation $\tau_r = \frac{1}{2\pi f_r}$, where f_r is the characteristic frequency minimum of the IMVS imaginary component. The collection (transport) time (τ_c) can be calculated from the equation $\tau_c = \frac{1}{2\pi f_c}$, where f_c is the characteristic frequency minimum of the IMPS imaginary component.

η_{cc} is strongly determined by competition between charge collection and recombination. $\eta_{cc} = 1 - \frac{\tau_c}{\tau_r}$ [39,40]. Experimental results demonstrate that when H-TiO$_2$ and P25-TiO$_2$ are mixed, the electron transport time can be reduced. It proves that after mixing, its connectivity becomes better, which can make the electron transfer path smoother. The values of τ_c and τ_r for the DSSC with the 3H7C scattering layer are 7.98 ms and 200.36 ms, respectively. The DSSC with the 3H7C scattering layer has the shortest collection time, causing the largest charge-collection efficiency (η_{cc} = 0.96) of our studied DSSCs. This result is consistent with our J-V measurement in this work. Consequently, the DSSC with the 3H7C scattering layer has the largest J_{sc} and η of the studied DSSCs.

4. Conclusions

In this work, hydrangea-shaped TiO$_2$ were hydrothermally synthesized and used in the scattering layer of DSSCs. Smaller particles of P25 TiO$_2$ were mixed with larger H-TiO$_2$ particles to form the scattering layer in the photoanode. When the (H-TiO$_2$): P25 TiO$_2$ ratio exceeded a particular value (3:7 in this study), as the H-TiO$_2$ content increased, the short-circuit current and conversion efficiency decreased, mainly because the reduction in dye adsorption capacity results in a reduction of short-circuit current. Experimental results demonstrate that a suitable mixture of small and large particles in light-scattering layers enhances the conversion efficiency of DSSCs. In our future work, we will add graphene to the structure of the studied DSSC and investigate its impact on device characteristics.

Author Contributions: Investigation, Y.-S.L.; Data curation, W.-H.C.; Writing—original draft, Y.-S.L. All authors have read and agreed to the published version of the manuscript.

Funding: This research was completed with funding from the National Science and Technology Council, Taiwan (Project No. NSTC 112-2221-E-259-014 and NSTC 113-2221-E-259-006).

Institutional Review Board Statement: Not applicable.

Informed Consent Statement: Not applicable.

Data Availability Statement: The original contributions presented in this study are included in the article. Further inquiries can be directed to the corresponding author.

Conflicts of Interest: The authors declare no conflicts of interest.

References

1. Gong, J.W.; Sumathy, K.; Qiao, Q.Q.; Zhou, Z.P. Review on dye-sensitized solar cells (DSSCs): Advanced techniques and research trends. *Renew. Sustain. Energy Rev.* **2017**, *68*, 234–246. [CrossRef]
2. Sekaran, P.D.; Marimuthu, R. An extensive analysis of dye-sensitized solar cell (DSSC). *Braz. J. Phys.* **2024**, *54*, 28. [CrossRef]
3. Di Carlo, V.; Prete, P.; Dubrovskii, V.G.; Berdnikov, Y.; Nico Lovergine, N. CdTe Nanowires by Au-Catalyzed Metalorganic Vapor Phase Epitaxy. *Nano Lett.* **2017**, *17*, 4075–4082. [CrossRef] [PubMed]
4. Lovergine, N.; Cingolani, R.; Mancini, A.M. Photoluminescence of CVD grown CdS epilayers on CdTe substrates. *J. Cryst. Growth* **1992**, *118*, 304–308. [CrossRef]
5. Kujovic, L.; Liu, X.L.; Abbas, A.; Jones, L.O.; Law, A.M.; Togay, M.; Curson, K.M.; Barth, K.L.; Bowers, J.W.; Walls, J.M.; et al. Achieving 21.4% Efficient CdSeTe/CdTe Solar Cells Using Highly Resistive Intrinsic ZnO Buffer Layers. *Adv. Funct. Mater.* **2024**, *34*, 2312528. [CrossRef]
6. Miccoli, I.; Prete, P.; Marzo, F.; Cannolletta1, D.; Lovergine, N. Synthesis of vertically-aligned GaAs nanowires on GaAs/(111)Si hetero-substrates by metalorganic vapour phase epitaxy. *Cryst. Res. Technol.* **2011**, *46*, 795–800. [CrossRef]
7. Raj, V.; Haggren, T.; Mayon, Y.O.; Jagadish, C.; Tan, H.H. 21.2% GaAs Solar Cell Using Bilayer Electron Selective Contact. *Sol. RRL* **2024**, *8*, 2300889. [CrossRef]
8. Pandikumar, A.; Jothivenkatachalam, K.; Bhojanaa, K.B. *Interfacial Engineering in Functional Materials for Dye-Sensitized Solar Cells*; John Wiley & Sons, Inc.: Hoboken, NJ, USA, 2020.
9. Lin, Y.-S.; Lu, C.-C. Improved AlGaN/GaN metal-oxide-semiconductor high-electron mobility transistors with TiO$_2$ gate dielectric annealed in nitrogen. *IEEE Trans. Electron Devices* **2018**, *65*, 783–787. [CrossRef]

10. Wang, Z.S.; Kawauchi, H.; Kashima, T.; Arakawa, H. Significant influence of TiO_2 photoelectrode morphology on the energy conversion efficiency of N719 dye-sensitized solar cell. *Coord. Chem. Rev.* **2004**, *248*, 1381–1389. [CrossRef]
11. Hore, S.; Vetter, C.; Kern, R.; Smit, H.; Hinsch, A. Influence of scattering layers on efficiency of dye-sensitized solar cells. *Sol. Energy Mater. Sol. Cells* **2006**, *90*, 1176–1188. [CrossRef]
12. Park, J.T.; Roh, D.K.; Chi, W.S.; Patel, R.; Kim, J.H. Fabrication of double layer photoelectrodes using hierarchical TiO_2 nanospheres for dye-sensitized solar cells. *J. Ind. Eng. Chem.* **2012**, *18*, 449–455. [CrossRef]
13. Shital, S.; Swami, S.K.; Barnes, P.; Dutta, V. Monte Carlo simulation for optimization of a simple and efficient bifacial DSSC with a scattering layer in the middle. *Sol. Energy* **2018**, *161*, 64–73. [CrossRef]
14. Nguyen, D.T.; Kurokawa, Y.; Taguchi, K. Enhancing DSSC photoanode performance by using Ni-doped TiO_2 to fabricate scattering layers. *J. Electron. Mater.* **2020**, *49*, 2578–2583. [CrossRef]
15. Chang, W.C.; Tang, B.H.; Lu, Y.W.; Yu, W.C.; Lin, L.Y.; Wu, R.J. Incorporating hydrangea-like titanium dioxide light scatterer with high dye-loading on the photoanode for dye-sensitized solar cells. *J. Power Sources* **2016**, *319*, 131–138. [CrossRef]
16. Chava, R.K.; Yu, Y.T.; Kang, M. Layer-by-layer deposition of hollow TiO_2 spheres with enhanced photoelectric conversion efficiency for dye-sensitized solar cell applications. *Nanomaterials* **2024**, *14*, 1782. [CrossRef]
17. Jiang, Y.J.; Zhao, F.Y. Application of bunchy TiO_2 hierarchical microspheres as a scattering layer for dye-sensitized solar cells. *Front. Energy Res.* **2021**, *9*, 682709. [CrossRef]
18. Chou, J.C.; Syu, R.H.; Lai, C.H.; Kuo, P.Y.; Yang, P.H.; Nien, Y.H.; Lin, Y.C.; Yong, Z.R.; Wu, Y.T. Optimization and application of TiO_2 hollow microsphere modified scattering layer for the photovoltaic conversion efficiency of dye-sensitized solar cell. *IEEE Trans. Semicond. Manuf.* **2022**, *35*, 363–371. [CrossRef]
19. Zhang, Z.; Cai, W.; Lv, Y.; Jin, Y.; Chen, K.; Wang, L.; Zhou, X. Anatase TiO_2 nanowires with nanoscale whiskers for the improved photovoltaic performance in dye-sensitized solar cells. *J. Mater. Sci. Mater. Electron.* **2019**, *30*, 14036–14044. [CrossRef]
20. Chen, Y.Z.; Wu, R.J.; Lin, L.Y.; Chang, W.C. Novel synthesis of popcorn-like TiO_2 light scatterers using a facile solution method for efficient dye-sensitized solar cells. *J. Power Sources* **2019**, *413*, 384–390. [CrossRef]
21. Ramakrishnan, V.M.; Sandberg, S.; Muthukumarasamy, N.; Kvamme, K.; Balraju, P.; Agilan, S.; Velauthapillai, D. Microwave-assisted solvothermal synthesis of worms-like TiO_2 nanostructures in submicron regime as light scattering layers for dye-sensitized solar cells. *Mater. Lett.* **2019**, *236*, 747–751. [CrossRef]
22. Cui, Y.; He, X.; Zhu, M.H.; Li, X. Preparation of anatase TiO_2 microspheres with high exposure (001) facets as the light-scattering layer for improving performance of dye-sensitized solar cells. *J. Alloys Compd.* **2017**, *694*, 568–573. [CrossRef]
23. Nemala, S.S.; Mokurala, K.; Bhargava, P.; Mallick, S. Titania nanobelts as a scattering layer with Cu_2ZnSnS_4 as a counter electrode for DSSC with improved efficiency. *Mater. Today Proc.* **2018**, *5*, 23351–23357. [CrossRef]
24. Chen, C.; Luo, F.; Li, Y.; Sewvandi, G.A.; Feng, Q. Single-crystalline anatase TiO_2 nanoleaf: Simple topochemical synthesis and light-scattering effect for dye-sensitized solar cells. *Mater. Lett.* **2017**, *196*, 50–53. [CrossRef]
25. Navarro-Pardo, F.; Benetti, D.; Benavides, J.; Zhao, H.G.; Cloutier, S.G.; Castano, V.M.; Vomiero, A.; Rosei, F. Nanofiber-structured TiO_2 nanocrystals as a scattering layer in dye-sensitized solar cells. *ECS J. Solid State Sci. Technol.* **2017**, *6*, N32–N37. [CrossRef]
26. Hu, J.H.; Tong, S.Q.; Yang, Y.P.; Cheng, J.J.; Zhao, L.; Duan, J.X. A composite photoanode based on $P25/TiO_2$ nanotube arrays/flower-like TiO_2 for high-efficiency dye-sensitized solar cells. *Acta Metall. Sin. (Engl. Lett.)* **2016**, *29*, 840–847. [CrossRef]
27. Mustafa, M.N.; Sulaiman, Y. Optimization of titanium dioxide decorated by graphene quantum dot as a light scatterer for enhanced dye-sensitized solar cell performance. *J. Electroanal. Chem.* **2020**, *876*, 114516. [CrossRef]
28. Xu, S.X.; Fang, D.; Xiong, F.M.; Ren, Y.X.; Bai, C.; Mi, B.X.; Gao, Z.Q. Electrophoretic deposition of double-layer ZnO porous films for DSSC photoanode. *J. Solid State Electrochem.* **2024**, *28*, 589–599. [CrossRef]
29. Min, K.W.; Chao, S.M.; Yu, M.T.; Ho, C.T.; Chen, P.R.; Wu, T.L. Graphene-TiO_2 for scattering layer in photoanodes of dye-sensitized solar cell. *Mod. Phys. Lett. B* **2021**, *35*, 2141005. [CrossRef]
30. Pang, B.L.; Shi, Y.T.; Lin, S.; Chen, Y.J.; Feng, J.G.; Dong, H.Z.; Yang, H.B.; Zhao, Z.X.; Yu, L.Y.; Dong, L.F. Triiodide reduction activity of hydrangea molybdenum sulfide/reduced graphene oxide composite for dye-sensitized solar cells. *Mater. Res. Bull.* **2019**, *117*, 78–83. [CrossRef]
31. Liu, T.C.; Wu, C.C.; Huang, C.H.; Chen, C.M. Effects of Ethyl Cellulose on Performance of Titania Photoanode for Dye-sensitized Solar Cells. *J. Electron. Mater.* **2016**, *45*, 6192–6199. [CrossRef]
32. Zhou, C.H.; Zhao, X.Z.; Yang, B.C.; Zhang, D.; Li, Z.Y.; Zhou, K.C. Effect of poly (ethylene glycol) on coarsening dynamics of titanium dioxide nanocrystallites in hydrothermal reaction and the application in dye sensitized solar cells. *J. Colloid Interface Sci.* **2012**, *374*, 9–17. [CrossRef] [PubMed]
33. Ju, K.Y.; Cho, J.M.; Cho, S.J.; Yun, J.J.; Mun, S.S.; Han, E.M. Enhanced efficiency of dye-sensitized solar cells with novel synthesized TiO_2. *J. Nanosci. Nanotechnol.* **2010**, *10*, 3623–3627. [CrossRef] [PubMed]
34. Choi, H.; Nahm, C.; Kim, J.; Moon, J.; Nam, S.; Jung, D.R.; Park, B. The effect of $TiCl_4$-treated TiO_2 compact layer on the performance of dye-sensitized solar cell. *Curr. Appl. Phys.* **2012**, *12*, 737–741. [CrossRef]

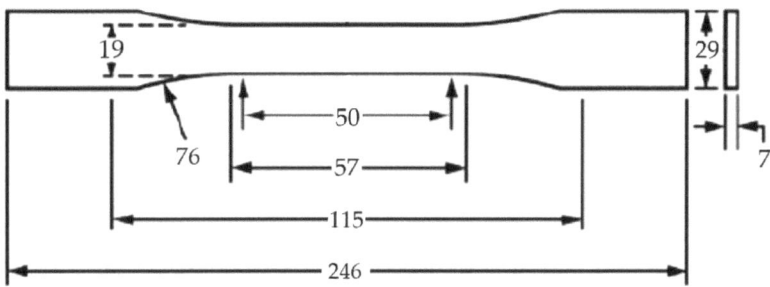

Figure 1. Dimensions of the tensile test specimen [27].

Rectangular samples measuring 200 mm × 200 mm × 3 mm for surface resistance testing were designed according to the appropriate standards [28].

The test samples were fabricated using the Material Extrusion (MEX) technique. The manufacturing process was carried out using a Prusa i3 MK3s 3D printer (Prusa Research, Prague, Czech Republic). The print codes were prepared in advance using the dedicated 3D printing software Prusa Slicer v2.5.2. The process parameters for all types of samples were kept consistent and are detailed in Table 1. Figure 2 shows printed samples.

Table 1. Printing parameters of the MEX.

Filament Diameter [mm]	Nozzle Diameter [mm]	Printing Temperature [°C]	Bed Temperature [°C]	Print Speed [mm/s]	Infill Pattern	Infill [%]	Number of Contours
1.75	0.4	230	80	70	linear	100	3

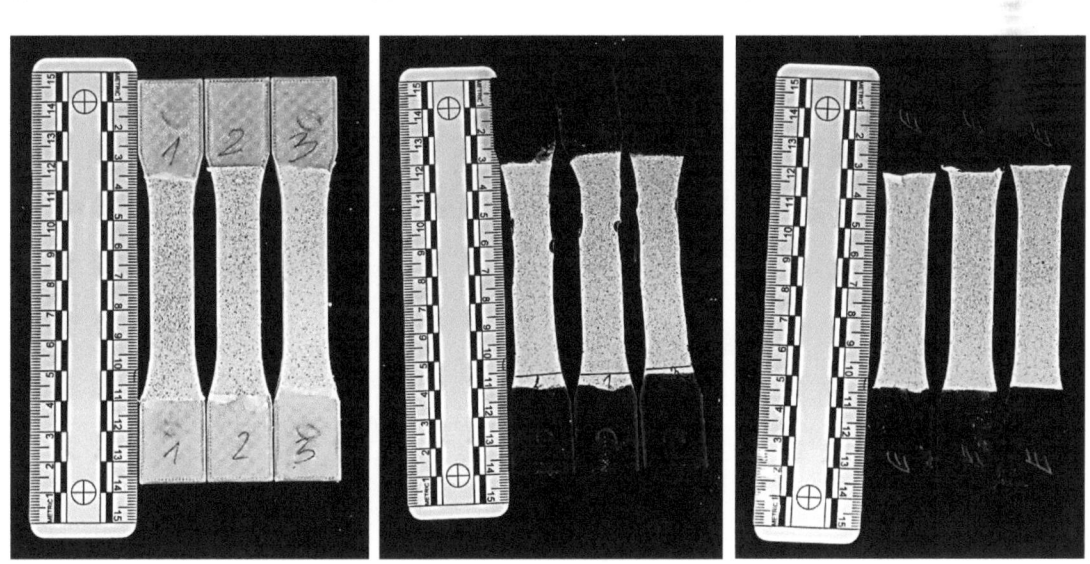

Figure 2. Printed models of samples, prepared for testing: (**a**). PET-G; (**b**). Carbon fiber PET-G; (**c**). ESD PET-G.

2.3. Testing Methods

The static tensile test of the fabricated samples was performed according to the standard ASTM D638-14:2022—Standard Test Method for Tensile Properties of Plastics [27]. Experiments were performed using a classic INSTRON 8802MTL tensile testing machine (Norwood, MA, USA) equipped with WaveMatrix software (version 2.0) (Instron, Norwood, MA, USA). Strain measurements were obtained using an extensometer (2620-604, INSTRON, Norwood, MA, USA) with a gauge length of 50 mm. Furthermore, the properties were examined by surface analysis using Digital Image Correlation (DIC).

Subsequently, the surface structure of the samples was analyzed after tensile testing using a Keyence VHX-7000 digital optical microscope (Keyence, Osaka, Japan). Representative samples of each material were selected for this analysis.

35. Lee, S.W.; Anh, K.S.; Zhu, K.; Neale, N.R.; Frank, A.J. Effects of TiCl$_4$ treatment of nanoporous TiO$_2$ films on morphology, Light harvesting, and charge-carrier dynamics in dye-sensitized solar cells. *J. Phys. Chem. C* **2012**, *116*, 21285–21290. [CrossRef]
36. Vu, H.H.T.; Atabaev, T.S.; Ahn, J.Y.; Dinh, N.N.; Kim, H.K.; Huang, Y.H. Dye-sensitized solar cells composed of photoactive composite photoelectrodes with enhanced solar energy conversion efficiency. *J. Mater. Chem. A* **2015**, *3*, 11130–11136. [CrossRef]
37. Xu, J.; Zhai, C.W.; Zheng, B.W.; Li, H.W.; Zhu, M.W.; Chen, Y.F. Large efficiency improvement in nanoporous dye-sensitized solar cells via vacuum assistant dye adsorption. *Vacuum* **2012**, *86*, 1161–1164. [CrossRef]
38. Hegazy, A.; Kinadjian, N.; Sadeghimakki, B.; Sivoththaman, S.; Allam, N.K.; Prouzet, E. TiO$_2$ nanoparticles optimized for photoanodes tested in large area dye-sensitized solar cells (DSSC). *Sol. Energy Mater. Sol. Cells* **2016**, *153*, 108–116. [CrossRef]
39. Zhu, K.; Neale, N.R.; Miedaner, A.; Frank, A.J. Enhanced charge-collection efficiencies and light scattering in dye-sensitized solar cells using oriented TiO$_2$ nanotubes arrays. *Nano Lett.* **2007**, *7*, 69–74. [CrossRef]
40. Ding, Y.; Mo, L.E.; Tao, L.; Ma, Y.M.; Hu, L.H.; Huang, Y.; Fang, X.Q.; Yao, J.X.; Xi, X.W.; Dai, S.Y. TiO$_2$ nanocrystalline layer as a bridge linking TiO$_2$ sub-microspheres layer and substrates for high-efficiency dye-sensitized solar cells. *J. Power Sources* **2014**, *272*, 1046–1052. [CrossRef]

Disclaimer/Publisher's Note: The statements, opinions and data contained in all publications are solely those of the individual author(s) and contributor(s) and not of MDPI and/or the editor(s). MDPI and/or the editor(s) disclaim responsibility for any injury to people or property resulting from any ideas, methods, instructions or products referred to in the content.

Article

Strength and Electrostatic Discharge Resistance Analysis of Additively Manufactured Polyethylene Terephthalate Glycol (PET-G) Parts for Potential Electronic Application

Julia Talecka [1], Janusz Kluczyński [2,*], Katarzyna Jasik [2], Ireneusz Szachogłuchowicz [2] and Janusz Torzewski [2]

1. Institute of Optoelectronics, Military University of Technology, Gen. S. Kaliskiego 2, 00-908 Warsaw, Poland; julia.talecka@student.wat.edu.pl
2. Institute of Robots & Machine Design, Faculty of Mechanical Engineering, Military University of Technology, Gen. S. Kaliskiego 2, 00-908 Warsaw, Poland; katarzyna.jasik@wat.edu.pl (K.J.); ireneusz.szachogluchowicz@wat.edu.pl (I.S.); janusz.torzewski@wat.edu.pl (J.T.)
* Correspondence: janusz.kluczynski@wat.edu.pl

Abstract: Optoelectronic components are crucial across various industries. They benefit greatly from advancements in 3D printing techniques that enable the fabrication of intricate parts. Among these techniques, Material Extrusion (MEX) stands out for its simplicity and cost-effectiveness. Integrating 3D printing into production processes offers the potential to create components with enhanced electrostatic discharge (ESD) resistance, a critical factor for ensuring the reliability and safety of optoelectronic devices. Polyethylene terephthalate glycol-modified (PET-G) is an amorphous copolymer renowned for its high transparency, excellent mechanical properties, and chemical resistance, which make it particularly suitable for 3D printing applications. This study focuses on analyzing the mechanical, structural, and electrostatic properties of pure PET-G as well as PET-G doped with additives to evaluate the effects of doping on its final properties. The findings highlight that pure PET-G exhibits superior mechanical strength compared to doped variants. Conversely, doped PET-G demonstrates enhanced resistance to electrostatic discharge, which is advantageous for applications requiring ESD mitigation. This research underscores the importance of material selection and optimization in 3D printing processes to achieve desired mechanical and electrical properties in optoelectronic components. By leveraging 3D printing technologies like MEX and exploring material modifications, industries can further innovate and enhance the production of optoelectronic devices, fostering their widespread adoption in specialized fields.

Keywords: additive manufacturing; material extrusion; PET-G; tensile test; digital image correlation; electrostatic discharge resistance

Citation: Talecka, J.; Kluczyński, J.; Jasik, K.; Szachogłuchowicz, I.; Torzewski, J. Strength and Electrostatic Discharge Resistance Analysis of Additively Manufactured Polyethylene Terephthalate Glycol (PET-G) Parts for Potential Electronic Application. *Materials* **2024**, *17*, 4095. https://doi.org/10.3390/ma17164095

Academic Editors: Jijun Feng and Shengli Pu

Received: 8 July 2024
Revised: 6 August 2024
Accepted: 16 August 2024
Published: 18 August 2024

Copyright: © 2024 by the authors. Licensee MDPI, Basel, Switzerland. This article is an open access article distributed under the terms and conditions of the Creative Commons Attribution (CC BY) license (https://creativecommons.org/licenses/by/4.0/).

1. Introduction

Optoelectronic components play a crucial role in various industrial sectors, facilitating advancements in telecommunications, medical technology, lighting, photography, and many other fields [1,2]. Given the complex structure of these components, the application of 3D printing techniques may be beneficial. Three-dimensional printing revolutionizes the production of small, intricate elements [3,4]. One of the most commonly used 3D printing methods is Material Extrusion (MEX), also known as Fused Deposition Modeling (FDM) or Fused Filament Fabrication (FFF) [5,6]. This technique uses thermoplastic polymer extrusion to create three-dimensional parts. Since its inception, MEX technology has become one of the most popular and recognizable additive manufacturing methods due to its ease of use and cost-effectiveness [7]. The use of this technology allows for the combination of Electrostatic Discharge Resistance (ESD) in polymer materials with appropriate additives while meeting the geometric requirements of optoelectronic components. Filament materials are readily available, and the extrusion process is precise, relatively inexpensive, and straightforward.

The authors of [8] produced diffusion tubes with Ion Mobility Spectrometry (IMS) using the MEX method. They demonstrated the possibility of single-run 3D printing of an IMS drift tube with integrated gate and aperture grids using a three-fiber 3D printing system, which provided better quality and less waste than traditional methods. The importance of testing printing parameters such as temperature and extrusion speed was also emphasized, as these affect the ESD conductivity of the filament and can yield different results on various 3D printers. Hauck et al. [9] designed a one-piece drift tube for IMS using 3D printing, consisting of alternating conductive and insulating layers. The goal was to increase the accuracy and repeatability of the measurements compared to traditional drift tubes assembled manually, which can introduce errors in calculating mobility. The results showed that the 3D printed drift tubes had uniform lengths and minimal weight differences, leading to accurate and repeatable calculations. It was confirmed that 3D printing could be an effective method for producing drift tubes of consistent length, thus improving measurement precision in IMS. Su et al. [10] developed a method for fully 3D printing flexible Organic Light-emitting Diode (OLED) displays. Through a multimodal printing approach that combines extrusion and spraying methods, they constructed devices with significantly improved uniformity of active layers and more stable polymer–metal connections. Spray printing, which was used to deposit active layers, improved their uniformity by reducing the directional mass transport. Additionally, mechanical reconfiguration of the liquid metal surface increased the contact area of the polymer–metal connections. The patent [11] describes an advanced LED lighting device that provides improved convergence of light beams and increased radiation range compared to traditional reflectors. The design includes a high thermal conductivity base with a curved upper portion, an internal container with a flat mounting surface, and a curved shield with an internally reflective surface. These components enable the focusing of LED light beams and minimize dispersion, significantly enhancing the device's lighting efficiency.

In the context of optoelectronic devices, there is a risk of ESD, which poses a threat to both operators and the devices themselves [12,13]. Phenomena such as separation and induction can generate excess electrostatic charges, which can lead to damage, especially in devices sensitive to ESD [14,15]. The electronics industry has numerous standards that define methods for controlling, testing, and taking preventive actions to ensure the safe handling of electronic components, including during production. Using materials with adequate surface resistance and good mechanical properties to manufacture structural components and enclosures of optoelectronic devices is crucial to ensuring the proper functioning and safety of these devices and their users [9]. The patent [16] presents innovative applications of electrostatic discharge-resistant enclosures, particularly for optoelectronic devices. One example of a patented enclosure includes a reflecting element designed to both reflect electromagnetic radiation emitted by the semiconductor device and absorb electromagnetic radiation directed toward the device. The reflecting part is covered with the same material as the rest of the enclosure, but the internal reflective part remains partially uncovered. The reflecting element is made of a different type of synthetic material than the rest of the enclosure, differing in at least one significant material property, such as thermal stability or resistance to electromagnetic radiation. A key aspect of the design is that the reflecting part is not connected to the rest of the enclosure using adhesives or mechanical macroscopic connections but, rather, joins directly through contact. The patent also presents a series of other solutions related to various materials and structural elements of enclosures used in different applications.

The polymers most commonly used in electronic applications are Acrylonitrile Butadiene Styrene (ABS), Polylactic Acid (PLA), Polyetheretherketone (PEEK), Polybutylene Terephthalate (PBT), and polyethylene terephthalate glycol-modified (PET-G) [17–24].

Polyethylene terephthalate glycol-modified (PET-G) is an amorphous copolymer with good transparency, mechanical properties, and chemical resistance, which is used in 3D printing. It is characterized by high mechanical strength and thermal stability. PET-G shows better thermal degradation and higher thermal stability compared to other materials used in 3D printing, making it suitable for producing components that require high strength and flexibility [25,26]. The patent [11] proposes an advanced optoelectronic device utilizing a housing made from two different types of plastics with specific properties. The first material, such as polybutylene terephthalate (PBT), PET-G, or polyetheretherketone (PEEK), is chosen for its high resistance to electromagnetic radiation and thermal stability. The second material, such as polyamides or polyphenylene sulfone, is selected for other structural aspects. The reflecting part of the housing is made from the first type of material, to which a white pigment like titanium dioxide is added to enhance its ability to reflect radiation. The enclosure provides a suitable environment for optoelectronic components such as light-emitting diodes (LEDs) or lasers that emit electromagnetic radiation. These elements are mounted in the housing in a manner that allows controlled reflection or scattering of radiation. The use of materials with increased resistance to electrostatic discharge, such as PET-G, enables the implementation of this solution using additive manufacturing methods, potentially increasing production flexibility and efficiency.

Given the benefits of PET-G compared to other widely used 3D printing materials, this study aims to investigate selected strength properties of PET-G and evaluate whether doping the material positively affects its final properties. There is a lack of available research results of this type; therefore, the novelty lies in compiling strength and structural analysis results with surface resistance data. This allows us to address whether the additively manufactured component will provide adequate strength while maintaining ESD resistance.

2. Materials and Methods

2.1. Materials for the Research

The first material selected for research was the ESD-resistant PET-G filament. This material is relatively easy to print with and does not require a heated chamber. The manufacturer claims that the filament is resistant to electrostatic discharge. The next material chosen was pure PET-G, which was selected to investigate whether the addition of compounds improves the properties of the final filament. The last material selected was PET-G doped with carbon fiber (up to 10% carbon fiber content). This material has high mechanical strength and is designed to operate in extreme temperatures. It is an industrial-grade filament that can be processed and used on 3D printers. It is safe for both humans and the environment, as it does not emit toxic fumes and exhibits minimal shrinkage. This selection of materials allowed us to compare and assess how doping affects the final mechanical and electrostatic properties of PET-G. All materials were supplied by Spectrum Filaments (Spectrum Filaments, Pęcice, Poland).

2.2. Manufacturing Process

The 3D models of the test samples were designed using SolidWorks CAD software (Dassault Systems; Waltham, QC, Canada) (version 2023). Dogbone-shaped samples were designed for static tensile strength according to the relevant standard ASTM D638-14:2022—Standard Test Method for Tensile Properties of Plastics [27] (Figure 1).

Following this, the resistance to electrostatic discharge was evaluated using the relevant standard. Resistance measurements were carried out using an Aijgo 61 resistance meter (Aijgo, Vác, Budapest). The tests were carried out in a laboratory setting after conditioning the samples for 48 h at a temperature of 23 ± 2 °C and a relative humidity of 12 ± 3% RH. Surface resistances were measured using various Concentric Ring Electrodes (CRE) following the standard guidelines. To reduce contact resistance, measurements were repeated using a Surface Resistivity Bar (SRB) electrode. The SRB electrode had a mass of 2900 g, a bar width of 3 mm, and a length of 50.8 mm, with a spacing of 25.4 mm between the bars. The contact surface was made of conductive rubber (3 mm thick, hardness A 60, $\rho V < 100$ Ωm) or copper. Dimensions were considered, and the results were expressed as resistance.

Material resistance was measured using an electrode with a mass of 2.3 kg and a diameter of 63 mm. Various types of conductive rubber were compared to determine the lowest possible contact resistance between the upper and lower electrodes and the sample. As a result of the unstable contact between the electrode and the samples, concentric ring electrodes (CREs) could not be used. Furthermore, DC resistance through the material was measured using an ESD contact probe with a rounded tip. Results were recorded after 15 s of electrification. The test voltage was 10 V for resistances below 1 MΩ or 100 V for resistances greater than 1 MΩ. All measurements were repeated ten times, with the results presented as minimum and maximum readings along with the geometric mean.

3. Results

All tests were performed on samples printed on three different materials. The descriptions of the results of all tests correspond to the classifications described in Table 2.

Table 2. Descriptions of the series of samples used during the research.

Material Condition	Specimen Description
PET-G	O
Carbon fiber PET-G	C
ESD PET-G	E

3.1. Static Tensile Test

To determine the tensile strength, five measurements were made for each type of sample. The results obtained are presented in Figures 3–5. While analyzing the graph for undoped PET-G presented in Figure 3, it was observed that the obtained samples are stable but lower in comparison to the results presented in articles such as [29]. Such a phenomenon could be related to the different suppliers of the material that could use different base materials for filament production. All the specimens also exhibited stress results below the standard, approximately 30 MPa. The likely cause of these differences is the use of the same manufacturing parameters for all material configurations. The curve selected for comparison was identified from all the tests in which the stress values were around 30 MPa, aligning with the results presented in other scientific studies on this material, such as [29]. Subsequently, tests were conducted on materials with additives enhancing electrostatic discharge (ESD) resistance. In the case of PET-G samples marked with the symbol O, one type of fracture mechanism occurred. All samples carried the highest load with a deformation in the range of 4–7%. This is due to the good cohesion of the external layer with the internal structure of the sample.

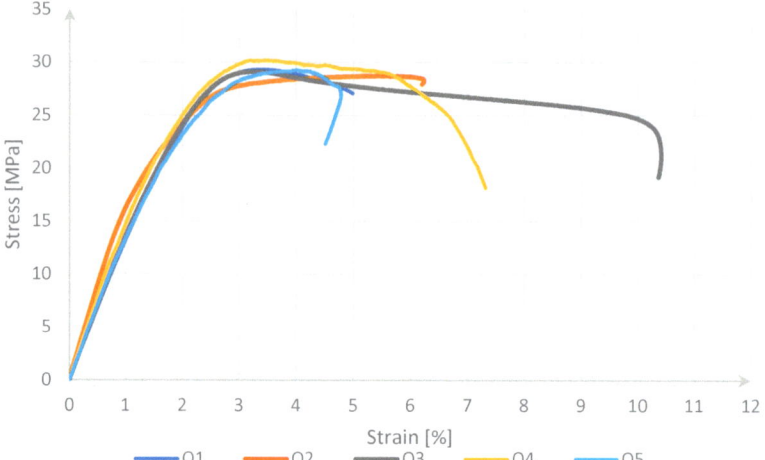

Figure 3. Stress–strain curve for PET-G.

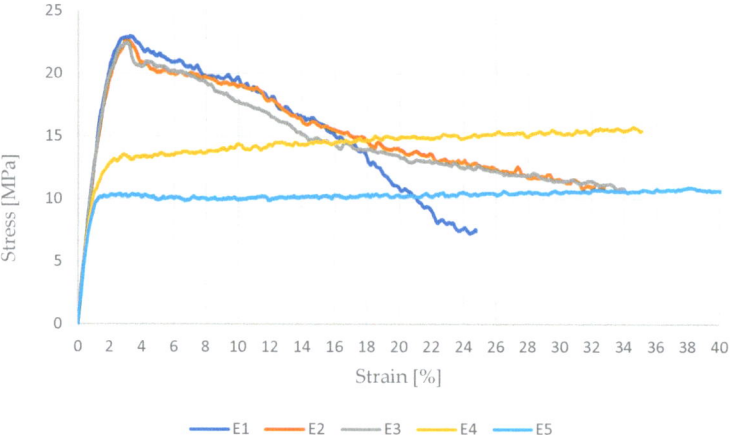

Figure 4. Stress–strain curve for ESD PET-G.

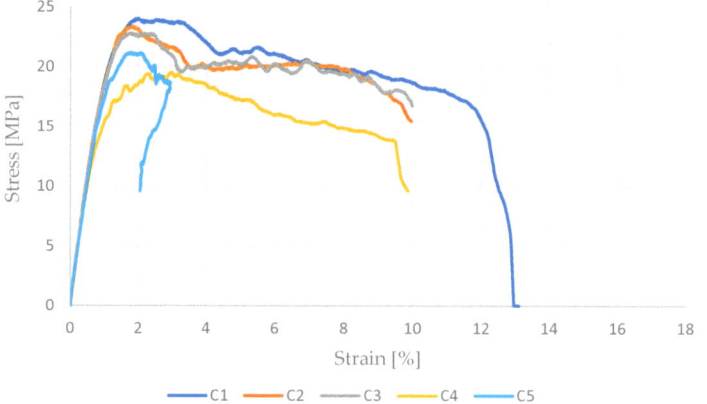

Figure 5. Stress–strain curve for PET-G carbon fiber.

The plotted graph (Figure 4) for the ESD material showed a significantly reduced maximum stress value compared to the reference material. The doping resulted in a decrease in strength to approximately 23 MPa. The graph showed two curves that significantly deviated from the rest. Hence, the first three samples exhibiting the highest tensile strength were taken as the standard. Considering the results presented in the article [30], it can be noted that they are significantly better, resembling more closely the values for undoped PET-G. The inferior properties of PETG ESD compared to the results of the study may be due to the use of two different manufacturers for this type of material. The material datasheet tested during the static tensile test does not provide information on the elements used during doping. Another reason for the reduced stress value could be the printing parameters. Additionally, an important factor is the fill pattern. The researchers in the article [30] applied values for several fill patterns, yielding different strength results ranging from approximately 34 to 51 MPa. The lowest result corresponded to the circular fill, while the highest was achieved with the linear fill pattern. In the case of samples marked "E", samples E1, E2, and E2 were considered representative and correctly made. The compact structure of the outer layer and the core ensured high strength and a gentle course of strength loss. In the case of samples E4 and E5, the internal structure mainly transferred elastic loads. After exceeding the yield point, the load was taken over by the outer layer. It gradually narrowed locally. This narrowing gradually propagated along the entire length of the sample.

Subsequently, samples made from carbon fiber-reinforced filament were tested. The graph in Figure 5 shows the stress–strain curves for PET-G reinforced with carbon fiber, which were the most consistent among all the tested samples in this series. The maximum values ranged from 21 MPa to 24 MPa. In studies published in the article [31], results around 30 MPa were reported, suggesting that the tests were conducted properly and the obtained results are comparable. For PET-G CF, the first three samples, whose curves almost overlapped, were considered representative. Given that all specimens, regardless of material type, were printed using the same parameters, it can be concluded that the applied additives significantly influence the strength properties of the printed elements. PET-G carbon fibre samples. Samples C1–C4 were characterized by good repeatability. Carbon fibers ensured a good connection of the elastic material and its cohesion in the range of plastic deformations. Sample C5 transferred the maximum load at an acceptable level, but the level of deformations was low. This resulted from the increased concentration of carbon fibers in a small area, which caused a dynamic fracture of the sample.

Figure 6 illustrates representative curves for all materials used in the strength tests. Due to the large scatter of results in mechanical tests, the authors selected curves for samples with the highest tensile strength. This criterion allows for the presentation of fracture mechanisms for samples carrying the highest load, during which the weakest links in the manufacturing process are clearly revealed.

The C2 sample achieves a maximum stress of about 23 MPa at a strain of around 6%, followed by a rapid decline in stress to near zero at a strain of about 10%. The E3 sample reaches a maximum stress of approximately 19 MPa at a strain of around 3%, and then the curve declines and remains relatively stable at around 15 MPa until a strain of about 35%. The O4 sample reaches a maximum stress of about 25 MPa at a strain of around 5%, followed by a gradual decline in stress to near zero at a strain of about 12%. Each of the tested materials exhibits different behavior and achieves different mechanical properties. The E3 sample shows the greatest ability to deform at relatively constant stress after reaching the maximum value, while the C2 and O4 samples exhibit distinct drops in stress after reaching their maximum values, with the C2 sample showing the fastest decline. The parameters used for printing in MEX technology significantly influenced the obtained results. Additionally, the storage conditions of the filaments may also be a factor contributing to lower strength results, with the addition of carbon fibers potentially increasing the material's water absorption.

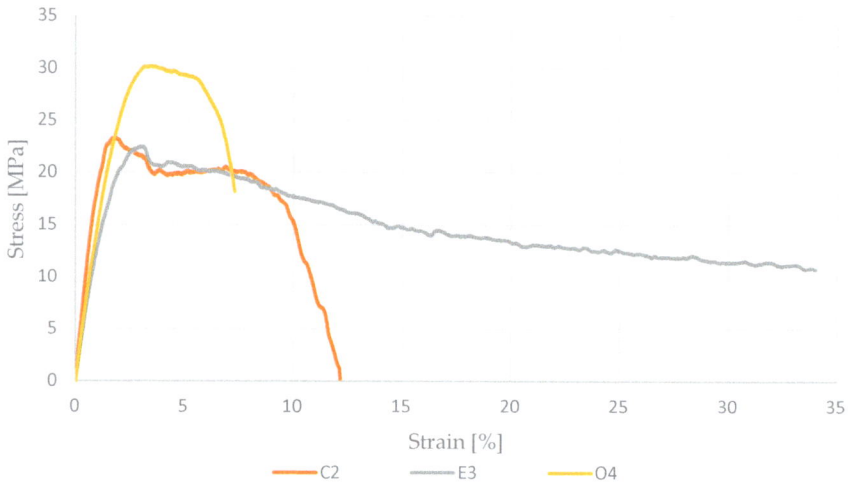

Figure 6. Representative stress–strain curves for all sample series.

3.2. Digital Image Correlation

During the static tensile test, strain analysis was conducted. The strain distributions are presented in Figures 7–9. The results obtained indicate a heterogeneous strain distribution on the surface of all the samples analyzed. This is particularly evident for pure PET-G (Figure 7), where two distinct areas of higher strain were observed after exceeding the yield point. This phenomenon may be due to inaccuracies in the print structure, where crack initiation likely occurred at these locations because of weak connections between the material's outline and the infill. Digital image correlation confirmed the tensile test results.

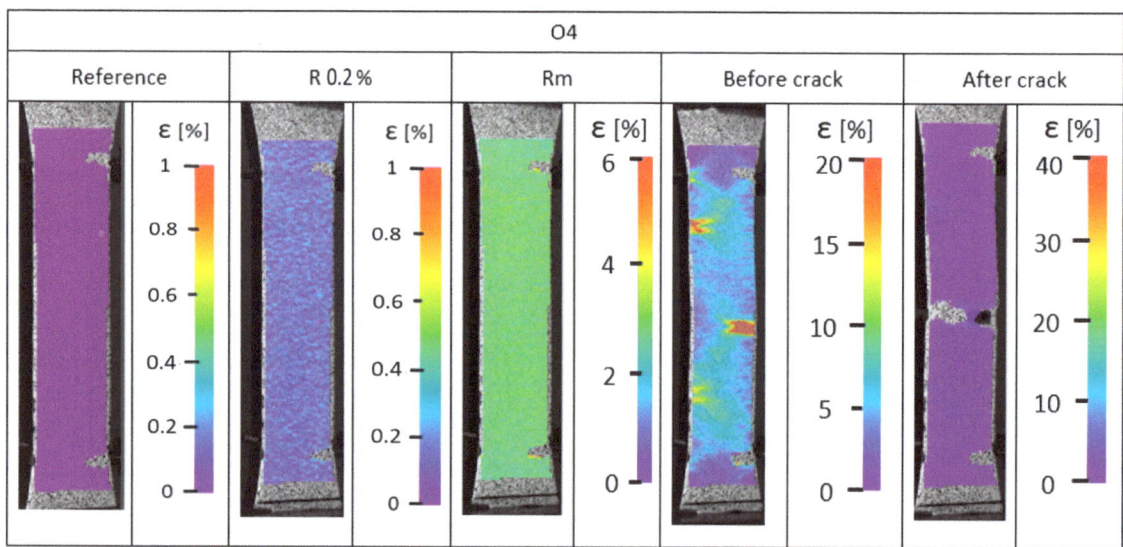

Figure 7. Strain distribution recorded using digital image correlation for PET-G.

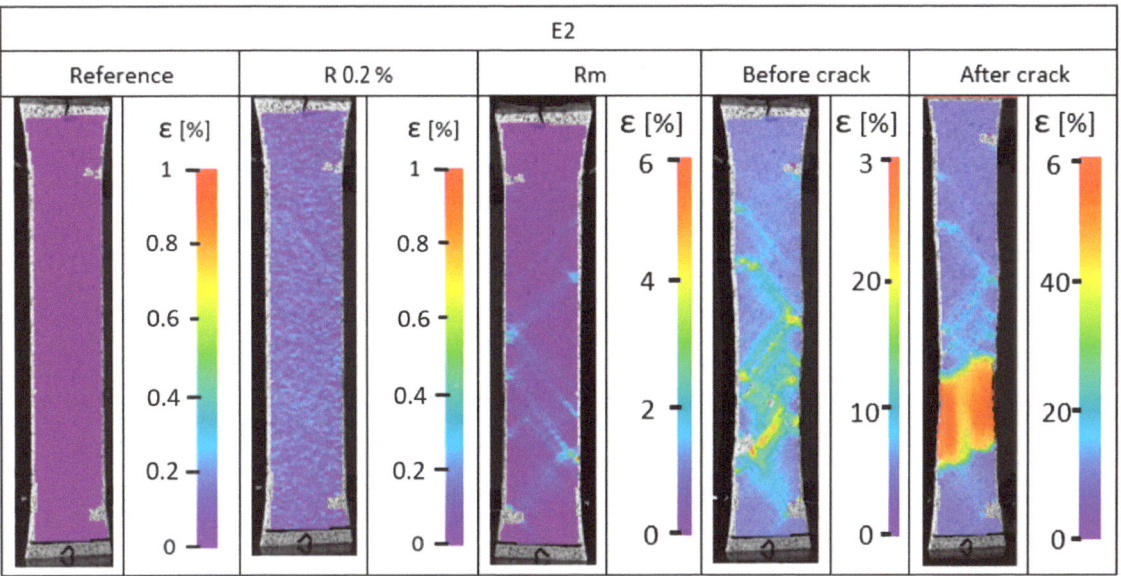

Figure 8. Strain distribution recorded using digital image correlation for EDS PET-G.

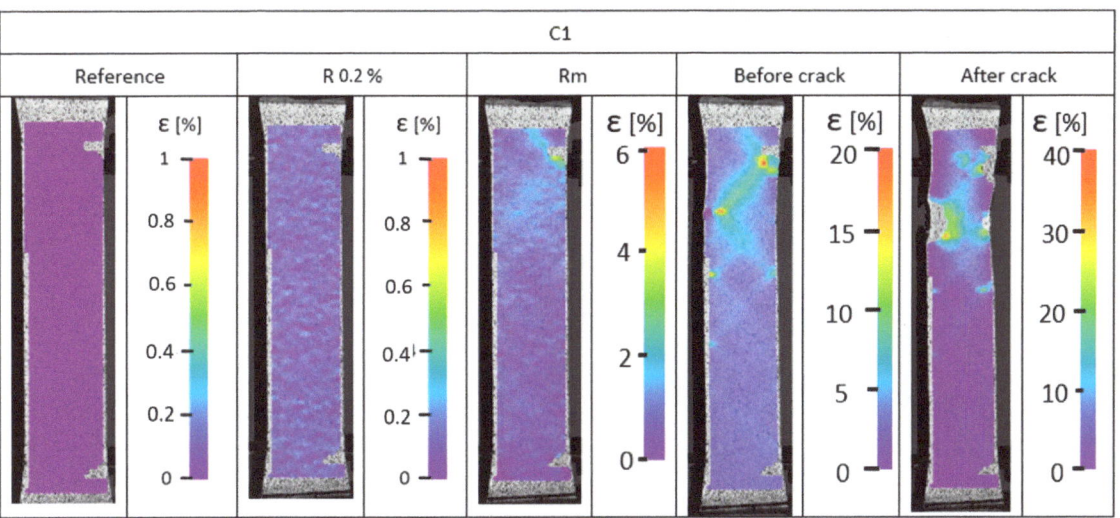

Figure 9. Strain distribution recorded using digital image correlation for carbon fiber PET-G.

The doped materials (Figures 8 and 9) exhibited much greater ductility, as measurements were taken only up to the end of the extensometer range rather than until the sample fracture. Consistently, for samples made of PC/PET-G (Figure 9) and PET-G ESD (Figure 8), areas of reduced tensile resistance were observed and identified as crack initiation points. For all samples, these regions are located near the boundary of the sample and propagate along the infill lines of the model. The propagation is attributed to the anisotropy of the mechanical properties, which indicates the variation of resistance in different directions, which is directly related to 3D printing technology.

For all samples tested, particularly just before fracture, linearly arranged areas (at an angle of approximately 45°) are observed with increased strain. This phenomenon is associated with the change in the orientation of layer deposition during the 3D printing process, which is another factor influencing the degradation of elements manufactured using PET-G-based materials.

3.3. Analysis of Fracture Surfaces after Static Tensile Testing

We show the results of the surface fracture analysis following the static tensile test, which was conducted for the cross-sections of the three types of samples examined (Figure 10A–C). The structure of pure PET-G (Figure 10A) is the most solid, with individual paths of deposited material fused together without distinct gaps. In the case of doped samples (Figure 10B,C), the individual paths of the printed material are visible, and single fibers used to dope the material can also be observed.

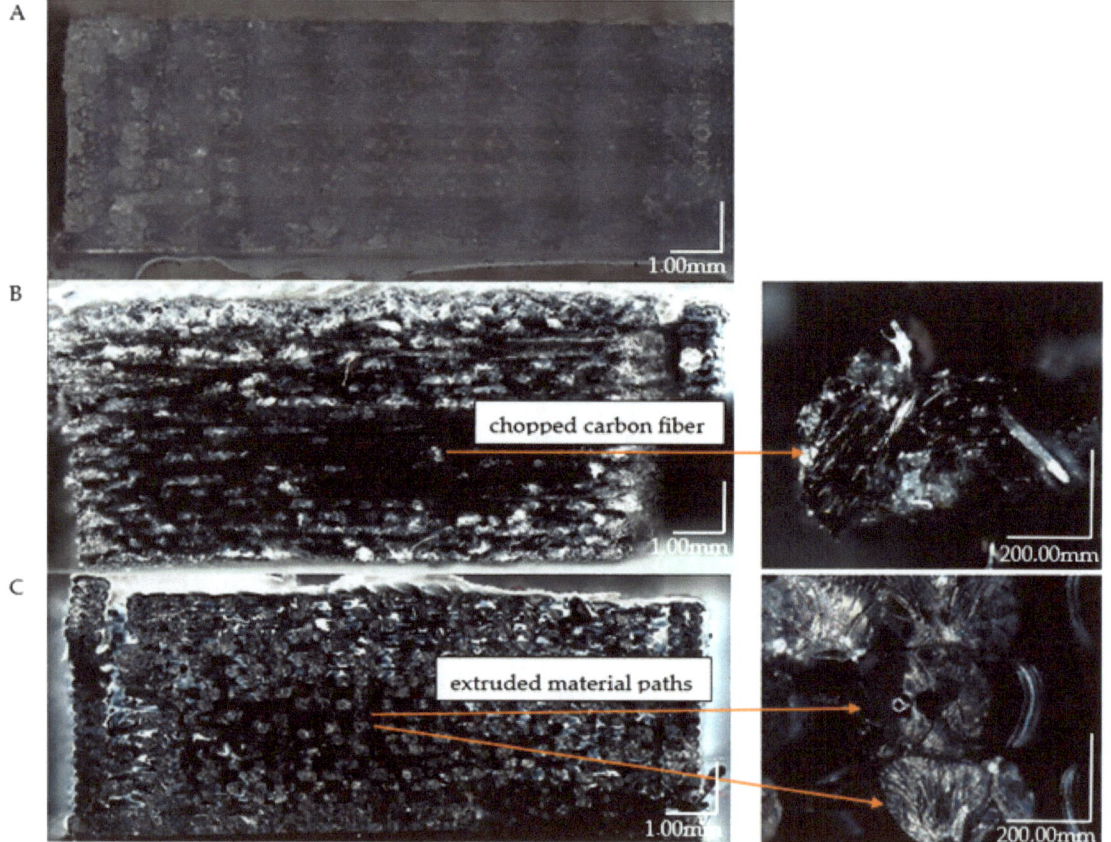

Figure 10. Sample fracture structures after static tensile testing for the following: (**A**) PET-G; (**B**) Carbon fiber PET-G; (**C**) EDS PET-G.

In the structures examined by means of a microscope for all samples, fractures predominantly exhibit brittle cracking. In some areas, plastic cracks are observed. The initiation points of material failure are also clearly visible. The first sample examined is pure PET-G (Figure 10A). The crack initiation point is marked in the close-up. The microscopic analysis also revealed that the surface in direct contact with the bed during printing has fibers much more tightly bonded than those of the upper layers. This phenomenon occurred in both pure PET-G and doped variants (Figure 10B,C). A difference in structure is visible between the wall and the filling. The second sample (Figure 10B) contained carbon fibers. It is noticeable that during the tensile test, the fibers were separated from the PET-G matrix, which has been confirmed in other studies [32]. Individual layers are not well bonded, and the broken carbon fibers create voids in the material structure. The strength of individual fibers affects the overall sample strength. In specimens with this additive, the direction in which the 3D printer places the infill negatively impacted the results. The best effects could be achieved with lines parallel to the tensile force. The destruction of material occurred over a large area of the sample. The course of the fracture process of the internal structure is unchangeable. Only the external structure has a random fracture course, and additionally, the authors wanted to present this in Figure 11.

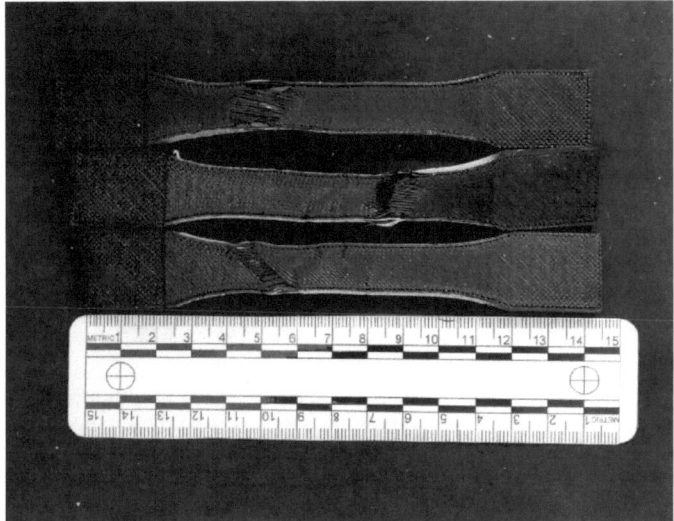

Figure 11. Samples after static tensile testing made of carbon fiber PET-G.

The last material tested was PET-G ESD (Figure 11), which exhibited the least cohesive fracture. In the microscopic image of the sample, large gaps in the central infill are noticeable. As with the other materials, the best bonding was observed between the layers printed on the bedside and those forming the walls. These structures influenced the final strength properties. The bottom surface of the sample also contacted the adhesive applied to the bed during printing. Due to the elevated temperature (approximately 60 °C), the adhesive likely bonded with the filament material. Close-ups reveal areas characteristic of plastic cracking. Unlike the other materials, the fibers in PET-G ESD mainly carried the load instead of breaking brittlely. Additionally, fibers with internal voids were observed, which could have significantly impacted the results obtained during the static tensile test.

3.4. Electrostatic Resistance

The results of the tests conducted on electrostatic discharge resistance for the printed samples are presented in Table 3.

The surface resistivity values for samples made of PET-G ESD confirm their resistance to electrostatic discharge. The resistance test results align with observations made under the microscope. A sample labeled E did not meet the ESD resistance requirements from the workbench side. The result within the limit is over 1000 times smaller. One possible reason for the difference between the bottom and top layers of the PETG ESD material sample could be the use of an additional adhesive substance. Due to the nature of 3D printing, the temperature on the bed during the process is 60 °C. The adhesive substance in direct contact with the first layer of the model may penetrate the material structure and alter its properties. For applications requiring ESD resistance, it would be advisable not to use such substances or to pre-treat the prototype surface to remove adhesive residues. In addition, parameters such as sample thickness and infill type can be significant. Depending on the application, it may be necessary to perform additional measurements to determine the ESD resistance values for the final models.

Table 3. Surface resistivity measurement results for the samples.

Tested Samples	Results on the Upper Surface		Results on the Lower Surface (Build Plate Side)	
	Voltage Load 100	Conclusion	Voltage Load 100	Conclusion
Samples O	over limit	material is an isolator	over limit	material is an isolator
Samples E	$8.7 \times 10^6 \ \Omega$	material exhibits dissipative properties	$17 \times 10^9 \ \Omega$	material does not exhibit dissipative properties—beyond ESD limit
Samples C	over limit	material is an isolator	over limit	material is an isolator

4. Conclusions

In this study, selected mechanical strength, structural, and electrostatic tests were performed on selected materials for optoelectronic applications. The aim was to understand the behavior of additively manufactured materials under mechanical load, including static tensile testing using DIC, fracture microstructure analysis, and their ESD. The results of these tests led to the following conclusions:

- PET-G without additives achieved the highest stress values (30 MPa), while ESD additives and carbon fibers reduced the strength to 23 MPa.
- PET-G ESD exhibited significant plasticity, reaching an elongation at a break of 32%, which is four times the breaking strain of pure PET-G.
- DIC analysis allowed for a detailed examination of surface deformations, showing greater plasticity in materials with additives and increased tensile resistance.
- Fracture microstructure analysis identified crack initiation sites and specific fracture features in doped materials.
- Surface resistance tests confirmed that PET-G ESD effectively disperses electrostatic discharges.

These findings underscore the dual benefits of PET-G in additive manufacturing: pure PET-G excels in mechanical strength, while doped PET-G formulations exhibit enhanced plasticity and superior ESD dissipative properties. This dual capability positions PET-G as a versatile material for a wide range of optoelectronic applications, from structural components requiring high mechanical integrity to devices demanding stringent ESD protection.

Author Contributions: Conceptualization, J.K. and J.T. (Julia Talecka); Methodology, K.J., J.K. and I.S.; Formal analysis, J.T. (Janusz Torzewski) and J.K.; Investigation, I.S., J.K. and J.T. (Julia Talecka); Data curation, K.J. and J.T. (Julia Talecka); Writing—original draft, J.T. (Julia Talecka), K.J. and J.K.; Writing—review and editing, J.K. and K.J.; Visualization, J.T. (Janusz Torzewski) and I.S.; Supervision, J.K.; Funding acquisition, J.K. All authors have read and agreed to the published version of the manuscript.

Funding: The research was funded by the Military University of Technology, grant number: 22-708.

Institutional Review Board Statement: Not applicable.

Informed Consent Statement: Not applicable.

Data Availability Statement: The raw data supporting the conclusions of this article will be made available by the authors on request.

Conflicts of Interest: The authors declare no conflicts of interest.

References

1. Goodfellow, R.C.; Debney, B.T.; Rees, G.J.; Buus, J. Optoelectronic Components for Multigigabit Systems. *IEEE Trans. Electron. Devices* **1985**, *32*, 2562–2571. [CrossRef]
2. Ostroverkhova, O. Organic Optoelectronic Materials: Mechanisms and Applications. *Chem. Rev.* **2016**, *116*, 13279–13412. [CrossRef] [PubMed]
3. Goh, G.L.; Zhang, H.; Chong, T.H.; Yeong, W.Y. 3D Printing of Multilayered and Multimaterial Electronics: A Review. *Adv. Electron. Mater.* **2021**, *7*, 2100445. [CrossRef]
4. Hu, T.; Zhang, M.; Mei, H.; Chang, P.; Wang, X.; Cheng, L. 3D Printing Technology Toward State-Of-The-Art Photoelectric Devices. *Adv. Mater. Technol.* **2023**, *8*, 2200827. [CrossRef]
5. Mazurkiewicz, M.; Kluczyński, J.; Jasik, K.; Sarzyński, B.; Szachogłuchowicz, I.; Łuszczek, J.; Torzewski, J.; Śnieżek, L.; Grzelak, K.; Małek, M. Bending Strength of Polyamide-Based Composites Obtained during the Fused Filament Fabrication (FFF) Process. *Materials* **2022**, *15*, 5079. [CrossRef] [PubMed]
6. Sawczuk, P.; Kluczyński, J.; Sarzyński, B.; Szachogłuchowicz, I.; Jasik, K.; Łuszczek, J.; Grzelak, K.; Płatek, P.; Torzewski, J.; Małek, M. Regeneration of the Damaged Parts with the Use of Metal Additive Manufacturing—Case Study. *Materials* **2023**, *16*, 3772. [CrossRef] [PubMed]
7. Bustos Seibert, M.; Mazzei Capote, G.A.; Gruber, M.; Volk, W.; Osswald, T.A. Manufacturing of a PET Filament from Recycled Material for Material Extrusion (MEX). *Recycling* **2022**, *7*, 69. [CrossRef]
8. Hauck, B.C.; Ruprecht, B.R.; Riley, P.C. Accurate and On-Demand Chemical Sensors: A Print-in-Place Ion Mobility Spectrometer. *Sens. Actuators B Chem.* **2022**, *362*, 131791. [CrossRef]
9. Hauck, B.C.; Ruprecht, B.R.; Riley, P.C.; Strauch, L.D. Reproducible 3D-Printed Unibody Drift Tubes for Ion Mobility Spectrometry. *Sens. Actuators B Chem.* **2020**, *323*, 128671. [CrossRef]
10. Su, R.; Park, S.H.; Ouyang, X.; Ahn, S.I.; McAlpine, M.C. 3D-Printed Flexible Organic Light-Emitting Diode Displays. *Sci. Adv.* **2022**, *8*, eabl8798. [CrossRef]
11. Gertrud, K.; Bernd, B. Patent: Method for Producing a Housing for an Optoelectronic Semiconductor Device, Housing, and Optoelectronic Semiconductor Device. Patent US20120280116A1, 14 July 2011. Available online: https://worldwide.espacenet.com/publicationDetails/biblio?FT=D&date=20110714&DB=EPODOC&CC=WO&NR=2011082876A1 (accessed on 5 February 2024).
12. Zheng, Z.; Chen, Z.; Xian, Y.; Fan, B.; Huang, S.; Jia, W.; Wu, Z.; Wang, G.; Jiang, H. Enhanced Electrostatic Discharge Properties of Nitride-Based Light-Emitting Diodes with Inserting Si-Delta-Doped Layers. *Appl. Phys. Lett.* **2011**, *99*, 111109. [CrossRef]
13. Neitzert, H.C.; Piccirillo, A. Sensitivity of Multimode Bidirectional Optoelectronic Modules to Electrostatic Discharges. *Microelectron. Reliab.* **1999**, *39*, 1863–1871. [CrossRef]
14. Chen, T.T.; Fu, H.K.; Dai, C.F.; Wang, C.P.; Chu, C.W.; Chou, P.T. Low-Frequency Noise Analysis of Electrostatic Discharge Tolerance of Ingan Light-Emitting Diodes. *IEEE Trans. Electron. Devices* **2013**, *60*, 3794–3798. [CrossRef]
15. Greason, W.D.; Kucerovsky, Z.; Flatley, M.W.; Bulach, S. Noninvasive Measurement of Electrostatic-Discharge-Induced Phenomena in Electronic Systems. *IEEE Trans. Ind. Appl.* **1998**, *34*, 571–579. [CrossRef]
16. Jack, H. Patent: Light Emitting Diode Lighting Device Having a Lens Connected to a Hood. Patent US7635206B2, 22 December 2009. Available online: https://worldwide.espacenet.com/publicationDetails/biblio?FT=D&date=20091222&DB=EPODOC&CC=US&NR=7635206B2 (accessed on 5 February 2024).
17. Dhawan, A.; Ghosh, T.K.; Seyam, A. Fiber-Based Electrical And Optical Devices And Systems. *Textile Progress* **2004**, *36*, 1–84. [CrossRef]
18. Letcher, T.; Rankouhi, B.; Javadpour, S. Experimental Study of Mechanical Properties of Additively Manufactured Abs Plastic as A Function of Layer Parameters. In Proceedings of the ASME 2015 International Mechanical Engineering Congress and Exposition, Houston, TX, USA, 13–19 November 2015.
19. Alafaghani, A.; Ablat, M.A.; Abedi, H.; Qattawi, A. Modeling the Influence of Fused Filament Fabrication Processing Parameters on the Mechanical Properties of ABS Parts. *J. Manuf. Process.* **2021**, *71*, 711–723. [CrossRef]
20. Hull, E.; Grove, W.; Zhang, M.; Song, X.; Pei, Z.J.; Cong, W. Effects of Process Variables on Extrusion of Carbon Fiber Reinforced ABS Filament for Additive Manufacturing. In Proceedings of the ASME 2015 International Manufacturing Science and Engineering Conference, Charlotte, NC, USA, 8–12 June 2015.
21. Boğa, C. Investigation of Mechanical and Fracture Behavior of Pure and Carbon Fiber Reinforced ABS Samples Processed by Fused Filament Fabrication Process. *Rapid Prototyp. J.* **2021**, *27*, 1220–1229. [CrossRef]
22. Olivera, S.; Muralidhara, H.B.; Venkatesh, K.; Gopalakrishna, K.; Vivek, C.S. Plating on Acrylonitrile–Butadiene–Styrene (ABS) Plastic: A Review. *J. Mater. Sci.* **2016**, *51*, 3657–3674. [CrossRef]
23. Andrzejewski, J.; Mohanty, A.K.; Misra, M. Development of Hybrid Composites Reinforced with Biocarbon/Carbon Fiber System. The Comparative Study for PC, ABS and PC/ABS Based Materials. *Compos. Part B Eng.* **2020**, *200*, 108319. [CrossRef]
24. Benfriha, K.; Ahmadifar, M.; Shirinbayan, M.; Tcharkhtchi, A. Effect of Process Parameters on Thermal and Mechanical Properties of Polymer-Based Composites Using Fused Filament Fabrication. *Polym. Compos.* **2021**, *42*, 6025–6037. [CrossRef]

25. Jia, Y.; He, H.; Peng, X.; Meng, S.; Chen, J.; Geng, Y. Preparation of a New Filament Based on Polyamide-6 for Three-Dimensional Printing. *Polym. Eng. Sci.* **2017**, *57*, 1322–1328. [CrossRef]
26. Vidakis, N.; Petousis, M.; Tzounis, L.; Grammatikos, S.A.; Porfyrakis, E.; Maniadi, A.; Mountakis, N. Sustainable Additive Manufacturing: Mechanical Response of Polyethylene Terephthalate Glycol over Multiple Recycling Processes. *Materials* **2021**, *14*, 1162. [CrossRef] [PubMed]
27. Hsueh, M.H.; Lai, C.J.; Wang, S.H.; Zeng, Y.S.; Hsieh, C.H.; Pan, C.Y.; Huang, W.C. Effect of Printing Parameters on the Thermal and Mechanical Properties of 3d-Printed Pla and Petg, Using Fused Deposition Modeling. *Polymers* **2021**, *13*, 1758. [CrossRef] [PubMed]
28. *ASTM D638-14*; 2022—Standard Test Method for Tensile Properties of Plastics. ASTM International: West Conshohocken, PA, USA, 2022.
29. *PN-EN 61340-2-3:2016-11/AC*; Elektryczność Statyczna Część 2-3: Metody Badań Stosowane Do Wyznaczania Rezystancji i Rezystywności Materiałów Stałych, Używanych Do Zapobiegania Gromadzeniu Się Ładunku Elektrostatycznego. Polish Committee for Standardization: Warsaw, Poland, 2024.
30. Stanciu, N. Effect of Low- and Extreme Low- Temperature on Mechanical Properties of 3D Printed Polyethylene Terephthalate Glycol Copolymer. *Rom. J. Tech. Sci. Appl. Mech.* **2019**, *64*, 21–41.
31. Paasi, J.; Viheriäkoski, T.; Tamminen, P.K.; Sutela, L. Electrostatic Discharge Attenuation Test for the Characterization of ESD Protective Materials. *J. Phys. Conf. Ser.* **2008**, *142*, 012038. [CrossRef]
32. Alarifi, I.M. PETG/Carbon Fiber Composites with Different Structures Produced by 3D Printing. *Polym. Test.* **2023**, *120*, 107949. [CrossRef]

Disclaimer/Publisher's Note: The statements, opinions and data contained in all publications are solely those of the individual author(s) and contributor(s) and not of MDPI and/or the editor(s). MDPI and/or the editor(s) disclaim responsibility for any injury to people or property resulting from any ideas, methods, instructions or products referred to in the content.

Article

Tungsten–SiO$_2$–Based Planar Field Emission Microtriodes with Different Electrode Topologies

Liga Avotina [1], Liga Bikse [2], Yuri Dekhtyar [3,*], Annija Elizabete Goldmane [1], Gunta Kizane [1], Aleksei Muhin [4], Marina Romanova [3], Krisjanis Smits [2], Hermanis Sorokins [3], Aleksandr Vilken [3] and Aleksandrs Zaslavskis [4]

1. Institute of Chemical Physics, University of Latvia, Jelgavas Street 1, LV-1004 Riga, Latvia; liga.avotina@lu.lv (L.A.); annija_elizabete.goldmane@lu.lv (A.E.G.); gunta.kizane@lu.lv (G.K.)
2. Institute of Solid State Physics, University of Latvia, Kengaraga Street 8, LV-1063 Riga, Latvia; lbikshe@cfi.lu.lv (L.B.); smits@cfi.lu.lv (K.S.)
3. Institute of Biomedical Engineering and Nanotechnologies, Riga Technical University, 6B Kipsalas Street, LV-1048 Riga, Latvia; marina.romanova@rtu.lv (M.R.); hermanis.sorokins@rtu.lv (H.S.); aleksandrs.vilkens@rtu.lv (A.V.)
4. Joint Stock Company "ALFA RPAR", 140 Ropazu Street, LV-1006 Riga, Latvia
* Correspondence: jurijs.dehtjars@rtu.lv

Abstract: This study examines the electrical properties and layer quality of field emission microtriodes that have planar electrode geometry and are based on tungsten (W) and silicon dioxide (SiO$_2$). Two types of microtriodes were analyzed: one with a multi-tip cathode fabricated using photolithography (PL) and the other with a single-tip cathode fabricated using a focused ion beam (FIB). Atomic force microscopy (AFM) analysis revealed surface roughness of the W layer in the order of several nanometers (Ra = 3.8 ± 0.5 nm). The work function values of the Si substrate, SiO$_2$ layer, and W layer were estimated using low-energy ultraviolet photoelectron emission (PE) spectroscopy and were 4.71 eV, 4.85 eV, and 4.67 eV, respectively. The homogeneity of the W layer and the absence of oxygen and silicon impurities were confirmed via X-ray photoelectron spectroscopy (XPS). The PL microtriode and the FIB microtriode exhibited turn-on voltages of 110 V and 50 V, respectively, both demonstrating a field emission current of 0.4 nA. The FIB microtriode showed significantly improved field emission efficiency compared to the PL microtriode, attributed to a higher local electric field near the cathode.

Keywords: planar field emission microtriode; tungsten; silicon dioxide; field emission; field emission cathode; electrical properties

1. Introduction

Advances in micro- and nanotechnology have led to a renewed interest in vacuum electronics [1–5]. Vacuum electronic devices have several advantages over semiconductor devices, making them the preferred choice for some applications. For example, vacuum devices are widely used in wireless communications and high-speed data transmission systems due to their ability to operate at high power levels and high frequencies [6–9]. Vacuum devices also have a high radiation tolerance, making them suitable for use in aerospace technology and other environments with high radiation activity levels, such as particle accelerators or radiation sources and detectors [10–14].

Microtriodes with field emission cathodes are one of the key components of vacuum electronic circuits and are used to amplify and control electrical signals. It is generally known that field emission cathodes have several advantages over thermionic cathodes, including longer life, reduced size and weight, higher power efficiency, and faster response time. However, they are more complex to manufacture because they require a very precise and controlled manufacturing process to achieve high tip sharpness. Field emitters for microelectronics are usually fabricated as vertically standing structures, such as sharp cones, tips, nanotubes, etc. [15–20]. To fabricate these structures, planar semiconductor technologies are often complemented by nanotechnologies. This complicates the design of field emission devices and increases their production costs. To reduce the costs, the planar geometry of the cathodes can be utilized.

This study compares the current and field emission characteristics of two types of microtriodes with the planar geometry of their electrodes. The first microtriode, referred to as a photolithography microtriode (PL), has a multi-tip cathode and is fabricated using planar semiconductor technologies and photolithography. The second microtriode, referred to as a focused ion beam microtriode (FIB), has a single-tip cathode and is fabricated using a focused ion beam. The choice of these two microtriodes allows us to study the influence of cathode design on the electrical characteristics of the microtriodes.

The choice of cathode type for the user will depend on the specific requirements of the application. Multi-tip cathodes have potentially higher-emission current density, lower threshold voltage, and greater resistance to damage. Multi-tip cathodes are generally more reliable than single-tip cathodes because there is a probability of electron emission from at least one of the tips, even if some of the tips are damaged or contaminated. On the other hand, the single-tip configuration allows for a more compact microtriode structure, which is an advantage in applications where size constraints are critical. Single-tip cathodes may also be preferred in applications where a high degree of special resolution is required, which is provided by the small size and sharpness of the emitting tip.

Furthermore, we evaluate the quality of the layers of the fabricated microtriodes by analyzing their surface roughness, elemental composition, and work function. The thermal properties and infrared spectroscopic analysis of the microtriode layers have been described in other works [21,22].

2. Materials and Methods

A schematic of the PL microtriode layers is shown in Figure 1a. To obtain the n+ gate layer, the surface of a p-type Si wafer was doped with phosphorus. Next, the oxidation process was performed at a temperature of 1130 °C to obtain a 0.6 µm thick thermal SiO_2 layer. A 0.2 µm thick tungsten (W) layer was then deposited on the thick thermal SiO_2 using DC magnetron sputtering. The deposition parameters included an argon atmosphere, a current of 150 mA, a pressure of 5×10^{-3} mBar, a temperature of 250 °C, and a deposition time of 3 min. The resulting resistance of the W layer was 3.8 ohm/square. The W layer was then etched to form the cathode and anode electrodes. Next, the thick oxide layer was etched to a thickness of 0.2 µm to obtain the gate oxide. Finally, windows were opened to make contact with the gate, and the aluminum (Al) wiring was formed.

The cathode of the PL microtriode has multiple tips oriented horizontally toward the anode, as shown in Figure 1b. The anode has a rectangular shape. The cathode tip angle is 22.6°, and the distance between the tips is 2.4 µm. There are 120 tips in total. The distance between the cathode and anode is 2 µm. Optical microscopy images of the fabricated PL microtriode are shown in Figure 2.

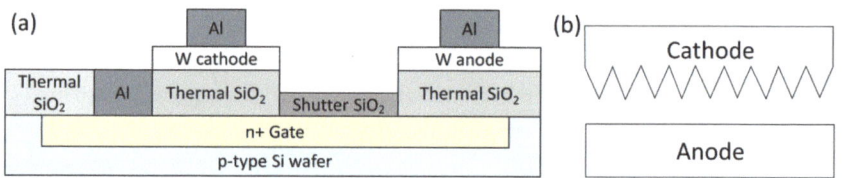

Figure 1. Schematic diagram of the PL microtriode: (**a**) microtriode layers; (**b**) mutual arrangement of the multi-tip cathode and rectangular anode, view from the top.

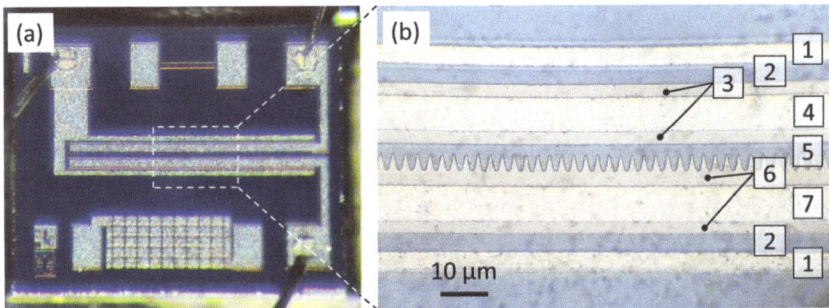

Figure 2. Optical microscope images of the PL microtriode: (**a**) image of the entire chip surface; (**b**) magnified view of the microtriode electrodes. Annotations in (**b**) correspond to the following layers: (1) Al connections to the gate; (2) thermal SiO_2; (3) anode; (4) Al connection to the anode; (5) shutter SiO_2; (6) cathode; (7) Al connection to the cathode.

Figure 3 shows scanning electron microscope (SEM) images of the second type of microtriode fabricated using a two-step FIB etching technique. The microtriode has a double-gate configuration with a cathode–anode distance of 100 nm, a gate-to-gate distance of 90 nm, and a cathode-to-gates distance of 15 nm. The gates, cathode, and anode electrodes have a taper angle of 30°. The radius of curvature of the anode electrode is 20 nm. A schematic diagram and the dimensions of the FIB microtriode are shown in Figure S1.

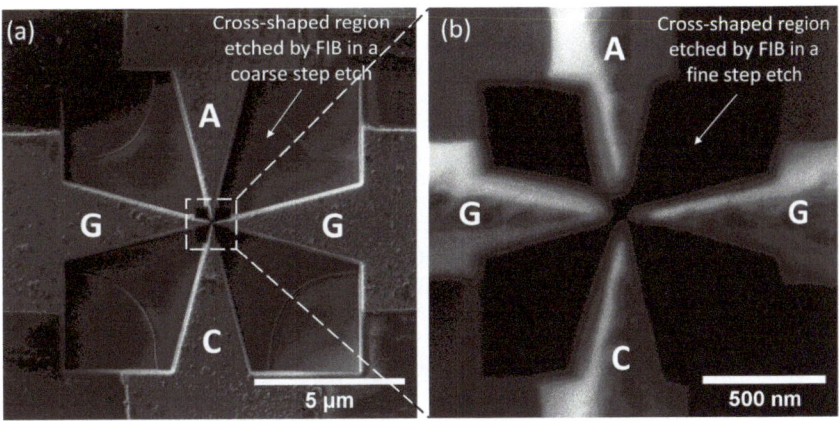

Figure 3. SEM images of the microtriode fabricated using FIB at different magnifications: (**a**) large-scale image; (**b**) close-up of the microtriode central area. Annotations: anode (A), cathode (C), two gate (G) electrodes.

To fabricate the microtriode, a cross-shaped blank was first prepared using planar semiconductor technologies to form W bridges that connect the cathode, anode, and gate electrodes. FIB was then used to cut these bridges and create a gap between the electrodes. To prepare the blank, a 1.5 μm thick SiO_2 layer was first grown on a p-type Si wafer through thermal oxidation at 1130 °C. A 0.2 μm thick layer of W was then deposited on the grown oxide using DC magnetron sputtering, using the same parameters as for the fabrication of the PL microtriode. To create the gap between the microtriode electrodes, a two-step FIB etching process was performed. In the first step, coarse structure etching was performed on a 10 × 10 μm area using a 30 kV and 1.2 nA ion beam, which reduced the connection area of the W layer. This etched region is shown in Figure 3a. Subsequently, a fine structure etching was performed on a 1.2 × 1.2 μm area using a 30 kV and 26 pA ion beam. The region etched in a second step is shown in Figure 3b. The fabrication of the microtriode was carried out using a Helios 5 UX dual-beam microscope (Thermo Scientific, Waltham, MA, USA).

To control the quality of the microtriode layers (W and SiO_2), reference samples were prepared simultaneously with the fabrication of the microtriodes. These reference samples consisted of similar layers deposited on p-type Si wafers. The quality of the layers was analyzed in terms of their elemental composition, surface roughness, and work function.

The elemental composition of the W layer was characterized using X-ray photoelectron emission spectroscopy (XPS). The measurements were performed with a ESCALAB Xi+ spectrometer (Thermo Scientific, Brno, Czech Republic). The base pressure in the analytical chamber was less than 2×10^{-7} Pa. Monoatomic Ar+ ions with an energy of 3000 eV were used to etch the surface for depth profiling. The raster size was 1×1 mm. The atomic concentrations of W4f, O1s, and Si2p were measured after every 10 s of etching, with an estimated etching rate of 13.77 nm/s (Ta_2O_5 equivalent).

The surface roughness of the W layer was characterized using atomic force microscopy (AFM). An Solver P-47 PRO microscope (NT-MDT, Zelenograd, Moscow, Russia) and NSG10/Pt AFM probes (TipsNano, Tallinn, Estonia) with a tip radius of 35 nm were used. AFM images were acquired with a scan size of 10×10 μm and processed using the Gwyddion software (version 2.63). Prior to the surface roughness analysis, the images were leveled using the mean plane subtraction method; then, the polynomial background was removed, and the minimum data value was shifted to zero.

The photoelectric work function of the fabricated layers was estimated using ultraviolet (UV) photoelectron emission (PE) spectroscopy. The measurements were performed in a vacuum of 10^{-3} Pa using a custom-made PE spectrometer. The PE was excited by a 30 W deuterium source (LOT-Oriel Europe, Darmstadt, Germany) emitting photons in an energy range of 4.13–6.20 eV (wavelengths from 295 to 200 nm). PE current was measured as a function of photon energy, and an MDR-2 UV monochromator (Lomophotonica, Saint Petersburg, Russia) with automatic scanning was used to select the wavelengths. The emitted photoelectrons were detected using an SEM-6M secondary electron multiplier (VTC Baspik, Vladikavkaz, North Ossetia-Alania, Russia), which was connected to a custom-made preamplifier, a Robotron 20046 radiometer (VEB Robotron-Meßelektronik, Dresden, Germany), and an M8784 counting board (Hamamatsu Photonics K.K., Shizuoka, Japan). The uncertainty in the photon energy measurement was within ±0.03 eV. To determine the work function, the low-energy region of the PE spectrum was analyzed by extrapolating the measured PE current to zero.

The electrical parameters of the fabricated microtriodes were measured in a custom-made vacuum chamber at a pressure of 5×10^{-5} Pa. A schematic diagram of the experimental setup for testing the electrical parameters is shown in Figure 4. The potentials were applied to the microtriode electrodes using a B5-50 DC power supply (JSC "Nizhny Novgorod plant RIAP", Nizhny Novgorod, Russia) and a C4840-02 high voltage power supply (Hamamatsu Photonics K.K., Shizuoka, Japan). The current flowing between the cathode and anode was measured using a Keithley 6485 picoammeter (Tektronix, Beaverton, OR, USA).

The field emission current passing through the vacuum gap between the cathode and anode was also detected using the electron counting method. In this case, the current was detected using an SEM-6M secondary electron multiplier (VTC Baspik, Vladikavkaz, North Ossetia-Alania, Russia), positioned above the microtriode in the vacuum chamber. The electron multiplier was connected to a custom-made preamplifier and a Robotron 20046 radiometer (VEB Robotron-Meßelektronik, Dresden, Germany). The accelerating potentials were applied to the electron multiplier using a T2DP-44 high voltage power supply (FAST ComTec Communication Technology GmbH, Oberhaching, Germany). The measurements were performed according to the setup shown in Figure 5.

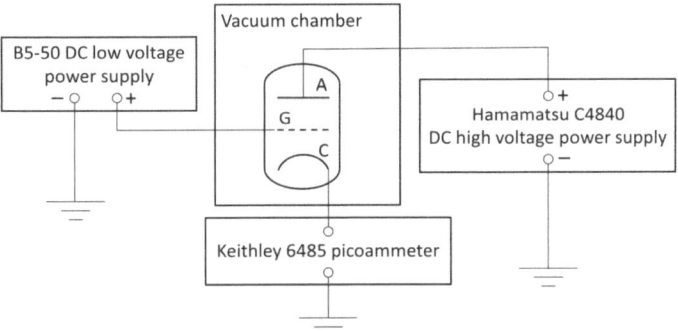

Figure 4. Schematic of the experimental setup for measuring the electrical parameters of the fabricated microtriodes.

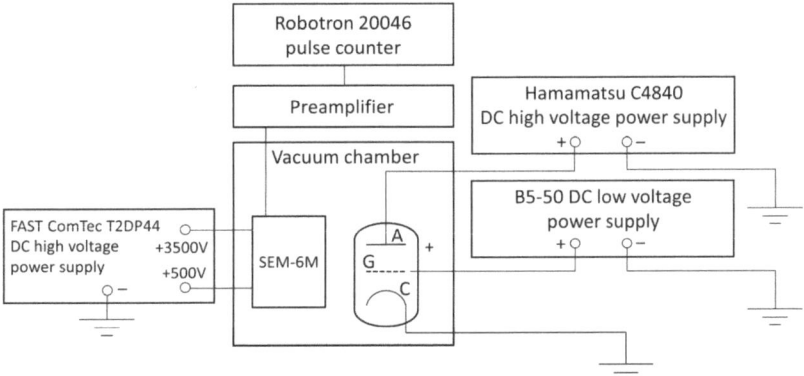

Figure 5. Schematic diagram of the setup for measuring field emission current through the cathode–anode vacuum gap using the electron counting method.

3. Results

The W layer had an average surface roughness (Ra) and root-mean-square roughness (RMS) of 3.8 ± 0.5 nm and 0.8 ± 0.1 nm, respectively, as measured using AFM. The low roughness of the emitting layer in the order of a few nanometers indicates a good fabrication quality, since the roughness of this layer should not exceed the size of the electron-emitting part of the cathode. In addition, the low roughness of the W layer reduces the occurrence of surface defects, thereby improving the electron emission properties and reducing the possibility of electron scattering [23,24].

The photoelectric work function was measured to be 4.71 ± 0.08 eV for the p-type Si substrate, 4.85 ± 0.11 eV for the SiO_2 layer, and 4.67 ± 0.06 eV for the W layer. The work function of the emitting layer must be lower compared to the materials of the other layers surrounding the cathode.

Figure 6 presents the XPS survey spectrum of the W layer. Prior to the measurements, the surface of the W layer was pre-etched with Ar+ ions for 10 s inside the XPS spectrometer chamber to remove possible surface carbon contamination. The XPS database from the reference [25] was used to identify the observed spectral features. The spectrum showed the presence of only W peaks, indicating the absence of O and Si contamination in the W layer. The binding energies where O1s and Si2p signals would be expected are also marked in Figure 6 for reference.

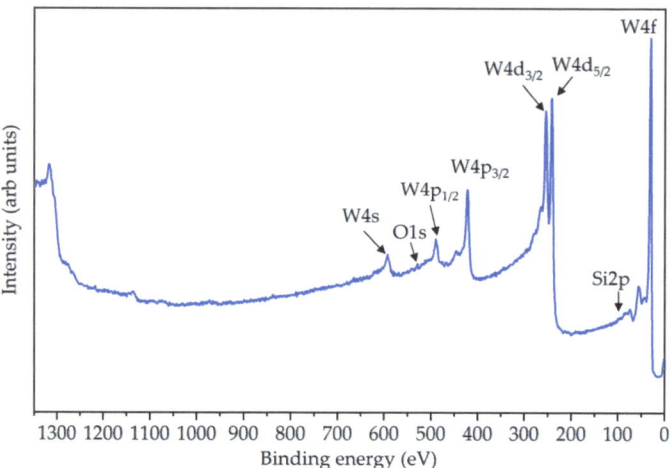

Figure 6. XPS survey spectrum of the W layer deposited on a Si/SiO$_2$ substrate; individual core levels of tungsten and the absence of oxygen and silicon signals are marked.

The XPS depth profiling results of the W layer for the presence of tungsten, oxygen, and silicon are shown in Figure S2. The increase in Si2p and O1s signals in the depth profiles indicated that the W layer was removed after about 1000 s of etching, and the underlying SiO$_2$ layer was reached. The presented depth profiling results demonstrated the absence of O and Si traces on both the surface and the bulk of the tungsten layer, indicating its homogeneity.

The high-resolution W4f spectrum is presented in Figure S3. To eliminate any probable impact from surface oxides, the spectrum was obtained after the W layer was etched for 400 s. The positions of the detected peaks were analyzed using databases from references [26,27]. The peak at 31.2 eV corresponds to W 4f7/2, the peak at 33.4 eV corresponds to W 4f5/2, and the peak at 36.7 eV corresponds to W 5p3/2. The binding energies of the peaks confirm the presence of metallic tungsten.

This section further presents the theoretical background and experimental results used to evaluate the electrical characteristics of the fabricated microtriodes.

When an external voltage is applied to a metal cathode at an electric field strength of 10^5 V/cm, the potential barrier height at the metal–vacuum interface decreases due to the Schottky effect [28]. If the field strength is further increased to 10^7–10^8 V/cm, the potential barrier height and width decrease so much that quantum mechanical tunneling becomes the dominant mechanism [29]. This leads to the emission of electrons into the vacuum, which is known as field electron emission. The relationship between the emission current density (J) and the electric field strength (E) between the electrodes in field electron emission is described by the Fowler–Nordheim equation [30]:

$$J = \frac{e^3 \cdot E^2}{8 \cdot \pi \cdot h \cdot \varphi \cdot t^2(E,\varphi)} \cdot \exp\left[-\frac{8 \cdot \pi \cdot (2m)^{1/2} \cdot \varphi^{3/2} \cdot \Theta(E,\varphi)}{3 \cdot h \cdot e \cdot E}\right] \quad (1)$$

where:

e—the charge of an electron,
φ—the work function of the cathode material,
m—the mass of an electron,
h—Planck's constant.

The functions $t(E,\varphi)$ and $\Theta(E,\varphi)$ are special functions that account for the influence of mirror image forces on the reduction in the triangular potential barrier, which affects the current value in field electron emission.

For practical purposes, the value of the function $t(E,\varphi)$ can be assumed to be equal to 1. The values of both $t(E,\varphi)$ and $\Theta(E,\varphi)$ have been tabulated in previous research [31].

Based on the experimental observation of field electron emission from metals, it is assumed that the electric field strength near the cathode surface is equal to or greater than 10^7 V/cm [28]. To achieve this, cathodes with non-uniform fields are commonly used, often employing tips with an extremely small radius of curvature. The electric field strength (E) at the apex of the tip is directly proportional to the applied voltage (U):

$$E = \beta \cdot U, \tag{2}$$

where β is the field enhancement factor [32]. The field enhancement factor is determined by solving the corresponding electrostatic problem and depends only on the geometry and dimensions of the cathode–anode system [33].

In an actual experimental setup, direct measurements of the current density (J) or the area of the electron-emitting surface (S) are not possible. Instead, the total current (I) is measured, which is the product of the current density and the electron-emitting surface area:

$$I = J \cdot S, \tag{3}$$

By utilizing Equations (2) and (3), substituting the values of the physical constants, and taking the logarithm of Equation (1), one can rewrite this equation in a form convenient for processing experimental data:

$$lg\left(\frac{I}{U^2}\right) = 10.188 + lg\left(\frac{S \cdot \beta^2}{\varphi \cdot t^2(E,\varphi)}\right) - \frac{0.297 \cdot \varphi^{3/2} \cdot \Theta(E,\varphi)}{\beta} \cdot \frac{1}{U} \tag{4}$$

where:
I—the field electron emission current in A,
U—the applied voltage in V,
E—the electric field strength in V/Å,
φ—the work function in eV,
β—the field enhancement factor in 1/Å,
S—the emitting surface area in cm^2.

The value of β can be determined by analyzing the slope of the linear part of the $lg(I/U^2)$ dependence on $1/U$. Also, the intersection point of this straight line with the $lg(I/U^2)$ axis gives the area of the electron-emitting surface (S).

Equation (4) shows that the field enhancement factor is inversely proportional to the slope:

$$\beta = \frac{0.297 \cdot \varphi^{3/2} \cdot \Theta(E,\varphi)}{\text{slope}} \tag{5}$$

Figure 7a shows the relationship between the anode–cathode current and the voltage across the anode and cathode. The current dependence on the applied voltage exhibits exponential behavior. The same relationship is represented in Figure 7b using Fowler–Nordheim coordinates. The turn-on voltage for the field emission was found to be 110 V, with a current of 0.4 nA. Beyond this voltage threshold, the dependence exhibits a linear pattern, indicating the onset of field emission current.

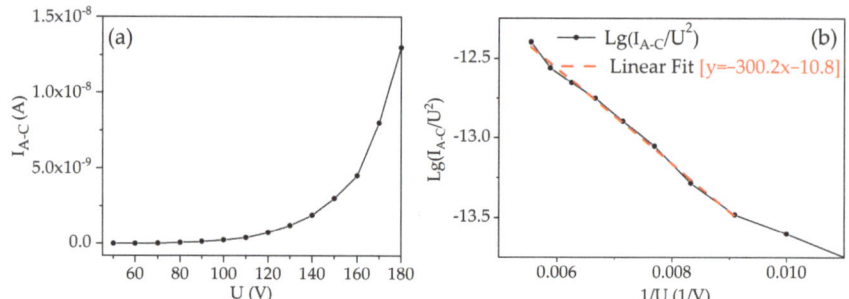

Figure 7. Anode–cathode current (I_{A-C}) characteristics for the PL microtriode: (**a**) I_{A-C} dependence on the anode–cathode voltage (U); (**b**) Fowler–Nordheim plot.

Figure 8 shows the relationship between the cathode current (I_C) and the potential of the gate electrode (U_G) at various anode potentials (U_A). As observed in the figure, increasing the anode potential leads to a corresponding increase in the cathode emission current. This can be attributed to the enhanced electric field strength in the anode–cathode gap.

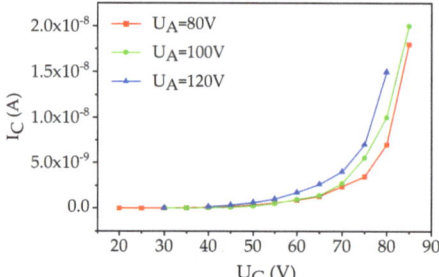

Figure 8. Dependence of the cathode current (I_C) on the gate electrode potential (U_G) at different anode potentials (U_A) for the PL microtriode.

To validate the occurrence of field emission current across the vacuum gap between the cathode and anode in the PL microtriode, electron current measurements were conducted using the electron counting method. The measurements were carried out following the setup shown in Figure 5. A voltage of 180 V was applied between the anode and cathode, and the current of 1000 electrons per second was recorded.

The electrical measurements were carried out on the FIB microtriodes according to the same procedure as for the PL microtriodes (schematics in Figure 4). The relationship between the anode–cathode current and the voltage across the anode and cathode is shown in Figure 9a, and the corresponding Fowler–Nordheim plot is shown in Figure 9b. The turn-on voltage for the FIB microtriode was found to be 50 V, with a current of 0.4 nA.

Figure 10 shows the relationship between the cathode current (I_C) and the potential of the gate electrode (U_G) at various anode potentials (U_A) for the FIB microtriode. When comparing Figures 8 and 10, it can be observed that the anode potential has a more pronounced effect on the dependence of $I_C(U_G)$ for the FIB microtriode.

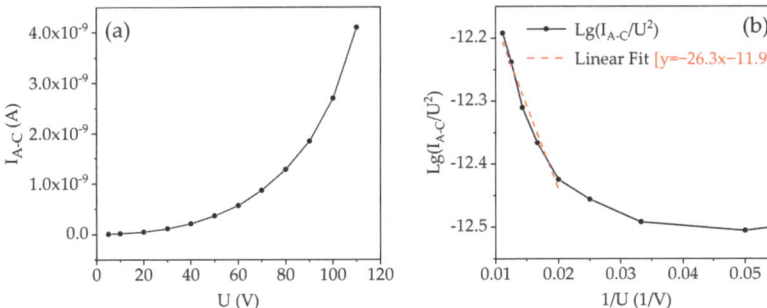

Figure 9. Anode–cathode current (I_{A-C}) characteristics for the FIB microtriode: (**a**) I_{A-C} dependence on the anode–cathode voltage (U); (**b**) Fowler–Nordheim plot.

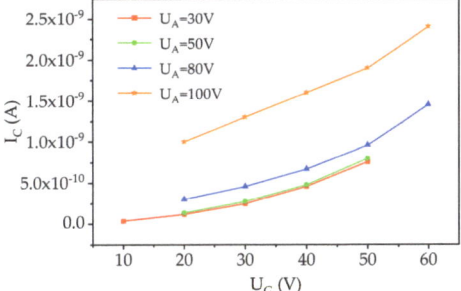

Figure 10. Dependence of the cathode current (I_C) on the gate electrode potential (U_G) at different anode potentials (U_A) for the FIB microtriode.

The local electric field strength near the cathode can be estimated using Equation (2), where the electric field strength is the product of the applied voltage and the field enhancement factor. The latter is inversely proportional to the slope according to Equation (5). The Fowler–Nordheim plots in Figures 7b and 9b show that the slope for the PL microtriode and the FIB microtriode was 300.2 and 26.3, respectively. Hence, it is inferred that the local electric field near the cathode of the FIB microtriode is one order of magnitude higher than that of the PL microtriode. This result can be attributed to the significantly reduced anode–cathode distance in the FIB microtriode, which is an order of magnitude smaller than that of the PL microtriode.

4. Conclusions

Two types of planar vacuum microtriodes and the fabrication quality of their layers were investigated. The PL microtriode had a multi-tip cathode and was fabricated using planar semiconductor technologies and photolithography. The FIB microtriode had a single cathode and was fabricated by FIB. Surface roughness analysis showed that the W layer of the microtriodes exhibited roughness values in the range of several nanometers, indicating good fabrication quality with reduced probability of surface defects. The photoelectric work function values of the Si substrate, SiO_2 layer, and W layer were estimated. XPS analysis of the W layer confirmed the absence of oxygen and silicon impurities, highlighting the homogeneity of the W layer.

Using the secondary electron multiplier and electron counting method, it was demonstrated that the current passes through the vacuum gap between the anode and cathode, confirming the occurrence of field electron emission. The Fowler–Nordheim equation was used to describe the relationship between the emission current density and the electric field strength. The analysis also showed that in the case of the FIB microtriodes, the local electric field near the cathode significantly exceeds that of the PL microtriodes, which was attributed to the smaller anode–cathode distance. This resulted in a significantly improved field electron emission efficiency for the FIB microtriodes. Also, the effect of the anode potential on the dependence of the cathode current on the gate electrode potential was more pronounced in the FIB microtriodes. The obtained results highlight the improved efficiency and performance of the FIB microtriodes in terms of field electron emission.

Supplementary Materials: The following supporting information can be downloaded at: https://www.mdpi.com/article/10.3390/ma16175781/s1, Figure S1: Configuration of the cathode (C), anode (A), and two gate electrodes (G) in a microtriode fabricated using FIB. Dimensions are in nanometers, not to scale; Figure S2: XPS depth profiling of a 200 nm thick W layer deposited on a Si/SiO$_2$ substrate, depth profiling spectra for tungsten, oxygen, and silicon are shown; Figure S3: W 4f5/2, W 4f7/2, and W 5p3/2 high-resolution XPS spectrum for tungsten metal.

Author Contributions: Methodology, Y.D. and A.Z.; Investigation, L.A., L.B., A.E.G., A.M., H.S. and A.V.; Resources, Y.D., G.K., K.S. and A.Z.; Writing—original draft, M.R. and A.V.; Writing—review & editing, Y.D. and M.R.; Supervision, Y.D.; Project administration, Y.D.; Funding acquisition, Y.D. and A.Z. All authors have read and agreed to the published version of the manuscript.

Funding: This research was supported by the European Regional Development Fund, Project No. 1.1.1.1/20/A/109 "Planar field emission microtriode structure". The Institute of Solid State Physics, University of Latvia, as a Center of Excellence, has received funding from the European Union's Horizon 2020 Framework Program H2020-WIDESPREAD-01-2016-2017-TeamingPhase2 under Grant Agreement No. 739508, Project CAMART2.

Institutional Review Board Statement: Not applicable.

Informed Consent Statement: Not applicable.

Data Availability Statement: The data presented in this study are available on request from the corresponding author.

Conflicts of Interest: The authors declare no conflict of interest.

References

1. Armstrong, C.M. The vitality of vacuum electronics. In Proceedings of the 2013 IEEE 14th International Vacuum Electronics Conference (IVEC), Paris, France, 21–23 May 2013.
2. She, J.; Huang, Z.; Huang, Y.; Huang, Z.; Chen, J.; Deng, S.; Xu, N. Introduction to the micro/nano-fabrication of modern vacuum electronic devices. In Proceedings of the 2017 30th International Vacuum Nanoelectronics Conference (IVNC), Regensburg, Germany, 10–14 July 2017.
3. Levush, B.; Abe, D.; Calame, J.; Danly, B.; Nguyen, K.; Dutkowski, E.J.; Abrams, R.; Parker, R. Vacuum Electronics: Status and Trends. *IEEE Aerosp. Electron. Syst. Mag.* **2007**, *22*, 28–34. [CrossRef]
4. Forati, E.; Dill, T.J.; Tao, A.R.; Sievenpiper, D. Photoemission-based microelectronic devices. *Nat. Commun.* **2016**, *7*, 13399. [CrossRef] [PubMed]
5. Han, P.; Li, X.; Cai, J.; Feng, J. Vertical Nanoscale Vacuum Channel Triodes Based on the Material System of Vacuum Electronics. *Micromachines* **2023**, *14*, 346. [CrossRef] [PubMed]
6. Srivastava, V. Vacuum microelectronic devices for THz communication systems. In Proceedings of the 2015 Annual IEEE India Conference (INDICON), New Delhi, India, 17–20 December 2015.
7. Rosker, M.J.; Wallace, H.B. Vacuum electronics and the world above 100 GHz. In Proceedings of the 2008 IEEE International Vacuum Electronics Conference, Monterey, CA, USA, 22–24 April 2008.
8. Hsu, S.-H.; Kang, W.P.; Raina, S.; Howell, M.; Huang, J.-H. Nanodiamond vacuum field emission microtriode. *J. Vac. Sci. Technol. B Nanotechnol. Microelectron. Mater. Process. Meas. Phenom.* **2017**, *35*, 032201. [CrossRef]
9. Han, S.-T.; Jeon, S.-G.; Shin, Y.-M.; Jang, K.-H.; So, J.-K.; Kim, J.-H.; Chang, S.-S.; Park, G.-S. Experimental investigations on miniaturized high-frequency vacuum electron devices. *IEEE Trans. Plasma Sci.* **2005**, *33*, 679–684.
10. Liu, X.; Li, Y.; Xiao, J.; Zhao, J.; Li, C.; Li, Z. Enhanced field emission stability of vertically aligned carbon nanotubes through anchoring for X-ray imaging applications. *J. Mater. Chem. C* **2023**, *11*, 2505–2513. [CrossRef]
11. Huang, W.; Huang, Y.; Liu, R.; Zhu, W.; Kang, S.; Qian, W.; Dong, C. A dual-functional micro-focus X-ray source based on carbon nanotube field emission. *Diamond Relat. Mater.* **2022**, *125*, 108970. [CrossRef]
12. Barysheva, M.M.; Zuev, S.Y.; Lopatin, A.Y.; Luchin, V.I.; Pestov, A.E.; Salashchenko, N.N.; Tsybin, N.N.; Chkhalo, N.I. Prospects for the use of X-ray tubes with a field-emission cathode and a through-type anode in the range of soft X-ray radiation. *Tech. Phys.* **2020**, *65*, 1726–1735. [CrossRef]

13. Harris, J.R.; Jensen, K.L.; Shiffler, D.A. Modelling field emitter arrays using line charge distributions. *J. Phys. D Appl. Phys.* **2015**, *48*, 385203. [CrossRef]
14. Kireeff Covo, M.; Albright, R.A.; Ninemire, B.F.; Johnson, M.B.; Hodgkinson, A.; Loew, T.; Benitez, J.Y.; Todd, D.S.; Xie, D.Z.; Perry, T.; et al. The 88-inch cyclotron: A one-stop facility for electronics radiation and detector testing. *Measurement* **2018**, *127*, 580–587. [CrossRef]
15. Kikukawa, R.; Ohkawa, Y.; Yamagiwa, Y. Effect of Xe plasma processing on characteristics of carbon nanotube-based field emission cathodes. *Diamond Relat. Mater.* **2022**, *122*, 108805. [CrossRef]
16. Schwoebel, P.R.; Spindt, C.A.; Holland, C.E. High current, high current density field emitter array cathodes. *J. Vac. Sci. Technol. B Microelectron. Nanometer Struct.* **2005**, *23*, 691–693. [CrossRef]
17. Laszczyk, K.U. Field emission cathodes to form an electron beam prepared from carbon nanotube suspensions. *Micromachines* **2020**, *11*, 260. [CrossRef] [PubMed]
18. Giubileo, F.; Grillo, A.; Passacantando, M.; Urban, F.; Iemmo, L.; Luongo, G.; Pelella, A.; Loveridge, M.; Lozzi, L.; Di Bartolomeo, A. Field Emission Characterization of MoS$_2$ Nanoflowers. *Nanomaterials* **2019**, *9*, 717. [CrossRef]
19. Zhang, Y.; Liu, X.; Zhao, L.; Li, Y.; Li, Z. Simulation and Optimization of CNTs Cold Cathode Emission Grid Structure. *Nanomaterials* **2023**, *13*, 50. [CrossRef]
20. Yu, Y.Y.; Rodiansyah, A.; Sawant, J.; Park, K.C. Patterning of Silicon Substrate with Self-Assembled Monolayers Using Vertically Aligned Carbon Nanotube Electron Sources. *Nanomaterials* **2022**, *12*, 4420. [CrossRef]
21. Goldmane, A.E.; Avotina, L.; Vanags, E.; Trimdale-Deksne, A.; Zaslavskis, A.; Kizane, G.; Dekhtyar, Y. Thermal oxidation of tungsten coatings for detection by infrared spectrometry method. *J. Phys. Conf. Ser.* **2023**, *2423*, 012022. [CrossRef]
22. Avotina, L.; Bumbure, L.; Goldmane, A.E.; Vanags, E.; Romanova, M.; Sorokins, H.; Zaslavskis, A.; Kizane, G.; Dekhtyar, Y. Thermal behavior of magnetron sputtered tungsten and tungsten-boride thin films. In Proceedings of the 2022 International Conference on Applied Electronics (A.E.), Pilsen, Czech Republic, 6–7 September 2022.
23. Kaser, A.; Gerlach, E. Scattering of conduction electrons by surface roughness in thin metal films. *Z. Phys. B Con. Mat.* **1995**, *97*, 139–146. [CrossRef]
24. Koch, J.F.; Murray, T.E. Electron scattering at a rough surface. *Phys. Rev.* **1969**, *186*, 722–727. [CrossRef]
25. Moulder, J.F.; Stichle, W.F.; Sobol, P.E.; Bomben, K.D. Tungsten. In *Handbook of X-ray Photoelectron Spectroscopy*; Chastain, J., Ed.; Perkin-Elmer Corporation, Physical Electronics Division: Eden Prairie, MN, USA, 1992; pp. 172–173.
26. NIST X-ray Photoelectron Spectroscopy Database, NIST Standard Reference Database 20, Version 4.1. Available online: https://srdata.nist.gov/xps/ (accessed on 14 August 2023).
27. X-ray Photoelectron Spectroscopy (XPS) Reference Pages, Tungsten. Available online: http://www.xpsfitting.com/search/label/Tungsten (accessed on 14 August 2023).
28. Farrall, G.A. Electrical breakdown in vacuum. In *Gas Discharge Closing Switches*; Schaefer, G., Kristiansen, M., Guenther, A., Eds.; Springer Science+Business Media: New York, NY, USA, 1990; pp. 196–197.
29. Gilmour, A.S. Cold cathodes. In *Microwave and Millimeter-Wave Vacuum Electron Devices: Inductive Output Tubes, Klystrons, Traveling-Wave Tubes, Magnetrons, Crossed-Field Amplifiers, and Gyrotrons*; Artech House: Norwood, MA, USA, 2020; pp. 120–121.
30. Fursey, G.N. Deviations from the Fowler–Nordheim theory and peculiarities of field electron emission from small-scale objects. *J. Vac. Sci. Technol. B Microelectron. Nanometer Struct.* **1998**, *16*, 910–915. [CrossRef]
31. Burgess, R.E.; Kroemer, H.; Houston, J.M. Corrected values of Fowler-Nordheim field emission functions v(y) and s(y). *Phys. Rev.* **1953**, *90*, 515. [CrossRef]
32. Lewis, P.A.; Alphenaar, B.W.; Ahmed, H. Measurements of geometric enhancement factors for silicon nanopillar cathodes using a scanning tunneling microscope. *Appl. Phys. Lett.* **2001**, *79*, 1348–1350. [CrossRef]
33. Bilici, M.A.; Haase, J.R.; Boyle, C.R.; Go, D.B.; Sankaran, R.M. The smooth transition from field emission to a self-sustained plasma in microscale electrode gaps at atmospheric pressure. *J. Appl. Phys.* **2016**, *119*, 223301. [CrossRef]

Disclaimer/Publisher's Note: The statements, opinions and data contained in all publications are solely those of the individual author(s) and contributor(s) and not of MDPI and/or the editor(s). MDPI and/or the editor(s) disclaim responsibility for any injury to people or property resulting from any ideas, methods, instructions or products referred to in the content.

Article

Application of Bis-Adducts of Phenyl-C$_{61}$ Butyric Acid Methyl Ester in Promoting the Open-Circuit Voltage of Indoor Organic Photovoltaics

Xueyan Hou [1], Xiaohan Duan [1], Mengnan Liang [2], Zixuan Wang [3,*] and Dong Yan [2,*]

1. International Collaborative Laboratory of 2D Materials for Optoelectronics Science and Technology of Ministry of Education, Institute of Microscale Optoelectronics, Shenzhen University, Shenzhen 518060, China
2. Guangdong-Hong Kong-Macao Joint Laboratory for Intelligent Micro-Nano Optoelectronic Technology, School of Physics and Optoelectronic Engineering, Foshan University, Foshan 528225, China
3. School of Mechanical Engineering and Automation, Northeastern University, Shenyang 110819, China
* Correspondence: wangzixuan_neu@163.com (Z.W.); yandong@fosu.edu.cn (D.Y.)

Abstract: Fullerene-based indoor OPVs, particularly phenyl-C$_{61}$ butyric acid methyl ester (PCBM), has been regarded as a prospective harvesting indoor light energy source to drive low-power consumption electronic devices such as sensors and IoTs. Due to the low tunability of its inherently spherical structure, the performance of the fullerene-based indoor OPVs seem to hit a bottleneck compared with the non-fullerene materials. Here, we explore the potential application of fullerene derivative bis-PCBM in indoor OPVs, which owns a higher the lowest unoccupied molecular orbital (LUMO) level than PCBM. The results show that when blended with PCDTBT, bis-PCBM devices yield a high V_{OC} of up to 1.05 V and 0.9 V under AM 1.5G illumination and 1000 lx indoor light, compared with the corresponding values of 0.93 V and 0.79 V for PCBM devices. Nevertheless, the disorders in bis-PCBM suppress the J_{SC} and FF and, therefore, result in a lower efficiency compared to PCBM devices. However, the efficiency and stability differences between the two kinds of cells were much reduced under indoor light conditions. After further optimization of the material composition and fabrication process, bis-PCBM could be an alternative to PCBM, offering great potential for indoor OPV with high performance.

Keywords: indoor photovoltaics; fullerene derivatives; open-circuit voltage; internet of things

Citation: Hou, X.; Duan, X.; Liang, M.; Wang, Z.; Yan, D. Application of Bis-Adducts of Phenyl-C$_{61}$ Butyric Acid Methyl Ester in Promoting the Open-Circuit Voltage of Indoor Organic Photovoltaics. *Materials* 2023, 16, 2613. https://doi.org/10.3390/ma16072613

Academic Editors: Shengli Pu and Jijun Feng

Received: 9 March 2023
Revised: 20 March 2023
Accepted: 23 March 2023
Published: 25 March 2023

Copyright: © 2023 by the authors. Licensee MDPI, Basel, Switzerland. This article is an open access article distributed under the terms and conditions of the Creative Commons Attribution (CC BY) license (https:// creativecommons.org/licenses/by/ 4.0/).

1. Introduction

Organic photovoltaics (OPVs) are considered to be a highly promising candidate for a large-area, flexible, and low-cost renewable energy source. Even though considerable progress has been made in the power conversion efficiencies (PCE) of OPVs; however, stability issues under harsh outdoor environments are a drawback for the commercialization of these devices [1–3]. The indoor environment has very different environmental stress factors associated with low light conditions, e.g., lack of elevated temperatures, intensive light socking, thermal cycling, and weathering, which can provide an alternative scenario for OPV commercialization. In the past few years, indoor PVs have attracted intense research attention due to their potential to harvest indoor light energy to drive low-power consumption electronic devices, such as indoor sensors and the internet of things (IoT) [4]. Among them, OPV-based indoor PVs have emerged as promising candidates for efficient light-harvesting technologies when employing highly efficient fullerene and non-fullerene materials as acceptors [5–8]. The highest efficiency of indoor OPVs based on fullerene materials is 28% under 1000 illuminance (lx) indoor light [6], and the optimized indoor OPVs based on non-fullerene materials exhibited a record PCE up to 33.2% under 19,500 lx LED light [9]. Due to the diverse structure of non-fullerene materials, as well as their remarkable optoelectronic characteristics, such as excellent light-harvesting ability and adjustable energy level, most of the current indoor OPV research is based on non-fullerene materials [5,9–11]. In contrast, fullerene materials have relatively low tunability due to their inherently spherical structure; thus, it seems that there is not much room for performance improvement for the fullerene-based, particularly PCBM, indoor OPVs.

However, fullerene materials could have greater application prospects than expected. PCBM and $PC_{71}BM$ have been utilized in indoor photovoltaic materials and achieved high efficiency [6,12,13]. Fullerene materials are usually adopted in ternary OPV devices, which include a third component in the original binary OPV devices [14]. This strategy can overcome the original devices' light absorption limitations and enhance the active layer's morphology and energy level alignments, thus increasing the PCE of OPVs [14]. The blue light emitted by indoor PVs can impair vision, and studies have shown that adding fullerene can lessen the effect [15]. Furthermore, commercial fullerene materials are cheaper than non-fullerene materials. Studies have shown that open-circuit voltage (V_{OC}) is important for the operation of indoor OPVs [7]. Fullerene derivatives, such as bis-PCBM, can increase the lowest unoccupied molecular orbital (LUMO), thereby increasing V_{OC} [16,17]. To our best knowledge, indoor OPVs based on bis-PCBM have not yet been studied. Thus, exploring the potential application of the fullerene derivative bis-PCBM in indoor OPVs is essential, as they own higher LUMO levels but have more disorders than PCBM.

The work in this paper employed bis-PCBM and the counterpart PCBM with PCDTBT in indoor OPV devices for the first time. The molecular structures and corresponding energy levels are shown in Figure 1 (the energies are from references [18–20]). The device performance indicated the positive effects of bis-PCBM on the V_{OC} under both AM 1.5G and indoor light. Although the PCE of bis-PCBM devices did not surpass that of the PCBM devices due to disorders and the amorphous property, the PCE and stability differences between the two kinds of devices were much reduced under indoor low light conditions. Bis-PCBM is expected to be a promising replacement for PCBM in high-efficient indoor OPVs with elevated V_{OC} and enhanced PCE via material synthesis and fabrication process optimization.

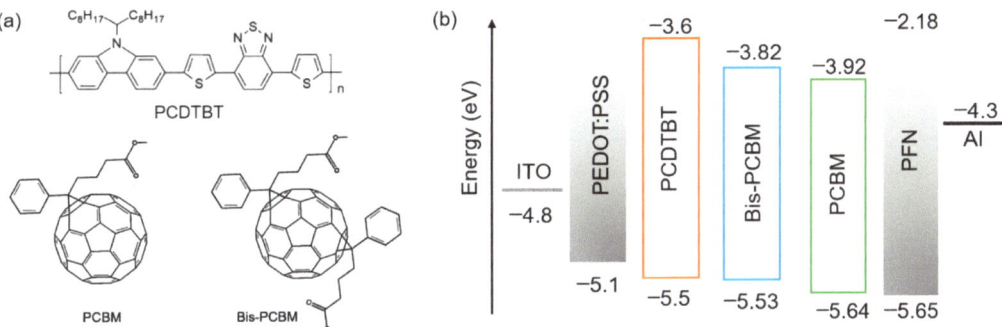

Figure 1. The chemical structure (**a**) and energy band diagram (**b**) of the polymer PCDTBT and fullerene acceptor PCBM and bis-PCBM. The chemical structure of bis-PCBM is a general one without considering its multi-isomers with the second side chain at different positions.

2. Materials and Methods

Materials: The regents and solvents were used as received from Sigma Aldrich (St. Louis, MI, USA). The materials were purchased from 1-materials.

Device fabrication and characterization: The schematic structure of the device used in this study is shown in Figure 2a. All layers were fabricated by spin coating from corresponding solutions. The metal electrode was thermally evaporated. The ITO glass (~1.1 mm) with an ITO thickness of ~100 nm and sheet resistance of ~15 Ω per square was prepared after cleaning and UV ozone treatment. The PEDOT:PSS electron transport layer was spin-coated on the ITO with a thickness of ~30 nm, followed by annealing at 150 °C for 20 min. The PCDTBT:fullerene blend solutions were prepared by dissolving solid PCDTBT and fullerene derivatives in chlorobenzene with a weight ratio of 1:2 and concentration of 20 mg/mL. The active layer was spun onto the ITO/PEDOT:PSS substrates with a thickness of ~85 nm under a nitrogen atmosphere. The film thicknesses were measured with a Dektak XT profilometer (Bruker, MA, USA). After the film was dry, the active layers were carried out with the solvent vapor annealing (SVA) treatment in a glass petri dish containing a THF atmosphere. The PFN was dissolved in methanol + 0.25 vol% acetic acid with a concentration of 0.2 mg/mL. Then the PFN was spin-cast on the active layer with a thickness of ~5 nm. Subsequently, the films were transferred to an evaporator for Al cathode deposition (100 nm). During the thermal evaporation of the Al electrode, a 6-pixel mask was used on the top of the substrate with a well-defined pixel area of 0.045 cm^2. The current density–voltage (J-V) characterization was carried out using a Keithley 2400 Source (Tektronix, OR, USA) measurement unit under the solar simulator (AM 1.5G standard light) or fluorescent lamps. The lux level was measured using a Luxmeter (Fluke, DC, USA). The EQE spectra were obtained using a grating spectrometer to create monochromatic light from a tungsten halogen light source in combination with a series of filters and a Stanford Research System SR380 lock-in amplifier to detect the photocurrent. A Silicon photodiode was used to calibrate the spectra.

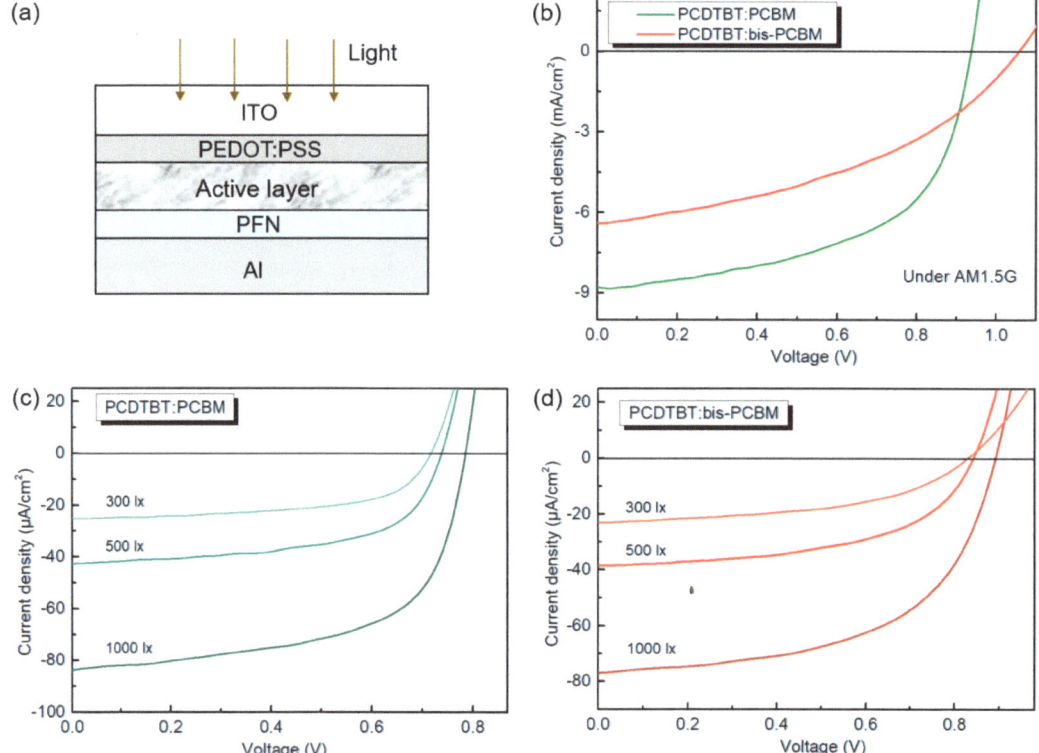

Figure 2. (a) Schematic diagram of the device structure. (b–d) *J-V* characteristics of the PCDTBT:PCBM and PCDTBT:bis-PCBM devices under AM 1.5G and indoor illuminations with the light intensity of 300 lx, 500 lx, and 1000 lx.

Degradation: Only the active layer coated ITO/PEDOT:PSS structures were used to perform the degradation treatment under AM 1.5G irradiation or 300 lx fluorescent light in air for different periods of time. After that, the interlayer and top electrode were fabricated to finish the devices.

3. Results and Discussion

3.1. Device Performance

OPV devices based on the conventional structure of ITO/PEDOT:PSS/active layer/PFN/Al (as shown in Figure 2a) were fabricated to investigate the effect of different cases of electron acceptors on the photovoltaic performance under both AM 1.5G and indoor lighting conditions. Figure 2b–d presents the experimentally measured typical J-V curves. The detailed device parameters are collected in Table 1. Under AM 1.5G light, PCDTBT:bis-PCBM cells exhibited a high V_{OC} of up to 1.05 eV, which was 120 meV higher than the V_{OC} value of the PCDTBT:PCBM devices (0.93 eV) due to the high-lying LUMO level of bis-PCBM. Nevertheless, the PCE of the PCDTBT:bis-PCBM cells (~2.9%) was much lower than that of the PCDTBT:PCBM cells (~4.7%). Of the performance parameters, the fill factor (FF) and short circuit current density (J_{SC}) were the short boards in the PCDTBT:bis-PCBM-based OPV devices, which were only 0.43 and 6.4 mA/cm^2, respectively (0.59 and 8.5 mA/cm^2 for PCDTBT:PCBM-based device). Earlier studies found that when using fullerene, higher adducts did not result in improved PCE due to a number of reasons: the addition of the second side chain on bis-PCBM inhibits the close packing of C$_{60}$ cage; bis-PCBM is amorphous; bis-PCBM is a mixture of multiple isomers which introduce disorders into both the electronic energy levels and molecular packing resulting in an adverse effect on electronic properties [21]; and the driving force for exciton dissociation at the PCDTBT/bis-PCBM interface is smaller than that of the PCDTBT/PCBM interface. These factors lead to insufficient charge transport (or electron mobility), charge injection and extraction, charge separation, and morphology problems [22,23].

Table 1. The photovoltaic parameters of the OPV cells under different light sources with different light intensities. The J_{SC} unit mA/cm^2 is for AM 1.5G, and μA/cm^2 is for indoor light.

Acceptor	Light	V_{OC}/V	FF	J_{SC}/mA cm^{-2} μA cm^{-2} (Indoor)	PCE/%
PCBM	AM 1.5G	0.93	0.59	8.5	4.7
	300 lx	0.72	0.61	25.2	13.2
	500 lx	0.74	0.61	42.1	13.6
	1000 lx	0.79	0.62	84	14.7
Bis-PCBM	AM 1.5G	1.05	0.43	6.4	2.9
	300 lx	0.83	0.52	23.3	12
	500 lx	0.85	0.53	38.9	12.6
	1000 lx	0.90	0.55	77.7	13.8

The light intensities under indoor environments are usually around 100–1000 lx [24]. Therefore, we studied the photovoltaic performance of the bis-PCBM and PCBM-based OPV cells under three light intensities in this range (e.g., 300 lx, 500 lx, and 1000 lx). As the light intensity increases, all the OPV cells show significantly increased J_{SC} and V_{OC} due to the increased input photon number. The FFs also increased slightly, leading to the overall increase in PCE from 13.2 to 14.7% (PCBM cells) and from 12 to 13.8% (bis-PCBM cells). Under the same light intensity, the J_{SC} values of bis-PCBM and PCBM-based OPV devices did not show as many differences as when under AM 1.5G illumination, which was around 25, 40, and 80 μA/cm^2 under 300, 500, and 1000 lx intensity, respectively. This may be due to the fact that all fullerene devices can effectively utilize indoor light. In comparison with the results obtained under AM 1.5G, the bis-PCBM-based devices also showed a higher V_{OC} value (110 meV) than the PCBM-based devices, while the FFs were still very low in the range of 0.52–0.55 compared to the FF range of PCBM (0.61–0.62).

Overall, the disadvantage in FF of the bis-PCBM cells offset their advantage in V_{OC}, leading to an overall lower PCE than seen in PCBM cells. The performance of bis-PCBM-based indoor PVs is still promising compared with other polymer:fullerene systems, such as the P3HT:PCBM, PBDB-T:PCBM, and PTB7:PCBM blends, which exhibited similar PCEs in the range of 9 to 15% but much lower V_{OC} (0.5 V, 0.67 V, and 0.57 V, respectively) under 1000 lx indoor light [25]. Replacing PCBM in these polymer:fullerene systems with bis-PCBM would improve the V_{OC} and potentially enhance the device performance in an indoor environment. Our work demonstrates that bis-PCBM has a higher indoor PV application prospect than mono-fullerene PCBM.

3.2. Device Degradation

Following the device performance study, we proceeded to investigate the degradation of the bis-PCBM and PCBM-based OPV devices after exposure of the active layer to AM 1.5G illumination or 300 lx indoor light, in air, for different periods. The degradation was carried out before the deposition of the top interlayer and electrode. The evolution of device performance parameters as a function of exposure time to light is summarized in Figure 3. All values were normalized to their respective initial values before degradation. It is obvious that the bis-PCBM and PCBM device parameters exhibited a range of degradation trends and the devices showed a slower degradation rate under low light intensity than that under AM 1.5G illumination. Under AM 1.5G, only the V_{OC} does not change much, maintaining >93% of the original value. The other parameters, J_{SC}, FF, and PCE, dropped faster in the initial 20 min and then leveled off until 60 min. The PCE remains 51% for PCBM and 35% for bis-PCBM after 60 min of degradation. This difference in degradation performance of bis-PCBM and PCBM-based blends may be due to crystalline PCBM having a lower tendency to take part in the oxidation reaction due to the denser molecular packing inhibiting the permeation of oxygen. Furthermore, the PCDTBT:fullerene blend degrades through a superoxide (O_2^-) degradation mechanism, whereby a higher LUMO of bis-PCBM speeds up the rate of electron transfer from fullerene to oxygen to form O_2^- [26]. Under irradiation at 300 lx, PCE of both PCDTBT:fullerene devices maintain over 80% of the original value after 60 min (85% for PCBM, 82% for bis-PCBM). Among the device parameters, the V_{OC} and FF of the two kinds of devices maintain very high stability under 300 lx and keep >97% of the initial values after continuous irradiation for 60 min. It is the reduced J_{SC} that accounts for the main PCE loss of the devices under low light.

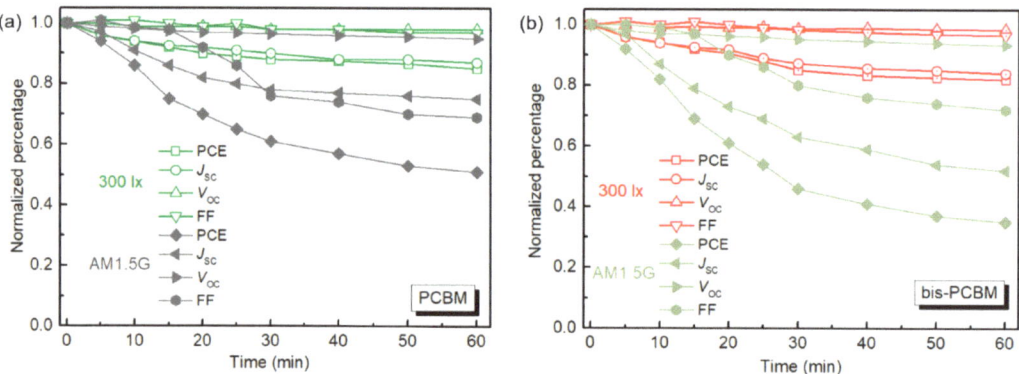

Figure 3. The normalized device performance parameters of the OPV devices based on PCDTBT blended with PCBM (**a**) and bis-PCBM (**b**) after degradation under AM 1.5G or 300 lx indoor light, in air, for a different time. All performance parameters are normalized to the value measured for the undegraded device.

To evaluate the photoresponse of the PCDTBT:fullerene blends after degradation, the corresponding EQE curves were measured. As illustrated in Figure 4, both undegraded devices exhibit a high photoresponse in the wavelength range between 400 and 600 nm, where the peak EQE is around 550 nm with a value of ~35% for bis-PCBM cells and ~49% for PCBM cells. The shape of EQE curves did not change after degradation under 300 lx indoor light for 60 min with some decrease in intensity, while under AM 1.5G the EQE curve of the bis-PCBM device flattened in the range of 400 to 600 nm, and the EQE peak value reduced to 18 and 33% for bis-PCBM and PCBM cells, respectively. The higher EQE value of PCBM devices in each case is consistent with the J_{SC} results of the J-V characterizations.

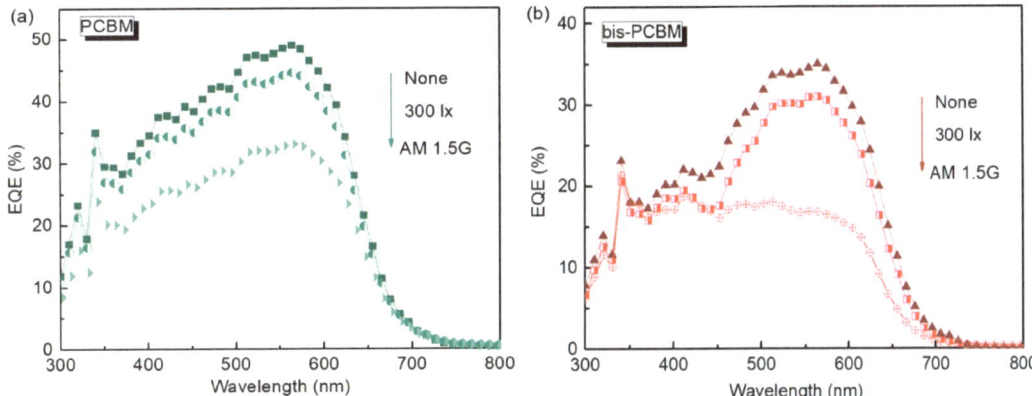

Figure 4. EQE of the OPV devices based on PCDTBT blended with PCBM (**a**) and bis-PCBM (**b**) after degradation under AM 1.5G or 300 lx indoor light, in air, for 60 min. EQE of the undegraded devices was used as references.

3.3. Voltage Loss Analysis

In addition to the device performance after degradation, it is also of great interest to analyze the voltage loss processes of the bis-PCBM OPV in order to understand its performance under indoor light and to maximize the V_{OC} by reducing the difference between the ideal and actual cases. Ideally, the OPV device can reach its maximum efficiency as indicated by the Shockley–Queisser (SQ) theory, which developed a theoretical framework to determine the limiting efficiency of a single-junction solar cell, based on the principle of detailed balance, equating the incoming and outgoing fluxes of photons for a device in open-circuit conditions [27]. For every above bandgap photon that is absorbed by the semiconductor, one electron-hole pair is generated, and all generated carriers are either collected as current in the leads or recombined, emitting a single photon per electron-hole pair. In other words, the SQ theory assumes that an ideal photovoltaic cell has a step-like absorptance with the band gap being the energy of the step, and all generated electron-hole pairs are collected with only radiative recombination occurring [28]. For the real device, it is necessary to consider the broadening of the absorption edge and the non-radiative recombination. Following previously reported methods [29–31], we investigated the voltage loss of PCDTBT:bis-PCBM OPVs before and after 60 min degradation under 300 lx indoor light. As illustrated in Figure 5, four loss components are summarized: the ideal maximum open-circuit voltage of SQ limit ($V_{OC,sq}$), which is typically 300 mV lower than the optical band gap (E_g) for the polymer:fullerene OPVs according to the previous studies in references [29]; the radiative open-circuit voltage ($V_{OC,rad}$), representing the actual light absorption spectrum; the difference between $V_{OC,sq}$ and $V_{OC,rad}$ ($\Delta V_{OC,abs}$), on behalf of the absorption edge; the difference between $V_{OC,rad}$ and V_{OC} ($\Delta V_{OC,nr}$), representing the non-radiative recombination. After degradation, the optical band gap E_g and $V_{OC,sq}$ increased slightly due to the change of absorption, while the difference between

E_g and $V_{OC,sq}$ (representing the unavoidable energy loss) did not change. The $V_{OC,rad}$ and V_{OC} decreased due to the increased loss of $\Delta V_{OC,abs}$ (from 0.18 to 0.23 V) and $\Delta V_{OC,nr}$ (from 0.25 to 0.29 V), as shown in the left part of Figure 5. The enhanced losses were due to the large shifts in energy between the main onset of absorption and the tail of absorption, and the increased non-radiative recombination after degradation. The results also indicate the increased disorders and trap states in the blend, which can slow down carrier collection and improve the likelihood of non-radiative recombination [32,33]. Reducing the intrinsic disorder of bis-PCBM and suppressing the production of new disorders during device operation can theoretically further improve the V_{OC} and device performance.

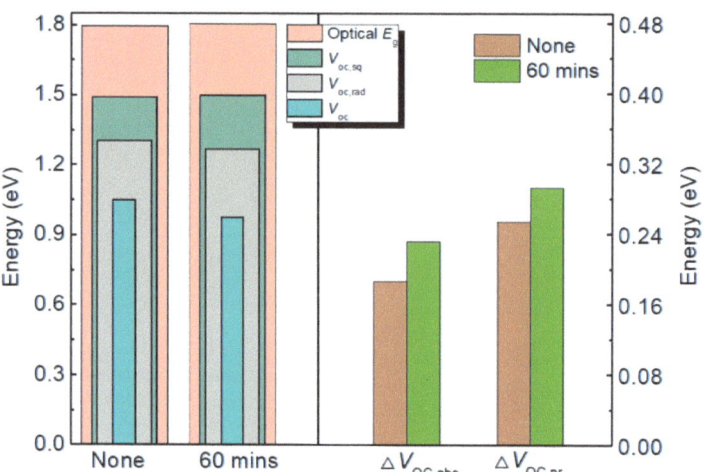

Figure 5. Comparison of the optical bandgap, SQ limit for V_{OC}, radiative V_{OC}, and V_{OC} for PCDTBT:bis-PCBM OPVs before and after degradation under 300 lx for 60 min. The $\Delta V_{OC,abs} = V_{OC,sq} - V_{OC,rad}$ and $\Delta V_{OC,nr} = V_{OC,rad} - V_{OC}$.

3.4. Discussion

The previous research found that the V_{OC} plays an important role in the OPVs under indoor light. Ensuring a high V_{OC} under one sun is important in achieving a high V_{OC} and, therefore, a high OPV device performance under low light conditions [7,34]. Bis-PCBM is a typical fullerene derivative that owns a 100 meV higher LUMO level than that of the usually used PCBM and a higher V_{OC} for the corresponding polymer:fullerene OPV devices under a one sun condition. Therefore, we carried out the indoor PVs study based on bis-PCBM to obtain an alternative material to PCBM and to achieve better indoor PV performance. The results show that the efficiency of the bis-PCBM devices is much worse than that of PCBM devices under strong illumination, even though the V_{OC} is 120 mV higher than PCBM. This is mainly due to the intrinsic disorders, low electron mobility, and amorphous property of bis-PCBM leading to adverse effects on the charge transport, charge injection and extraction, charge separation, and morphology. Under indoor light, the PCDTBT:bis-PCBM cells exhibit an expected high V_{OC} of up to 0.9 V under 1000 lx indoor light compared to the V_{OC} value of 0.79 V of PCBM-based cells. The voltage loss analysis of the bis-PCBM devices indicates a potentially higher voltage with reducing the disorders in bis-PCBM. Even though the PCE of bis-PCBM cells is still lower than that of PCBM cells (13.8 vs. 14.7%), the PCE differences between the two kinds of devices are reduced. The reason for the difference in performance between the two devices under low light illumination should be similar to that under strong illumination.

Although, at present, our bis-PCBM device does not surpass the PCBM under low light, bis-PCBM still has great application prospects in fullerene-based indoor PV. Firstly, bis-PCBM has lower electron mobility and higher molecular weight compared to PCBM, the optimal composition of bis-PCBM in the blend should be higher than PCBM, and the PCDTBT:bis-PCBM devices should have better device performance after further optimization of the fabrication process [35,36]. Secondly, bis-PCBM is a mixture of multiple isomers with energetic and structural disorders, which can be purified as isolated isomer groups or pure isomers with fewer or no disorders. The PV cells based on these isomer groups or pure isomers should have improved performance [19,37]. In terms of stability, bis-PCBM has a similar degradation rate to that of PCBM under low illumination. The top electrode and interlayer can suppress the oxygen and water permeation and decrease the degradation rate [38]. For practical application, encapsulation is also needed to protect the devices. These factors may together contribute to stable bis-PCBM OPV under an indoor environment. Moreover, the studies found that bis-PCBM is more stable than PCBM under an anaerobic environment due to the less opportunity to form dimer [26]. In addition, the OPVs may require a much shorter device lifetime considering the different target applications under indoor light (e.g., integration with consumer electronics or wireless sensors, which have a typical lifespan of less than 3–5 years) [7]. We will focus on the optimized bis-PCBM and continue to study its application in indoor OPVs in the future.

4. Conclusions

In summary, we carried out the performance investigation of indoor OPV devices comprised of bis-PCBM and its counterpart PCBM as the acceptor, respectively. With the advantage of high LUMO level, the PCDTBT:bis-PCBM devices yield a high V_{OC} up to 1.05 V and 0.9 V under AM 1.5G illumination and 1000 lx indoor light compared with the corresponding values of 0.93 V and 0.79 V of PCBM. Under AM 1.5G, the PCE of the PCDTBT:bis-PCBM cells is much lower than that of the PCDTBT:PCBM cells (2.9 vs. 4.7%), and the PCE drops faster remaining 35% of the initial value after 60 min degradation in air (remained 51% for PCBM cells). Under indoor light, the bis-PCBM and PCBM cells exhibit comparable PCE (13.8 vs. 14.7%) and device stability (maintained over 80% of the original value). The results prove that the application of bis-PCBM is an efficient and feasible approach to improve the V_{OC} of polymer–fullerene OPVs under both one sun and indoor light conditions. Because of the drawback of disorders and the amorphous property, the J_{SC} and FF of bis-PCBM devices are suppressed, leading to a lower PCE compared with PCBM devices. However, the PCE differences, as well as the stability differences between the two kinds of devices, are much reduced under indoor light. The voltage loss analysis of the bis-PCBM devices indicates a potentially higher voltage with reducing the disorders in bis-PCBM. After further optimization of the material composition and device fabrication process, bis-PCBM could be an alternative to typically used PCBM and offer great potential for high-performance OPV with enhanced V_{OC} and improved PCE.

Author Contributions: Conceptualization, X.H. and D.Y.; methodology, X.H.; software, X.D. and M.L.; validation, X.H., X.D. and D.Y.; formal analysis, X.H.; investigation, M.L. and D.Y.; resources, D.Y.; data curation, X.H. and D.Y.; writing—original draft preparation, X.D. and M.L.; writing—review and editing, X.H., D.Y. and Z.W.; visualization, X.H.; supervision, X.H. and D.Y.; project administration, X.H., D.Y. and Z.W.; funding acquisition, X.H., D.Y. and Z.W. All authors have read and agreed to the published version of the manuscript.

Funding: This research was funded by the National Natural Science Foundation of China (61704027), the Research Fund of Guangdong–Hong Kong–Macao Joint Laboratory for Intelligent Micro-Nano Optoelectronic Technology (No. 2020B1212030010), the Foshan Science and Technology Innovation Project (No. 2020001004898), the China Postdoctoral Science Foundation (No. 2021M700717), and the Fundamental Research Funds for the Education Department of Liaoning Province (No. LJKMZ20220348).

Institutional Review Board Statement: Not applicable.

Informed Consent Statement: Not applicable.

Data Availability Statement: The data presented in this study are available on request from the corresponding author. The data are not publicly available due to privacy.

Acknowledgments: This research was supported by the National Natural Science Foundation of China (61704027), the Research Fund of Guangdong–Hong Kong–Macao Joint Laboratory for Intelligent Micro-Nano Optoelectronic Technology (No. 2020B1212030010), the Foshan Science and Technology Innovation Project (No. 2020001004898), the China Postdoctoral Science Foundation (No. 2021M700717), and the Fundamental Research Funds for the Education Department of Liaoning Province (No. LJKMZ20220348).

Conflicts of Interest: The authors declare no conflict of interest.

References

1. Zhu, L.; Zhang, M.; Xu, J.; Li, C.; Yan, J.; Zhou, G.; Zhong, W.; Hao, T.; Song, J.; Xue, X.; et al. Single-junction organic solar cells with over 19% efficiency enabled by a refined double-fibril network morphology. *Nat. Mater.* **2022**, *21*, 656–663. [CrossRef] [PubMed]
2. Wang, Y.; Lee, J.; Hou, X.; Labanti, C.; Yan, J.; Mazzolini, E.; Parhar, A.; Nelson, J.; Kim, J.S.; Li, Z. Recent Progress and Challenges toward Highly Stable Nonfullerene Acceptor-Based Organic Solar Cells. *Adv. Energy Mater.* **2021**, *11*, 2003002. [CrossRef]
3. Speller, E.M.; Clarke, A.J.; Luke, J.; Lee, H.K.H.; Durrant, J.R.; Li, N.; Kim, J.-S.; Wang, T.; Wong, H.C.; Tsoi, W.C.; et al. From Fullerene Acceptors to Non-Fullerene Acceptors: Prospects and Challenges in the Stability of Organic Solar Cells. *J. Mater. Chem. A* **2019**, *7*, 23361–23377. [CrossRef]
4. Pecunia, V.; Occhipinti, L.G.; Hoye, R.L.Z. Emerging Indoor Photovoltaic Technologies for Sustainable Internet of Things. *Adv. Energy Mater.* **2021**, *11*, 2100698. [CrossRef]
5. Zhou, X.; Wu, H.; Bothra, U.; Chen, X.; Lu, G.; Zhao, H.; Zhao, C.; Luo, Q.; Lu, G.; Zhou, K.; et al. Over 31% efficient indoor organic photovoltaics enabled by simultaneously reduced trap-assisted recombination and non-radiative recombination voltage loss. *Mater. Horiz.* **2023**, *10*, 566–575. [CrossRef] [PubMed]
6. Lee, H.K.H.; Wu, J.; Barbé, J.; Jain, S.M.; Wood, S.; Speller, E.M.; Li, Z.; Castro, F.A.; Durrant, J.R.; Tsoi, W.C. Organic photovoltaic cells—Promising indoor light harvesters for self-sustainable electronics. *J. Mater. Chem. A* **2018**, *6*, 5618–5626. [CrossRef]
7. Hou, X.; Wang, Y.; Lee, H.K.H.; Datt, R.; Uslar Miano, N.; Yan, D.; Li, M.; Zhu, F.; Hou, B.; Tsoi, W.C.; et al. Indoor application of emerging photovoltaics—Progress, challenges and perspectives. *J. Mater. Chem. A* **2020**, *8*, 21503–21525. [CrossRef]
8. Nam, M.; Lee, C; Ko, D.-H. Sequentially processed quaternary blends for high-performance indoor organic photovoltaic applications. *Chem. Eng. J.* **2022**, *438*, 135576. [CrossRef]
9. Hyuk Kim, T.; Jin Chung, J.; Ahsan Saeed, M.; Youn Lee, S.; Won Shim, J. High-efficiency (over 33%) indoor organic photovoltaics with band-aligned and defect-suppressed interlayers. *Appl. Surf. Sci.* **2023**, *610*, 155558. [CrossRef]
10. Cui, Y.; Wang, Y.; Bergqvist, J.; Yao, H.; Xu, Y.; Gao, B.; Yang, C.; Zhang, S.; Inganäs, O.; Gao, F. Wide-gap non-fullerene acceptor enabling high-performance organic photovoltaic cells for indoor applications. *Nat. Energy* **2019**, *4*, 768–775. [CrossRef]
11. Wang, Z.; Tang, A.; Wang, H.; Guo, Q.; Guo, Q.; Sun, X.; Xiao, Z.; Ding, L.; Zhou, E. Organic photovoltaic cells offer ultrahigh VOC of ~1.2 V under AM 1.5 G light and a high efficiency of 21.2% under indoor light. *Chem. Eng. J.* **2023**, *451*, 139080. [CrossRef]
12. Qian, X.; She, L.; Li, Z.; Kang, X.; Ying, L. High-performance indoor organic photovoltaics enabled by screening multiple cases of electron acceptors. *Org. Electron.* **2023**, *113*, 106721. [CrossRef]
13. Jahandar, M.; Kim, S.; Kim, Y.H.; Lim, D.C. Large-Area Wide Bandgap Indoor Organic Photovoltaics for Self-Sustainable IoT Applications. *Adv. Energy Sustain. Res.* **2023**, *4*, 2200117. [CrossRef]
14. Chang, L.; Sheng, M.; Duan, L.; Uddin, A. Ternary organic solar cells based on non-fullerene acceptors: A review. *Org. Electron.* **2021**, *90*, 106063. [CrossRef]
15. Wu, Z.; Shi, R.; Chen, T.; Liu, J.; Du, X.; Ji, Z.; Hao, X.; Yin, H. Fullerene derivatives—Promising blue light absorbers suppressing visual hazards for efficient indoor light harvesters. *Appl. Phys. Lett.* **2022**, *121*, 133905. [CrossRef]
16. Zhang, F.; Shi, W.; Luo, J.; Pellet, N.; Yi, C.; Li, X.; Zhao, X.; Dennis, T.J.S.; Li, X.; Wang, S.; et al. Isomer-Pure Bis-PCBM-Assisted Crystal Engineering of Perovskite Solar Cells Showing Excellent Efficiency and Stability. *Adv. Mater.* **2017**, *29*, 1606806. [CrossRef] [PubMed]
17. Nasiri, S.; Dashti, A.; Hosseinnezhad, M.; Rabiei, M.; Palevicius, A.; Doustmohammadi, A.; Janusas, G. Mechanochromic and thermally activated delayed fluorescence dyes obtained from D–A–D′ type, consisted of xanthen and carbazole derivatives as an emitter layer in organic light emitting diodes. *Chem. Eng. J.* **2022**, *430*, 131877. [CrossRef]
18. Huang, F.; Wu, H.; Wang, D.; Yang, W.; Cao, Y. Novel Electroluminescent Conjugated Polyelectrolytes Based on Polyfluorene. *Chem. Mater.* **2004**, *16*, 708–716. [CrossRef]
19. Shi, W.; Hou, X.; Liu, T.; Zhao, X.; Sieval, A.B.; Hummelen, J.C.; Dennis, T.J.S. Purification and electronic characterisation of 18 isomers of the OPV acceptor material bis-60 PCBM. *Chem. Commun.* **2017**, *53*, 975–978. [CrossRef]
20. Sun, Y.; Takacs, C.J.; Cowan, S.R.; Seo, J.H.; Gong, X.; Roy, A.; Heeger, A.J. Efficient, Air-Stable Bulk Heterojunction Polymer Solar Cells Using MoOx as the Anode Interfacial Layer. *Adv. Mater.* **2011**, *23*, 2226–2230. [CrossRef]
21. Frost, J.M.; Faist, M.A.; Nelson, J. Energetic disorder in higher fullerene adducts: A quantum chemical and voltammetric study. *Adv. Mater.* **2010**, *22*, 4881–4884. [CrossRef] [PubMed]

22. Guilbert, A.A.; Reynolds, L.X.; Bruno, A.; MacLachlan, A.; King, S.P.; Faist, M.A.; Pires, E.; Macdonald, J.E.; Stingelin, N.; Haque, S.A.; et al. Effect of multiple adduct fullerenes on microstructure and phase behavior of P3HT: Fullerene blend films for organic solar cells. *Acs Nano* **2012**, *6*, 3868–3875. [CrossRef]
23. Faist, M.A.; Keivanidis, P.E.; Foster, S.; Wöbkenberg, P.H.; Anthopoulos, T.D.; Bradley, D.D.; Durrant, J.R.; Nelson, J. Effect of multiple adduct fullerenes on charge generation and transport in photovoltaic blends with poly (3-hexylthiophene-2,5-diyl). *J. Polym. Sci. Part B Polym. Phys.* **2011**, *49*, 45–51. [CrossRef]
24. Lee, H.K.H.; Barbé, J.; Tsoi, W.C. Organic and Perovskite Photovoltaics for Indoor Applications. In *Solar Cells and Light Management*; Elsevier: Amsterdam, The Netherlands, 2020; pp. 355–388.
25. You, Y.J.; Song, C.E.; Hoang, Q.V.; Kang, Y.; Goo, J.S.; Ko, D.H.; Lee, J.J.; Shin, W.S.; Shim, J.W. Highly Efficient Indoor Organic Photovoltaics with Spectrally Matched Fluorinated Phenylene-Alkoxybenzothiadiazole-Based Wide Bandgap Polymers. *Adv. Funct. Mater.* **2019**, *29*, 1901171. [CrossRef]
26. Hou, X.; Clarke, A.J.; Azzouzi, M.; Yan, J.; Eisner, F.; Shi, X.; Wyatt, M.F.; Dennis, T.J.S.; Li, Z.; Nelson, J. Relationship between molecular properties and degradation mechanisms of organic solar cells based on bis-adducts of phenyl-C61 butyric acid methyl ester. *J. Mater. Chem. C* **2022**, *10*, 7875–7885. [CrossRef] [PubMed]
27. Xu, Y.; Gong, T.; Munday, J.N. The generalized Shockley-Queisser limit for nanostructured solar cells. *Sci. Rep.* **2015**, *5*, 1–9. [CrossRef] [PubMed]
28. Shockley, W.; Queisser, H.J. Detailed balance limit of efficiency of p-n junction solar cells. *J. Appl. Phys.* **1961**, *32*, 510–519. [CrossRef]
29. Tuladhar, S.M.; Azzouzi, M.; Delval, F.; Yao, J.; Guilbert, A.A.; Kirchartz, T.; Montcada, N.F.; Dominguez, R.; Langa, F.; Palomares, E. Low open-circuit voltage loss in solution-processed small-molecule organic solar cells. *ACS Energy Lett.* **2016**, *1*, 302–308. [CrossRef]
30. Azzouzi, M.; Yan, J.; Kirchartz, T.; Liu, K.; Wang, J.; Wu, H.; Nelson, J. Nonradiative energy losses in bulk-heterojunction organic photovoltaics. *Phys. Rev. X* **2018**, *8*, 031055. [CrossRef]
31. Yao, J.; Kirchartz, T.; Vezie, M.S.; Faist, M.A.; Gong, W.; He, Z.; Wu, H.; Troughton, J.; Watson, T.; Bryant, D.; et al. Quantifying losses in open-circuit voltage in solution-processable solar cells. *Phys. Rev. Appl.* **2015**, *4*, 014020. [CrossRef]
32. Lee, H.K.H.; Telford, A.M.; Röhr, J.A.; Wyatt, M.F.; Rice, B.; Wu, J.; Maciel, A.C.; Tuladhar, S.M.; Speller, E.; McGettrick, J.; et al. The Role of Fullerenes in The Environmental Stability of Polymer: Fullerene Solar Cells. *Energy Environ. Sci.* **2018**, *11*, 417–428. [CrossRef]
33. Street, R.A.; Hawks, S.A.; Khlyabich, P.P.; Li, G.; Schwartz, B.J.; Thompson, B.C.; Yang, Y. Electronic Structure and Transition Energies in Polymer–Fullerene Bulk Heterojunctions. *J. Phys. Chem. C* **2014**, *118*, 21873–21883. [CrossRef]
34. Lee, H.K.; Li, Z.; Durrant, J.R.; Tsoi, W.C. Is organic photovoltaics promising for indoor applications? *Appl. Phys. Lett.* **2016**, *108*, 253301. [CrossRef]
35. Ye, L.; Zhang, S.; Qian, D.; Wang, Q.; Hou, J. Application of bis-PCBM in polymer solar cells with improved voltage. *J. Phys. Chem. C* **2013**, *117*, 25360–25366. [CrossRef]
36. Lenes, M.; Wetzelaer, G.J.A.; Kooistra, F.B.; Veenstra, S.C.; Hummelen, J.C.; Blom, P.W. Fullerene bisadducts for enhanced open-circuit voltages and efficiencies in polymer solar cells. *Adv. Mater.* **2008**, *20*, 2116–2119. [CrossRef]
37. Bouwer, R.K.; Wetzelaer, G.-J.A.; Blom, P.W.; Hummelen, J.C. Influence of the isomeric composition of the acceptor on the performance of organic bulk heterojunction P3HT: Bis-PCBM solar cells. *J. Mater. Chem.* **2012**, *22*, 15412–15417. [CrossRef]
38. Mateker, W.R.; McGehee, M.D. Progress in understanding degradation mechanisms and improving stability in organic photovoltaics. *Adv. Mater.* **2017**, *29*, 1603940. [CrossRef]

Disclaimer/Publisher's Note: The statements, opinions and data contained in all publications are solely those of the individual author(s) and contributor(s) and not of MDPI and/or the editor(s). MDPI and/or the editor(s) disclaim responsibility for any injury to people or property resulting from any ideas, methods, instructions or products referred to in the content.

Article

Simultaneous Measurement of Magnetic Field and Temperature Utilizing Magnetofluid-Coated SMF-UHCF-SMF Fiber Structure

Ronghui Xu [1,2], Yipu Xue [1,2], Minmin Xue [1,2], Chengran Ke [1,2], Jingfu Ye [1,2], Ming Chen [1,2], Houquan Liu [1,2,*] and Libo Yuan [1,2]

[1] Photonics Research Center, School of Optoelectronic Engineering, Guilin University of Electronic Technology, Guilin 541004, China
[2] Guangxi Key Laboratory of Optoelectronic Information Processing, Guilin University of Electronics Technology, Guilin 541004, China
* Correspondence: houquanliu@163.com

Abstract: We have proposed and experimentally demonstrated a dual-parameter optical fiber sensor for simultaneous measurement of magnetic field and temperature. The sensor is a magnetofluid-coated single-mode fiber (SMF)-U-shaped hollow-core fiber (UHCF)-single-mode fiber (SMF) (SMF-UHCF-SMF) fiber structure. Combined with the intermodal interference and the macro-bending loss of the U-shaped fiber structure, the U-shaped fiber sensor with different bend diameters was investigated. In our experiments, the transmission spectra of the sensor varied with magnetic field strength and temperature around the sensing structure, respectively. The dip wavelengths of the interference spectra of the proposed sensor exhibit red shifts with magnetic field strength and temperature, and the maximum sensitivity of magnetic field strength and temperature were 1.0898 nm/mT and 0.324 nm/°C, respectively.

Citation: Xu, R.; Xue, Y.; Xue, M.; Ke, C.; Ye, J.; Chen, M.; Liu, H.; Yuan, L. Simultaneous Measurement of Magnetic Field and Temperature Utilizing Magnetofluid-Coated SMF-UHCF-SMF Fiber Structure. *Materials* **2022**, *15*, 7966. https://doi.org/10.3390/ma15227966

Academic Editor: Marcel Poulain

Received: 30 September 2022
Accepted: 8 November 2022
Published: 11 November 2022

Publisher's Note: MDPI stays neutral with regard to jurisdictional claims in published maps and institutional affiliations.

Copyright: © 2022 by the authors. Licensee MDPI, Basel, Switzerland. This article is an open access article distributed under the terms and conditions of the Creative Commons Attribution (CC BY) license (https://creativecommons.org/licenses/by/4.0/).

Keywords: dual-parameter optical fiber sensor; magnetofluid; SMF-UHCF-SMF fiber structure

1. Introduction

Magnetic field sensors have found numerous potential applications in aerospace, underwater military, power grid, geological exploration, medical instrument and precision measurement [1–5]. There are many conventional magnetic field sensors, such as those based on the Hall effect, those using magneto-transistors or those that are magneto-resistive, that have some disadvantages, such as complicated structure, expensive cost, difficulty withstanding harsh environment conditions, being not easy to multiplex and to monitor remotely and so on [6,7]. Compared to conventional magnetic field sensors, fiber-optic-based magnetic field sensors have many benefits such as compact size, immunity to electromagnetic interference, remote monitoring, multiplexing capability, high sensitivity and resolution, good insulation and so on [8,9]. Therefore, optical fiber magnetic field sensors have aroused more and more attention during the past four decades [10,11].

Magnetic fluid (MF), also known as magnetofluid or magnetic liquid, is a stable colloidal liquid consisting of magnetic nanoparticles (Fe_3O_4, $CoFe_2O_4$ or $MnFe_2O_4$) coated with surfactant (oleic acid) and dispersed in a suitable liquid carrier (water, organic solvent or oil) [12,13]. MF possesses a unique magnetic-field-dependent refractive index (RI) as an external magnetic field is applied. Thus, MF exhibits some abundant magneto-optical characteristics, such as the Faraday effect, the birefringence effect, dichroism and tunable refractive index (RI), and can be used to design various optical devices such as wavelength filters, magneto-optical modulators, optical switches, magnetic field sensors and so on [14,15].

In recent years, many optic fiber magnetic field sensors combined with MF have been reported, which have different structures or principles, such as those based on tapered fiber [16,17], microstructure fiber [18,19], D-shaped fiber [20,21] or those based on interference mechanism [22–25], surface plasmon resonance (SPR) [26–28], photonic band gap effect [29,30] and so on. In 2016, Zhengyong Li et al. [24] reported an ultrasensitive magnetic field sensor based on a compact in-fiber Mach-Zehnder interferometer (MZI) created in twin-core fiber. This sensor has an ultrahigh magnetic field sensitivity of 20.8 nm/mT from 5 mT to 9.5 mT. In 2020, Qianyu Lin et al. [31] developed the half-side-gold-coated MSM structure magnetic field sensor, and the sensor magnetic field intensity is 10.08 nm/mT at the magnetic field range of 2–12 mT. These sensors have high sensitivity. Nevertheless, the properties of the MF are also sensitive to temperature. Therefore, several fiber-optic sensors have also been reported to achieve the simultaneous measurement of magnetic field and temperature in recent years. An effective and simple method is cascaded with a fiber Bragg grating (FBG) [32–34]. In 2016, Yaofei Chen et al. [32] realized a macro-bending SMF cascaded with a FBG structure with the sensitivities of 1.426 nm/mT and 0.329 nm/°C. The macro-bending SMF and FBG induce two types of dips in the transmission spectrum of the sensor, which can be realized in dual-parameter measurement by monitoring the wavelength shifts of the two types of dips. In 2022, Yuxiu Zhang et al. [35] proposed a nonadiabatic tapered microfiber cascaded with FBG structure, and this sensor obtained high sensitivity of magnetic field and temperature that are 1.159 nm/mT and −1.737 nm/°C, respectively. This sensor can achieve high sensitivity, but sensors based on microfiber-nanofiber may be fragile. In addition to the method of cascading FBGs, some special structures can also measure the magnetic field and temperature at the same time. A single-mode-D-shaped-single-mode fiber structure was proposed by Yue Dong et al. in 2018 [36]. The D-shaped fiber is a side-polished fiber structure, and the fabrication of the sensor structure is time-consuming. The magnetic field intensity and temperature of this sensor are 0.997 nm/mT and 0.0775 nm/°C, respectively, and this sensor has high magnetic field sensitivity. In 2022, Dongying Wang et al. [37] designed a two-channel photonic crystal fiber (PCF)-SPR structure applied in detecting the magnetic field and temperature. It requires gold film or other special materials in the film-plating process in different channels and is filled with polydimethylsiloxane (PDMS) in channel 2 to achieve a response to the temperature. This sensor's maximum sensitivities of magnetic field intensity and temperature are 0.65 nm/mT and 0.52 nm/°C. In the same year, Bing Sun et al. [22] reported a SMF-(Mn_3O_4-PDMS-Air) FP cavity-SMF structure that detected magnetic field intensity and temperature, and this sensor has a high sensitivity of temperatures up to 1.16 nm/°C, and the maximum sensitivity of magnetic field intensity is 0.563 nm/mT. This sensor mainly focused on the measurement of small magnetic fields, and the fabrication methods of the sensor are relatively complex.

In this work, we propose a dual-parameter fiber sensor based on HCF and MF for simultaneously measuring magnetic field intensity and temperature. The proposed sensor probe consists mainly of a SMF-UHCF-SMF structure, with a UHCF sandwiched between two SMFs. The UHCF section is totally sealed in a U-shaped vessel filled with MF, with UV glue at both ends preventing the fluid from leaking out. In our experiment, the maximum magnetic field intensity sensitivity of 1.0898 nm/mT and temperature sensitivity of 0.324 nm/°C were obtained, respectively. Unlike other dual parameter sensors, our sensor is very simple to fabricate because only fusion splicing and encapsulation are involved. Our designed sensing structure has good sensitivity, good linearity of response, low cost, no additional FBG and good mechanical performance.

2. Preparation and Principle

The HCF in the experiment is silica capillary with an inner diameter of 10 μm and a cladding diameter is 125 μm. The cross-section and parameters of the HCF are shown in Figure 1a. The SMF-UHCF-SMF structure is fabricated by sandwiching a length of HCF between two segments of SMFs. The fabrication process of the SMF-UHCF-SMF fiber structure is as follows:

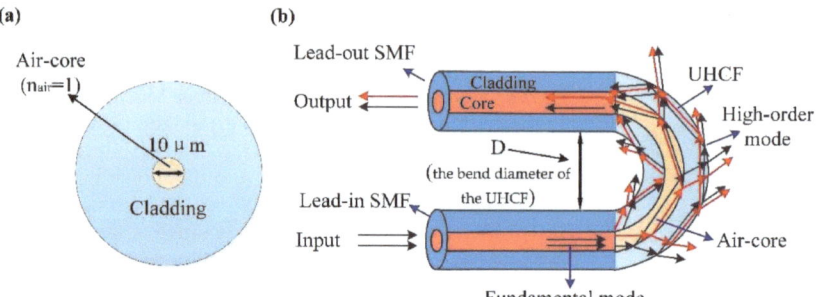

Figure 1. (a) Cross-section of the HCF. (b) Schematic diagram of the SMF-UHCF-SMF structure with optical path.

Firstly, a segment of the HCF without a coating layer is spliced with a section of SMF with the coated layer stripped. We must ensure that the length of the HCF is 5 mm by fixed length cutting, and then another section of SMF with coated layer stripped is spliced at the other end of the HCF to form the SMF-HCF-SMF structure. Then, a regular U-shape vessel (the diameter of the vessel is 10 mm) is used as a simple bending guide, and the straight SMF-HCF-SMF fiber structure is bent to the SMF-UHCF-SMF structure. Finally, the SMF-UHCF-SMF structure is in a U-shaped vessel, and the vessel is filled with MF. To keep the optical fiber stable, both ends of the U-shaped vessel are sealed with UV glue.

Two sections of SMFs act as beam splitter and optical coupler, respectively. The high-order modes are excited at the spliced region of the lead-in SMF and the UHCF. The optical path of SMF-UHCF-SMF structure is shown in Figure 1b. When the lightwave guided in the lead-in SMF core is transmitted to the UHCF, due to the collapsed splicing region between the SMF and the UHCF, the incident lightwave diffracts and spreads out into the collapsed splicing region and excites high-order cladding modes. The high-order cladding modes propagate along the capillary wall and the air-core of the UHCF. In addition, there is a portion of incident lightwave that still propagates within the air-core of the UHCF and enters the capillary wall of the UHCF. When light in bent fibers leaks from the fiber core to the cladding, it causes a portion of ray radiation to penetrate the fibers and reflect onto the interface between the fiber cladding and the air [38]. At the spliced region of the lead-in SMF and the UHCF, the fundamental mode in the lead-in SMF is split into the air-core modes and the high-order cladding modes. The cladding modes and the air-core mode propagate along the UHCF. These mixing modes experienced different bending attenuation; meanwhile, they have different responses to the environmental RI (ERI). The air-core modes and the cladding modes mutually interfere at the splicing region of the lead-out SMF and the UHCF. The interference lightwave of the sensor is led-out from the lead-out SMF and observed by an OSA. In addition, the losses mainly happen where a fiber's straight section transitions to a curved section (i.e., where the bend radius changes) [39]. Through conformal transformation, the SMF-UHCF-SMF structure with bending diameter D can be regarded as an equivalent straight waveguide (ESW). The effective refractive index (ERI) of ESW can be expressed as:

$$n_{esw} = n(x,y)\left(1 + \frac{x}{D}\right) \quad (1)$$

where x is the diameter of the fiber, and D is the bend diameter of the UHCF. In addition, $n(x,y)$ is the original RI of the straight fiber and n_{esw} is the RI of the outer portion of the fiber. At the lead-out SMF, the cladding modes and the core mode couple back to the SMF core; thus, the multiple-mode interference effect will produce. According to the interference theory, the transmission spectra of the SMF-UHCF-SMF structure can be described as [40–42]:

$$I_{out} = I_{core} + I_{clad} + 2\sqrt{I_{core}I_{clad}}\cos\phi \quad (2)$$

where I_{out} is the output light intensity, and I_{core} and I_{clad} represent the light intensity of fundamental mode and the cladding modes of the HCF participating in the interference, respectively. In addition, ϕ is the phase difference between the fundamental mode and the cladding modes, which can be expressed as follows:

$$\phi = \frac{(n_{core}^{eff} - n_{clad}^{eff})}{\lambda} L \qquad (3)$$

λ is the wavelength of the incident light, L is the length of the UHCF and n_{core}^{eff} and n_{clad}^{eff} are the effective RIs of the core mode and the cladding mode, respectively. Then the interference spectrum shift wavelength is given by:

$$\lambda_{dip} = \frac{2}{2m+1} \Delta n_{eff} L \qquad (4)$$

where Δn_{eff} is the ERI difference between the core mode and the cladding mode, $\Delta n_{eff} = n_{core}^{eff} - n_{clad}^{eff}$, because of the change in the RI of the external environment, so from Equation (4), it can be reasoned that the RI changes in the ambient environment will affect the resonance wavelength of the dip.

According to Equation (4), we can calculate that the position of the interference dip depends on the effective refractive index difference between the fundamental mode and the cladding mode when the interference path length is fixed. The effective RI of the cladding mode changes with the external environment RI, while the effective RI of the core mode does not change, which makes the position of the interference dip change. Therefore, the external environment variables can be detected and measured by monitoring the changes of the wavelength of the interference dip. Compared to the SMF-HCF-SMF structure, the SMF-UHCF-SMF structure has a bending oscillation interference and intermodal interference form similar to the MZI effect.

The RI of magnetic fluid is affected not only by magnetic field intensity but also temperature, and the relationship can be described as [43]:

$$n(H) = \begin{cases} n_0, H \leq H_c \\ n_s - n_0 \left[\coth\left(\frac{H - H_{c,n}}{T}\right) - \frac{T}{\alpha(H - H_{c,n})} \right] + n_0, H > H_{c,n} \end{cases} \qquad (5)$$

where $H_{c,n}$ is the critical magnetic field intensity, n_s is the saturation RI of MF, n_0 is the RI when the magnetic field intensity is less than $H_{c,n}$, α is the fitting parameter and H and T are the external magnetic field intensity and temperature, respectively.

3. Experimental Results and Discussion

The experimental setup of the magnetic field measurement is illustrated in Figure 2. In our experiment, the SMF-UHCF-SMF structure or the sensor head was fixed in the center of an adjustable permanent magnet (YP Magnetic Technology Development Co., Changchu, China), which was used to supply a uniform magnetic field source. The magnetic field intensity of the magnetic field source was monitored by a Gauss meter with resolution of 0.1 mT. The continuous light from the supercontinuum broadband light source (SBS, Wuhan Yangtze Soton Laser Co., Ltd., Wuhan, China) was emitted into the magnetic field sensor by the lead-in SMF. The SBS wavelength range was 600 nm–1700 nm. The transmission spectrum of the sensor was outputted from the lead-out SMF and observed by an optical spectrum analyzer (OSA, Yokogawa, AQ6370C, Tokyo, Japan) with a spectral resolution of 0.02 nm.

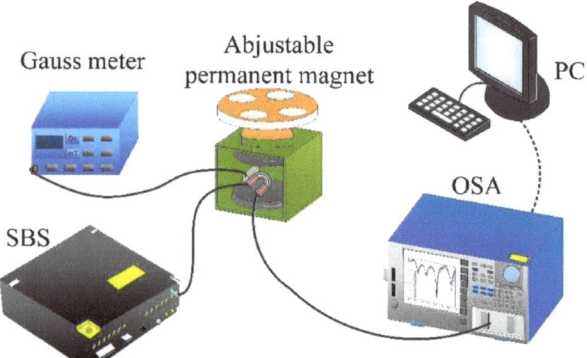

Figure 2. Experimental setup of the magnetic field measurement.

3.1. Refractive Index Measurement

Since the prepared sensor is based on the adjustable RI feature of the MF to measure magnetic field intensity, we needed to explore the RI sensitivity of the sensor firstly. In order to select an optimal bending curvature and prevent the sensing structure (SMF-UHCF-SMF structure) to break, the bending diameter of the UHCF could not be selected too short. Additionally, the bending diameter of the UHCF could not be too large for obtaining a high sensitivity. Thus, we chose three sensor samples with bend diameter of the UHCF around 47.0 mm, 50.0 mm and 54.0 mm, respectively, to verify and choose an optimize sensing structure.

When the room temperature was maintained at 25 °C, the response of the sensor to external RI was studied by completely immersing the sensor in a glycerol solution. RI solutions consisted of glycerol mixed with water in various proportions. The RI of the solution was calibrated by an Abbe refractometer. The actually measured corresponding RIs were 1.340, 1.350, 1.360, 1.370, 1.380, 1.390, 1.400 and 1.410, respectively. The transmission spectra and the wavelength shift of three sensor samples in the different RI are shown in Figure 3.

Figure 3a–c show the transmission spectra of the three selected sensing structure parameters or the bending diameter of the HCF with the refractive index. From Figure 3a–c, we can see the changes of the three sensor samples in transmission spectra are similar. The interference spectra shift toward longer wavelength (red shift), and the transmission loss also change as the glycerol RI increases. In the range of 1500~1600 nm, the fitting results show that the RI shift is linear, the RI sensitivity of D = 47 mm is 184.17143 nm/RIU, the RI sensitivity of D = 50 mm is 113.0856 nm/RIU and the RI sensitivity of D = 54 mm is 284.503 nm/RIU. To represent the accuracy of the data, we added error bar analysis to the discrete data. As shown in Figure 3d, these red vertical lines represent data errors, and the error bars (red vertical lines) indicate that the RI sensing error is little.

Actually, in the experimental results, the sensing structure with 50 mm exhibits lower RI sensitivity than that of ones with 47 and 54 mm. The RI sensitivity does not show an obviously regularity with the diameter or curvature of the UHCF among the three sensing samples. This may be related to the different high-order cladding modes involved in the intermodal interference.

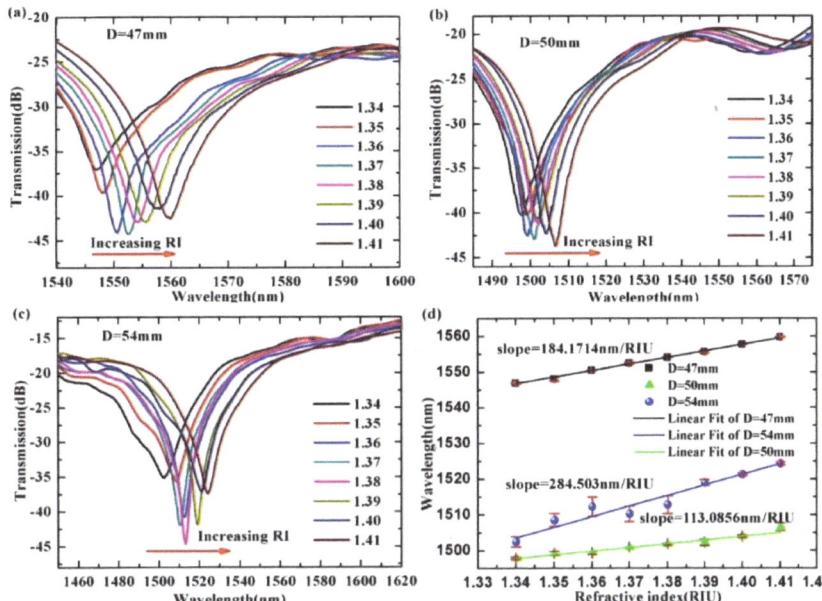

Figure 3. Transmission spectra and the wavelength shift trend of the corresponding valley of the SMF-UHCF-SMF structure immersed in solution with different RIs. (**a**) the bending diameter of the UHCF is 47mm; (**b**) the bending diameter of the UHCF is 50mm; (**c**) the bending diameter of the UHCF is 47mm; (**d**) the wavelength shift trend of the corresponding valley with different RIs.

Figure 4 shows the spatial frequency spectra of three different transmission spectra, which corresponds to different sensor bending diameter (D = 47.0 mm, 50.0 mm, 54.0 mm). Figure 4 shows the high-order modes involved in the intermodal interference for the interference valleys of the three sensor samples. We believe that the sensing structures with different bending diameters excite different high-order cladding modes. In addition, due to these different high-order modes, we achieve individual transmission attenuation. Moreover, these high-order modes have different sensitivities to the ERI. Because the SMF-UHCF-SMF structure with a bending diameter of 54 mm has higher RI sensitivity, we choose this bending diameter as the sensor structure parameter.

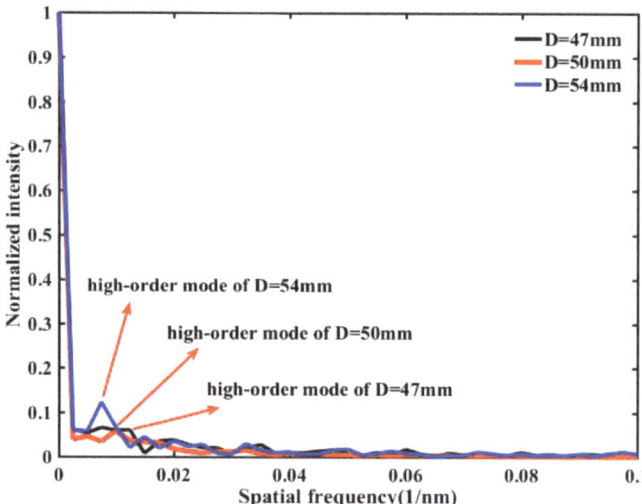

Figure 4. Spatial frequency distribution of transmission spectra of the proposed sensor samples with different bending diameter (D = 47.0 mm, 50.0 mm, 54.0 mm).

3.2. Magnetic Field Intensity Measurement

To investigate the transmission spectrum of the SMF-UHCF-SMF sensor structure to magnetic field intensity, we kept the sensor structure in the center of the adjustable permanent magnet and then gradually changed the intensity of the magnetic field by adjusting the distance between the poles. In addition, the magnetic field direction was perpendicular to the position of the U-shaped surface of the SMF-UHCF-SMF structure. From Figure 5, one can find that the transmission spectrum has two adjacent interference dips (the dip1 and the dip2), and the transmission spectra shift toward a long wavelength (or red shift) in the measurement range from 38 mT to 62 mT.

From Figure 5a, one can find that the wavelength of the dip1 shifts monotonically from 1378.48 nm to 1401.33 nm with magnetic field intensity. Similarly, we can observe that the wavelength of the dip2 shifts monotonically from 1587.19 nm to 1607.76 nm in Figure 5c. Additionally, when the magnetic field intensity increases to 58 mT or a larger value, the wavelength shifts of the transmission spectra are hardly varied. We understand that the saturation magnetic field strength of the magnetic fluid is reached. The two dips in the transmission spectrum of the sensing structure correspond different interference modes. The different modes propagate in the sensing structure experience different loss. Thus, different interference modes exhibit different response to the magnetic field intensity or the RI of the magnetic field around the sensing structure. The wavelength sensitivity to magnetic field of the dip1 and the dip2 are shown in Figure 5b,d, respectively. The fitting results in Figure 5b indicate that the response of the absolute wavelength shift as a function of magnetic field intensity is generally linear. Furthermore, the magnetic field intensity response has good linearity ($R^2 = 0.99$) from 38 mT to 58 mT. The maximum sensitivity of wavelength shift is 1.0898 nm/mT. Figure 5d shows that maximum wavelength sensitivity to magnetic field of the dip2 is 1.05406 nm/mT, and the linearity is 0.95 in the same measurement range of magnetic field intensity. Similarly, we added error bars (red vertical lines) to represent the accuracy of the data, and the error bars (red vertical lines) indicate that the magnetic field sensing error is little.

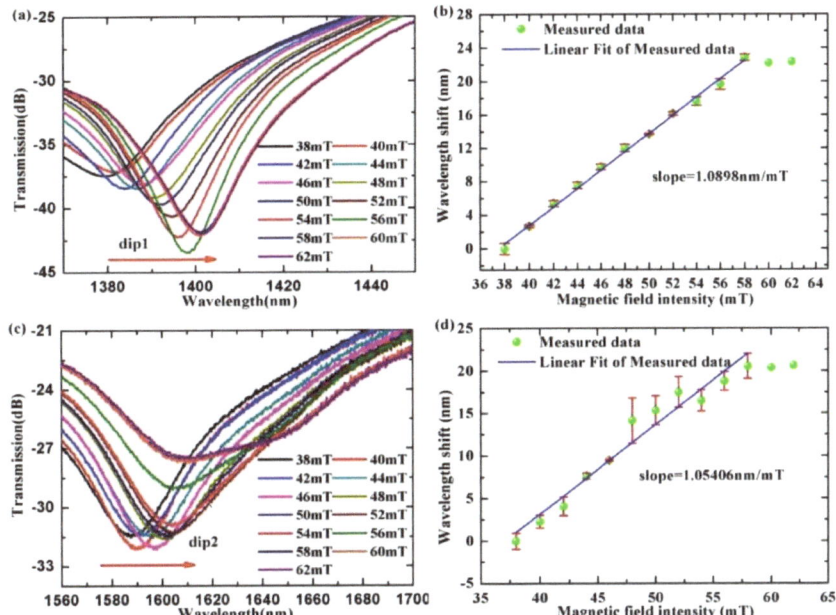

Figure 5. (**a,c**) Wavelength shifts of the dip1 and the dip2 of the interference spectra with the magnetic field intensity. (**b,d**) The wavelength shift trends of the dip1 and the dip2 of the interference spectra with magnetic field intensity.

3.3. Temperature Measurement

In order to investigate the temperature sensitivity of the sensor, we measured different temperatures and monitored wavelength shift of the dips of the transmission spectra by the OSA. The sensor was placed on a heating platform to ensure that the optical fiber structure was not stretched by external axial tension. We placed a thermometer probe near the sensing structure to monitor the temperature. We measured temperature in steps of 5 °C, within the temperature range of 25 °C to 65 °C. After reaching the set temperature, it remained steady for 20 min per step to reduce any measurement error. Figure 6a,c show the wavelength shift of the dip1 and the dip2 under the range of 25 °C to 65 °C. The data fitting results in Figure 6b,d indicate that the response of the absolute wavelength shift as a function of temperature is generally linear. Temperature sensitivities of the dip1 and the dip2 are 0.324 nm/°C and 0.069 nm/°C, respectively. The error bars (red vertical lines) in the fitted lines imply that the errors are very low.

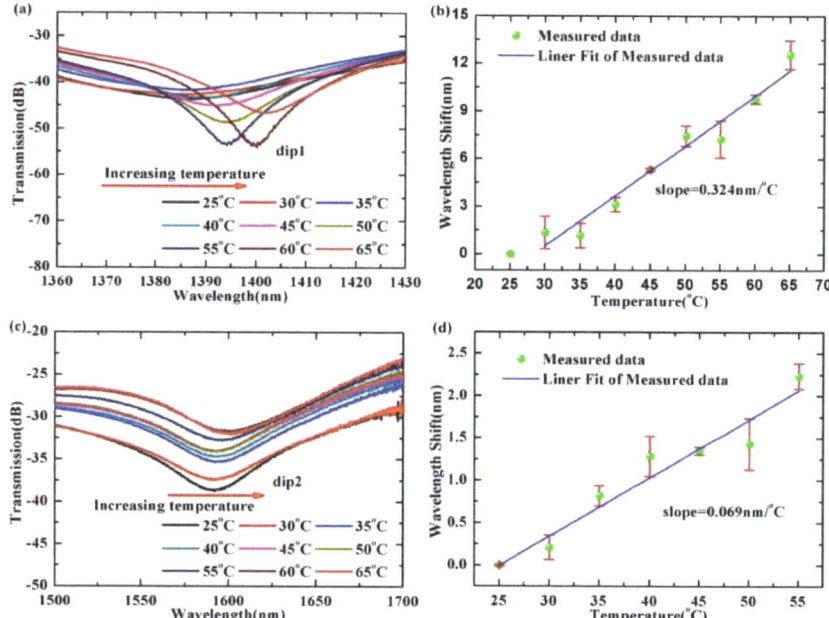

Figure 6. (a,c) Wavelength shifts of the dip1 and the dip2 of the interference spectra with the temperature. (b,d) The wavelength shift trends of the dip1 and the dip2 of the interference spectra with temperature.

3.4. Analysis of Experimental Results

According to the above experimental results, we ascertained that the interference of different order modes leads to the difference of magnetic field strength and temperature sensitivity. By monitoring the wavelength shifts of the dip1 and the dip2 of the transmission spectra, magnetic field intensity and temperature can be measured simultaneously. The relationship between wavelength shift and magnetic field intensity and the relationship between wavelength shift and temperature are described in the following equation:

$$\begin{bmatrix} \Delta\lambda_{dip1} \\ \Delta\lambda_{dip2} \end{bmatrix} = \begin{bmatrix} K_{H,dip1} & K_{T,dip1} \\ K_{H,dip2} & K_{T,dip2} \end{bmatrix} \begin{bmatrix} \Delta H \\ \Delta T \end{bmatrix} \qquad (6)$$

where $\Delta\lambda_{dip1}$ and $\Delta\lambda_{dip2}$ represent the changes of wavelength shifts at the dip1 and the dip2, respectively, $K_{H,dip1}$ and $K_{T,dip1}$ denote the sensitivities of the magnetic field intensity and temperature for the dip1. Similarly, $K_{H,dip2}$ and $K_{T,dip2}$ are those for the dip2. ΔH expresses the change of external magnetic field intensity, and ΔT is the temperature change around the sensor structure.

By taking the inverse matrix of the coefficient matrix in Equation (7), the changes in magnetic field intensity and temperature can be calculated from the wavelength shifts of the transmission spectra of the dip1 and the dip2 and are expressed as follows:

$$\begin{bmatrix} \Delta H \\ \Delta T \end{bmatrix} = \begin{bmatrix} K_{H,dip1} & K_{T,dip1} \\ K_{H,dip2} & K_{T,dip2} \end{bmatrix}^{-1} \begin{bmatrix} \Delta\lambda_{dip1} \\ \Delta\lambda_{dip2} \end{bmatrix} \qquad (7)$$

Based on the above experimental results, $K_{H,dip1} = 1.0898$, $K_{H,dip2} = 1.0506$, $K_{T,dip1} = 0.324$ and $K_{T,dip2} = 0.069$. When we substitute these into Equation (7), we can achieve:

$$\begin{bmatrix} \Delta H \\ \Delta T \end{bmatrix} = -\frac{1}{0.2652} \begin{bmatrix} 0.324 & -0.069 \\ -1.0506 & 1.0898 \end{bmatrix} \begin{bmatrix} \Delta \lambda_{dip1} \\ \Delta \lambda_{dip2} \end{bmatrix} \qquad (8)$$

With this matrix, the magnetic field intensity and temperature variation can be calculated out after measuring the resonant wavelength shifts of the dip1, and the dip2.

3.5. Comparison with Other Sensing Structure

In comparison with some other dual-parameter sensors of magnetic field and temperature, we list the reported works in Table 1. From the table, one can find easily that, all in all, the proposed dual-parameter fiber sensor has a good sensitivity. Additionally, compared with some reported works, such as those based on side-polished fiber structure, taped-fiber structure, MF-infiltrated PCF structure, gold film-coated fiber SPR structure, Mn_3O_4-PDMS cap structure, optical fiber cascaded with a FBG structure and so on, our designed sensing structure is simple, low-cost, easy to fabricate, requires no additional FBG and has good mechanical performance.

Table 1. Comparison with some other dual-parameter sensors of magnetic field and temperature.

Sensing Structure	Sensitivity	Ref.
SMF-(Mn_3O_4-PDMS-Air) FP cavity-SMF Structure	0.563 nm/mT (0–4 mT) 1.16 nm/°C (30–70 °C)	[22]
SMF-MF infiltrated PCF-SMF	0.72 nm/mT (0–6.66 mT); −0.08 nm/°C (20–60 °C)	[29]
Macro-bending SMF cascaded with a FBG	1.426 nm/mT (3–10 mT) 0.329 nm/°C (28.6–57.2 °C)	[32]
MF-infiltrated PCF Cascaded with a FBG	0.925 nm/mT (0–10 mT); 0.165 nm/°C (20–70 °C)	[33]
Two-taped Fiber Joints Cascaded with a FBG	0.408 nm/mT (0–25 mT); −0.363 nm/°C (25–55 °C)	[34]
A nonadiabatic tapered microfiber Cascaded with a FBG	1.159 nm/mT (0–18 mT) −1.737 nm/°C (25–50 °C)	[35]
SMF-D-shaped Fiber-SMF	0.997 nm/mT (0–2.1 mT); 0.0775 nm/°C (30–55 °C)	[36]
SMF-exposed core fiber (ECF)-SMF	−0.18 nm/mT (0–7.3 mT); −0.16 nm/°C (26–50 °C)	[44]
SMF-no-core fiber (NCF)-SMF	0.074 nm/mT (2–16 mT); −0.28667 nm/°C (20–70 °C)	[45]
Two-channel PCF-SPR Structure	0.65 nm/mT (5–13 mT); −0.52 nm/°C (17.5–27.5 °C)	[37]
SMF-UHCF-SMF	1.0898 nm/mT (38–62 mT); 0.324 nm/°C (25–65 °C)	This work

4. Conclusions

In conclusion, by using magnetofluid encapsulated a U-shaped HCF sandwiched between two segments of SMFs, we have achieved a dual-parameter optical fiber sensor of magnetic field and temperature. At the spliced region of the lead-in SMF and the UHCF, the fundamental mode in the lead-in SMF is split into the air-core modes and the high-order cladding modes. The cladding modes and the air-core mode propagate along the UHCF. These mixing modes experienced different bending attenuation; meanwhile, they have different responses to the environmental RI (ERI). The air-core modes and the cladding modes mutually interfere at the splicing region of the lead-out SMF and the UHCF. The interference lightwave of the sensor is led-out from the lead-out SMF and observed by an OSA. The experimental results show that interference dips of the transmission spectra are sensitive to magnetic field intensity and temperature around the sensor structure. The maximum magnetic field intensity and temperature sensitivities are 1.0898 nm/mT and 0.324 nm/°C, respectively. The proposed optical fiber magnetic field sensor has the advantages of simple structure, low cost, convenient to fabrication and good mechanical performance and may provide a potential candidate for industry applications.

Author Contributions: Conceptualization, R.X. and Y.X.; software, Y.X.; validation, Y.X., J.Y., M.X., C.K. and M.C.; data curation, Y.X. and R.X.; writing—original draft preparation, Y.X.; writing—review and editing, R.X.; project administration, R.X., H.L. and L.Y; funding acquisition, R.X., M.C., H.L. and L.Y. All authors have read and agreed to the published version of the manuscript.

Funding: This research was supported in part by the National Key Research and Development Program of China under Grant 2019YFB2203903; in part by the National Major Scientific Research Instrument Development Project of China under Grant 61827819; in part by the National Natural Science Foundation of China under Grant 62075047, Grant 62175044, Grant 62065006, Grant 12164010 and Grant 61965006; in part by the Guangxi Key Research and Development Program under Grant AA18242043 and Grant AB18221033; in part by the Guangxi Natural Science Foundation under Grant 2020GXNSFDA297019, Grant 2020GXNSFAA238040 and Grant 2018GXNSFAA281272; and in part by the Science and Technology Project of Guangxi under Grant AD21220078.

Institutional Review Board Statement: Not applicable.

Informed Consent Statement: Not applicable.

Data Availability Statement: The original contributions presented in the study are included in the article. Further inquiries can be directed to the corresponding author.

Conflicts of Interest: The authors declare no conflict of interest.

References

1. Heremans, J. Solid-state magnetic-field sensors and applications. *J. Phys. D Appl. Phys.* **1993**, *26*, 1149–1168. [CrossRef]
2. Li, G.X.; Joshi, V.; White, R.L.; Wang, S.X.; Kemp, J.T.; Webb, C.; Davis, R.W.; Sun, S.H. Detection of single micron-sized magnetic bead and magnetic nanoparticles using spin valve sensors for biological applications. *J. Appl. Phys.* **2003**, *93*, 7557–7559. [CrossRef]
3. Chen, L.X.; Huang, X.G.; Zhu, J.H.; Li, G.C.; Lan, S. Fiber magnetic-field sensor based on nanoparticle magnetic fluid and Fresnel reflection. *Opt. Lett.* **2011**, *36*, 2761–2763. [CrossRef] [PubMed]
4. Herrera-May, A.L.; Soler-Balcazar, J.C.; Vazquez-Leal, H.; Martinez-Castillo, J.; Vigueras-Zuniga, M.O.; Aguilera-Cortes, L.A. Recent Advances of MEMS Resonators for Lorentz Force Based Magnetic Field Sensors: Design, Applications and Challenges. *Sensors* **2016**, *16*, 1359. [CrossRef] [PubMed]
5. Murzin, D.; Mapps, D.J.; Levada, K.; Belyaev, V.; Omelyanchik, A.; Panina, L.; Rodionova, V. Ultrasensitive Magnetic Field Sensors for Biomedical Applications. *Sensors* **2020**, *20*, 1569. [CrossRef]
6. Bichurin, M.; Petrov, R.; Sokolov, O.; Leontiev, V.; Kuts, V.; Kiselev, D.; Wang, Y.J. Magnetoelectric Magnetic Field Sensors: A Review. *Sensors* **2021**, *21*, 6232. [CrossRef] [PubMed]
7. Reig, C.; Cubells-Beltran, M.D.; Munoz, D.R. Magnetic Field Sensors Based on Giant Magnetoresistance (GMR) Technology: Applications in Electrical Current Sensing. *Sensors* **2009**, *9*, 7919–7942. [CrossRef]
8. Rashleigh, S.C. Magnetic-field sensing with a single-mode fiber. *Opt. Lett.* **1981**, *6*, 19–21. [CrossRef]
9. Zhao, Y.; Wu, D.; Lv, R.Q.; Ying, Y. Tunable Characteristics and Mechanism Analysis of the Magnetic Fluid Refractive Index with Applied Magnetic Field. *IEEE Trans. Magn.* **2014**, *50*, 4600205. [CrossRef]
10. Alberto, N.; Domingues, M.F.; Marques, C.; Andre, P.; Antunes, P. Optical Fiber Magnetic Field Sensors Based on Magnetic Fluid: A Review. *Sensors* **2018**, *18*, 4325. [CrossRef]
11. Liu, C.; Shen, T.; Wu, H.B.; Feng, Y.; Chen, J.J. Applications of magneto-strictive, magneto-optical, magnetic fluid materials in optical fiber current sensors and optical fiber magnetic field sensors: A review. *Opt. Fiber Technol.* **2021**, *65*, 102634. [CrossRef]
12. Odenbach, S. MAGNETIC FLUIDS. *Adv. Colloid. Interfac.* **1993**, *46*, 263–282. [CrossRef]
13. Nikolaev, V.I.; Shipilin, A.M.; Shkolnicov, E.N.; Zaharova, I.N. "Elastic properties" of magnetic fluids. *J. Appl. Phys.* **1999**, *86*, 576–577. [CrossRef]
14. Odenbach, S. Recent progress in magnetic fluid research. *J. Phys. Condens. Matter* **2004**, *16*, R1135–R1150. [CrossRef]
15. Martinez, L.; Cecelja, F.; Rakowski, R. A novel magneto-optic ferrofluid material for sensor applications. *Sens. Actuators A Phys.* **2005**, *123*, 438–443. [CrossRef]

16. Luo, L.F.; Pu, S.L.; Tang, J.L.; Zeng, X.L.; Lahoubi, M. Reflective all-fiber magnetic field sensor based on microfiber and magnetic fluid. *Opt. Express* **2015**, *23*, 18133–18142. [CrossRef]
17. Lu, L.; Miao, Y.P.; Zhang, H.M.; Li, B.; Fei, C.W.; Zhang, K.L. Magnetic sensor based on serial-tilted-tapered optical fiber for weak-magnetic-field measurement. *Appl. Opt.* **2020**, *59*, 2791–2796. [CrossRef]
18. Gao, R.; Jiang, Y.; Abdelaziz, S. All-fiber magnetic field sensors based on magnetic fluid-filled photonic crystal fibers. *Opt. Lett.* **2013**, *38*, 1539–1541. [CrossRef]
19. Candiani, A.; Argyros, A.; Leon-Saval, S.G.; Lwin, R.; Selleri, S.; Pissadakis, S. A loss-based, magnetic field sensor im-plemented in a ferrofluid infiltrated microstructured polymer optical fiber. *Appl. Phys. Lett.* **2014**, *104*, 111106. [CrossRef]
20. Violakis, G.; Korakas, N.; Pissadakis, S. Differential loss magnetic field sensor using a ferrofluid encapsulated D-shaped optical fiber. *Opt. Lett.* **2018**, *43*, 142–145. [CrossRef]
21. Chen, Y.F.; Hu, Y.C.; Cheng, H.D.; Yan, F.; Lin, Q.Y.; Chen, Y.; Wu, P.J.; Chen, L.; Liu, G.S.; Peng, G.D.; et al. Side-Polished Single-Mode-Multimode-Single-Mode Fiber Structure for the Vector Magnetic Field Sensing. *J. Lightwave Technol.* **2020**, *38*, 5837–5843. [CrossRef]
22. Sun, B.; Bai, M.; Ma, X.; Wang, X.; Zhang, Z.; Zhang, L. Magnetic-Based Polydimethylsiloxane Cap for Simultaneous Measurement of Magnetic Field and Temperature. *J. Lightwave Technol.* **2022**, *40*, 2625–2630. [CrossRef]
23. Song, B.B.; Miao, Y.P.; Lin, W.; Zhang, H.; Liu, B.; Wu, J.X.; Liu, H.F.; Yan, D.L. Loss-Based Magnetic Field Sensor Employing Hollow Core Fiber and Magnetic Fluid. *IEEE Photonics Technol. Lett.* **2014**, *26*, 2283–2286. [CrossRef]
24. Li, Z.; Liao, C.; Song, J.; Wang, Y.; Zhu, F.; Wang, Y.; Dong, X. Ultrasensitive magnetic field sensor based on an in-fiber Mach–Zehnder interferometer with a magnetic fluid component. *Photonics Res.* **2016**, *4*, 197–201. [CrossRef]
25. Zhang, S.; Li, X.; Liu, Y.; Chen, W.H.; Niu, H.W.; Yan, Q.; Wang, S.J.; Li, S.; Geng, T.; Sun, W.M.; et al. A MMF-TSMF-MMF Structure Coated Magnetic Fluid for Magnetic Field Measurement. *IEEE Photonics Technol. Lett.* **2021**, *33*, 1105–1108. [CrossRef]
26. Zhou, X.; Li, X.; Li, S.; An, G.; Cheng, T. Magnetic Field Sensing Based on SPR Optical Fiber Sensor Interacting with Magnetic Fluid. *IEEE Trans. Instrum. Meas.* **2019**, *68*, 234–239. [CrossRef]
27. Zhu, L.; Zhao, N.; Lin, Q.; Zhao, L.; Jiang, Z. Optical fiber SPR magnetic field sensor based on photonic crystal fiber with the magnetic fluid as cladding. *Meas. Sci. Technol.* **2021**, *32*, 075106. [CrossRef]
28. Rodriguez-Schwendtner, E.; Navarrete, M.C.; Diaz-Herrera, N.; Gonzalez-Cano, A.; Esteban, O. Advanced Plasmonic Fiber-Optic Sensor for High Sensitivity Measurement of Magnetic Field. *IEEE Sens. J.* **2019**, *19*, 7355–7364. [CrossRef]
29. Li, X.G.; Zhou, X.; Zhao, Y.; Lv, R.Q. Multi-modes interferometer for magnetic field and temperature measurement using Photonic crystal fiber filled with magnetic fluid. *Opt. Fiber Technol.* **2018**, *41*, 1–6. [CrossRef]
30. Wang, E.; Cheng, P.; Li, J.; Cheng, Q.; Zhou, X.; Jiang, H. High-sensitivity temperature and magnetic sensor based on magnetic fluid and liquid ethanol filled micro-structured optical fiber. *Opt. Fiber Technol.* **2020**, *55*, 102161. [CrossRef]
31. Lin, Q.Y.; Hu, Y.C.; Yan, F.; Hu, S.Q.; Chen, Y.; Liu, G.S.; Chen, L.; Xiao, Y.; Chen, Y.F.; Luo, Y.H.; et al. Half-side gold-coated hetero-core fiber for highly sensitive measurement of a vector magnetic field. *Opt. Lett.* **2020**, *45*, 4746–4749. [CrossRef] [PubMed]
32. Chen, Y.F.; Han, Q.; Yan, W.C.; Yao, Y.Z.; Liu, T.G. Magnetic Field and Temperature Sensing Based on a Macro-Bending Fiber Structure and an FBG. *IEEE Sens. J.* **2016**, *16*, 7659–7662. [CrossRef]
33. Zhao, Y.; Zhang, Y.Y.; Wu, D.; Wang, Q. Magnetic field and temperature measurements with a magnetic fluid-filled photonic crystal fiber bragg grating. *Instrum. Sci. Technol.* **2013**, *41*, 463–472. [CrossRef]
34. Zhang, R.; Pu, S.L.; Li, Y.Q.; Zhao, Y.L.; Jia, Z.X.; Yao, J.L.; Li, Y.X. Mach-Zehnder Interferometer Cascaded with FBG for Simultaneous Measurement of Magnetic Field and Temperature. *IEEE Sens. J.* **2019**, *19*, 4079–4083. [CrossRef]
35. Zhang, Y.X.; Pu, S.L.; Li, Y.X.; Hao, Z.J.; Li, D.H.; Yan, S.K.; Yuan, M.; Zhang, C.C. Magnetic Field and Temperature Dual-Parameter Sensor Based on Nonadiabatic Tapered Microfiber Cascaded with FBG. *IEEE Access* **2022**, *10*, 15478–15486. [CrossRef]
36. Dong, Y.; Wu, B.L.; Wang, M.G.; Xiao, H.; Xiao, S.Y.; Sun, C.R.; Li, H.S.; Jian, S.S. Magnetic field and temperature sensor based on D-shaped fiber modal interferometer and magnetic fluid. *Opt. Laser Technol.* **2018**, *107*, 169–173. [CrossRef]
37. Wang, D.Y.; Yi, Z.; Ma, G.L.; Dai, B.; Yang, J.B.; Zhang, J.F.; Yu, Y.; Liu, C.; Wu, X.W.; Bian, Q. Two-channel photonic crystal fiber based on surface plasmon resonance for magnetic field and temperature dual-parameter sensing. *Phys. Chem. Chem. Phys.* **2022**, *24*, 21233–21241. [CrossRef]
38. Wu, C.W.; Chen, C.T.; Chiang, C.C. A novel U-shaped, packaged, and microchanneled optical fiber strain sensor based on macro-bending induced whispering gallery mode. *Sens. Actuators A Phys.* **2019**, *288*, 86–91. [CrossRef]
39. Chou, Y.L.; Wen, H.Y.; Weng, Y.Q.; Liu, Y.C.; Wu, C.W.; Hsu, H.C.; Chiang, C.C. A U-Shaped Optical Fiber Temperature Sensor Coated with Electrospinning Polyvinyl Alcohol Nanofibers: Simulation and Experiment. *Polymers* **2022**, *14*, 2110. [CrossRef]
40. Geng, Y.F.; Li, X.J.; Tan, X.L.; Deng, Y.L.; Yu, Y.Q. High-Sensitivity Mach-Zehnder Interferometric Temperature Fiber Sensor Based on a Waist-Enlarged Fusion Bitaper. *IEEE Sens. J.* **2011**, *11*, 2891–2894. [CrossRef]
41. Tang, J.; Pu, S.; Dong, S.; Luo, L. Magnetic Field Sensing Based on Magnetic-Fluid-Clad Multimode-Singlemode-Multimode Fiber Structures. *Sensors* **2014**, *14*, 19086–19094. [CrossRef] [PubMed]
42. Chen, L.; Zhang, W.G.; Wang, L.; Zhou, Q.; Sieg, J.; Zhao, D.L.; Wang, B.; Yan, T.Y.; Wang, S. Fiber refractive index sensor based on dual polarized Mach-Zehnder interference caused by a single-mode fiber loop. *Appl. Opt.* **2016**, *55*, 63–69. [CrossRef] [PubMed]
43. Hong, C.Y.; Horng, H.E.; Yang, S.Y. Tunable refractive index of magnetic fluids and its applications. *Proc. Phys. Status Solidi C Magn. Supercond. Mater. Proc.* **2004**, *1*, 1604–1609. [CrossRef]

44. Yu, Q.; Li, X.G.; Zhou, X.; Chen, N.; Wang, S.K.; Li, F.; Lv, R.Q.; Nguyen, L.V.; Warren-Smith, S.C.; Zhao, Y. Temperature Compensated Magnetic Field Sensor Using Magnetic Fluid Filled Exposed Core Microstructure Fiber. *IEEE Trans. Instrum. Meas.* **2022**, *71*, 7004408. [CrossRef]
45. Su, G.H.; Shi, J.; Xu, D.G.; Zhang, H.W.; Xu, W.; Wang, Y.Y.; Feng, J.C.; Yao, J.Q. Simultaneous Magnetic Field and Temperature Measurement Based on No-Core Fiber Coated with Magnetic Fluid. *IEEE Sens. J.* **2016**, *16*, 8489–8493. [CrossRef]

Article

Fiber-Optic Vector-Magnetic-Field Sensor Based on Gold-Clad Bent Multimode Fiber and Magnetic Fluid Materials

Weinan Liu [1], Shengli Pu [1,2,*], Zijian Hao [1], Jia Wang [1], Yuanyuan Fan [1], Chencheng Zhang [1] and Jingyue Wang [1]

1. College of Science, University of Shanghai for Science and Technology, Shanghai 200093, China
2. Shanghai Key Laboratory of Modern Optical System, University of Shanghai for Science and Technology, Shanghai 200093, China
* Correspondence: shlpu@usst.edu.cn; Tel.: +86-21-65667034; Fax: +86-21-55271663

Abstract: A kind of bent multimode fiber (MMF) vector magnetic sensor based on surface plasmon resonance (SPR) was proposed. By plating gold film on the curved part of the bent multimode fiber, the surface plasmon mode (SPM) was excited via a whispering gallery mode (WGM). Fabricating the structure only required bending the fiber and plating it with gold, which perfectly ensured the integrity of the fiber and made it more robust compared with other structures. The sensor used magnetic fluid (MF) as the magnetically sensitive material. Through monitoring the shift of the surface plasmon resonance dip, the as-fabricated sensor not only had a high magnetic field intensity sensitivity of 9749 pm/mT but could also measure the direction of a magnetic field with a high sensitivity of 546.5 pm/°. The additional advantages of the proposed sensor lay in its easy fabrication and good integrity, which make it attractive in the field of vector-magnetic-field sensing.

Keywords: vector-magnetic-field sensing; surface plasmon resonance; magnetic fluid

Citation: Liu, W.; Pu, S.; Hao, Z.; Wang, J.; Fan, Y.; Zhang, C.; Wang, J. Fiber-Optic Vector-Magnetic-Field Sensor Based on Gold-Clad Bent Multimode Fiber and Magnetic Fluid Materials. *Materials* **2022**, *15*, 7208. https://doi.org/10.3390/ma15207208

Academic Editor: Simone Pisana

Received: 12 September 2022
Accepted: 12 October 2022
Published: 16 October 2022

Publisher's Note: MDPI stays neutral with regard to jurisdictional claims in published maps and institutional affiliations.

Copyright: © 2022 by the authors. Licensee MDPI, Basel, Switzerland. This article is an open access article distributed under the terms and conditions of the Creative Commons Attribution (CC BY) license (https://creativecommons.org/licenses/by/4.0/).

1. Introduction

Magnetic fluid is a kind of stable colloidal material, which is composed of magnetic nanoparticles, surfactant coated on the surface of the particles and carrier liquid. It has the magnetic properties of solid magnetic materials and unique optical properties. Due to its tunable refractive index (RI) characteristics, optical anisotropy and good compatibility with optical fiber, MF has been widely used in optical fiber sensors [1–5]. Recently, many fiber sensors combined with MF have been proposed for magnetic field, temperature and curvature measurement. The employed structures include tapered fibers [6–8], Fabry–Perot interferometers [9,10], photonic crystal fibers [11,12], bent fibers [13,14], fiber multimode interferometers [15,16], microfiber couplers [17,18] and Sagnac interferometer [19].

Compared with other structures, the bending structure does not require complex processes of manufacturing (such as mechanical polishing, chemical etching or tapering), which ensures the mechanical integrity of the optical fiber. In addition, it intrinsically meets the prerequisite for vector-magnetic-field measurement, i.e., the optical fiber structure is geometrically non-centrosymmetric [20–22]. In addition, the bending structure automatically enables the reflective-type configuration and can be regarded as a sensor probe, which is more pragmatic compared with other structures. Moreover, the small size of the structure and the reflection-like probe are convenient for magnetic field detection in a narrow space.

The bending structure can also easily stimulate whispering gallery modes (WGMs) within the cladding [23–26], which makes it possible to combine the bending structure with a surface-plasmon-resonance (SPR)-sensing system. It is well known that sensors based on SPR have relatively higher sensitivity [27–30], which can measure very tiny changes in the environmental RI around the structure. Though many sensors based on bent fiber have been proposed and demonstrated, the SPR effect has not attracted much attention from researchers. Most of the sensing mechanisms of those sensors have been based on Mach–Zehnder interference via coupling between core mode (CM) and WGM [31,32]. As the structures belong to the interference type, there is a free spectral range (FSR), which leads to a limited detection range of the sensor (i.e., the main deficiency of this kind of device). This is due to the very small interval between the two neighboring interference dips [33].

In view of the above-mentioned facts, we proposed and experimentally demonstrated a kind of vector-magnetic-field sensor based on a MF-clad bending fiber-optic structure with SPR, which possessed the benefits of both bent fiber and SPR such as easy fabrication, good integrity and high sensitivity.

2. Fabrication and Sensing Principle

The structure of the proposed sensor is shown in Figure 1, which was made of multi-mode fiber (MMF, SI 105/125-15/250, Yangtze Optical Fibre and Cable Joint Stock Limited Company, Wuhan, China) with cladding and with core diameters of 125 μm and 105 μm, respectively. It has been reported that a too large or too small bending radius is not conducive to producing a clear SPR resonant dip [34]. On the other hand, it is necessary to avoid fiber fracturing caused by bending when the diameter is too small. After a series of pre-experiments, we made three sensors with nominal bending diameters of 8.0 mm, 9.0 mm and 10.0 mm (the actual measured bending diameters were 7.8 mm, 8.9 mm and 10.1 mm, respectively), which are hereafter referred to as sensor 1, sensor 2 and sensor 3, respectively.

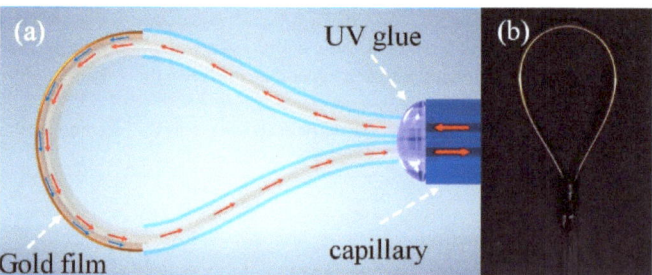

Figure 1. (a) Sensing structure of gold-clad bent multimode fiber; (b) Photograph of the sensor sample.

To fabricate the sensing structure, the coating of a piece of MMF was stripped to obtain the uncoated fiber with lengths of 12.6 mm, 14.1 mm and 15.8 mm, respectively. Then, the coating-removed fiber was bent and inserted into the capillary of an appropriate length. The fiber at the end of the capillary was stretched to control the bending radius and the position of the fiber. After reaching the appropriate position, UV glue was dropped into the capillary and fixed by UV-irradiation. By this simple method, the coating-stripped fiber was bent into a semicircle and the length of the coating-stripped fiber was equal to the circumference of the semicircle. So, it was convenient and simple to control the bending diameter of the sensor probe. Then, gold film was plated on the surface of the bent MMF using an ion-sputtering apparatus (ETD-900M, Vision Precision Instruments, Beijing, China) with a 10 mA sputtering current for 90 s. The thickness of the deposited gold film was about 17.3 nm. It should be pointed out that the thickness and roughness of gold film will influence the SPR effect and thus the sensing performance. Generally, the sensitivity of the sensor increases with the thickness of the film. However, the normalized depth of the resonance dip decreases at the same time, which is caused by the enhanced evanescent field energy absorption in the gold film [35]. Moreover, the full width at half-maximum of the resonance dip will increase with the gold film thickness, which will reduce the measurement accuracy of the sensor.

Before being packaged in a capsule filled with MF, the RI response of the as-prepared sensors was measured by placing the sensor in liquids with different concentrations. A mixture of glycerol and water with different proportions were used as the liquids with different RIs. Then, the MF with an appropriate RI was chosen, which guaranteed that the magnetic-field-modulated RI range lay within the linear response interval of the SPR dip wavelength. The typical SPR spectra of the as-fabricated three sensors clad with liquids of different RIs (1.33–1.39) are shown in Figure 2a–c. The SPR resonance dip was very obvious. Three tests were performed for each RI to verify their robustness and repeatability. When the RI of the external solution increased, the dip redshifted. In order to facilitate the application of sensing, linear fitting in two limited variation ranges was employed. A very good linearity and a very small error were obtained. The three sensors had high sensitivities in the RI range from 1.35 to 1.38, which were 3171.65 nm/RIU, 3030.85 nm/RIU and 3244.71 nm/RIU, respectively (see Figure 2d–f).

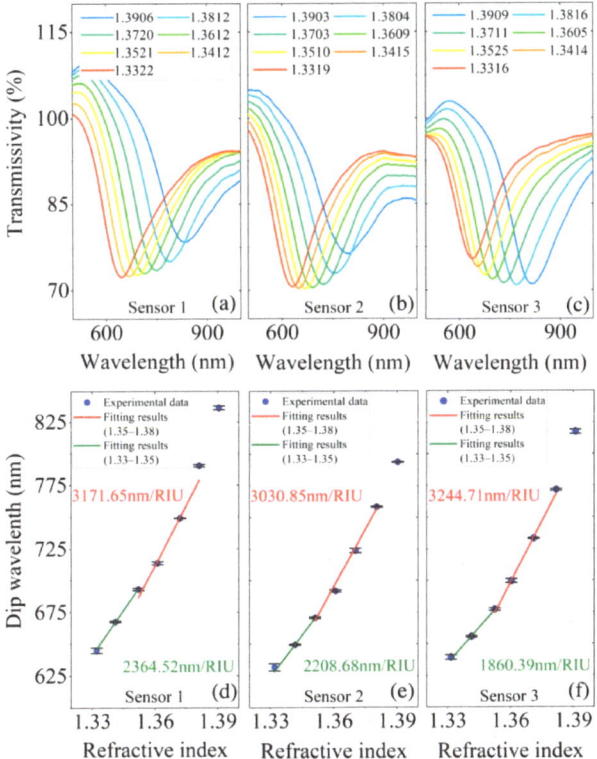

Figure 2. Spectra of the as-fabricated sensing probes clad with liquids of different RIs: (**a**–**c**). Relationship between resonance wavelength and RI: (**d**–**f**). (**a**,**d**), (**b**,**e**) and (**c**,**f**) correspond to sensor 1, sensor 2 and sensor 3, respectively.

It is obvious from Figure 2a–c that the contrast of SPR dip decreased with the RI, which may be assigned to the increased loss when the ambient RI approached that of the fiber. This also increased the sensitivity in a high RI environment, as shown in Figure 2d–f.

After the RI measurement, the structure was packaged in a capsule filled with MF. Water-based MF with surfactant-coated Fe_3O_4 nanoparticles (EMG 605) was employed in this work. Although the larger the RI of the MF is, the better the sensing performance of the structure will be, the resonant dip broadened under a higher magnetic field (i.e., a higher RI), which was counterintuitive. In our experiments, the MF was diluted with deionized water. The achieved RI was about 1.360 (measured by a refractometer A670, Jinan, China), which was within the linear range of the response of the SPR resonance wavelength.

Due to the bending effect, the WGM can be excited within the cladding, and then surface plasmon mode (SPM) can be excited by the evanescent field of the WGM. The coupling between the CM within the core and the SPM within the gold film is mediated by the WGM. So, coupling among the CM, cladding of the WGM and symmetric SPM can be realized simultaneously. Since the effective refractive index (ERI) of SPM depends on the surrounding medium RI (n_s), the ERI of the WGM-SPM hybrid mode also depends on n_s. Meanwhile, resonance loss will occur when the CM and WGM-SPM hybrid mode are coupled. This leads to an environment-dependent resonant dip. So, the drifting of the coupling wavelength of CM-WGM-SPM is closely related with n_s [34].

The evanescent field of WGM can penetrate the metal film and then excite the SPR if the following phase-matching condition is fulfilled [27]:

$$\frac{2\pi}{\lambda} n_c \sin(\phi) = \text{Re}\left[\frac{2\pi}{\lambda}\left(\frac{\varepsilon_m n_s^2}{\varepsilon_m + n_s^2}\right)^{\frac{1}{2}}\right], \quad (1)$$

where λ represents the wavelength of the incident light, n_c is the RI of the fiber cladding, ϕ is the incidence angle, ε_m is the dielectric constant of the metal film and n_s is the RI of the surrounding medium. The sensitivity of an SPR-based sensor is generally defined as [36]

$$S = \frac{d\lambda}{dn_s} = \frac{n_g^4(\omega_p\lambda)^2 n_s}{\pi c\left(n_s^2 - n_g^2(1+n_s^2)\right)\sqrt{-\Gamma^2 + 4\omega_p^2 \frac{n_s^2 - n_g^2}{n_s^2 - n_g^2(1+n_s^2)}}}, \quad (2)$$

where n_g is the effective RI of the guided mode propagating in the optical fiber, Γ is the electron relaxation rate given by $\Gamma = 1/\tau$, τ is the electron relaxation time, ω_p is the plasma frequency and c is the light velocity.

According to Equation (2), the theoretical sensitivity S as a function of λ was calculated and the corresponding results are in plotted Figure 3. The following parameters were taken for the calculation: $\omega_p = 9$ eV/$\hbar$ ($\hbar = 6.582119514 \times 10^{-16}$ eV s is the reduced Planck's constant), $\tau = 14$ fs and $n_s = 1.360$. At the typical resonance wavelength (700 nm), the theoretical sensitivity was 4878 nm/RIU, which was slightly larger than the experimental one.

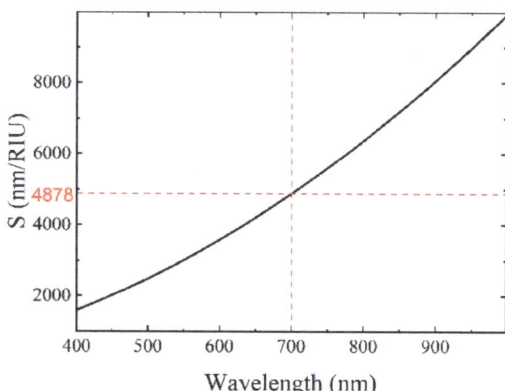

Figure 3. Relationship between theoretical sensitivity S and SPR resonance wavelength.

Magnetic nanoparticles in MF will interact with each other and move under the action of a magnetic field. They will form nanochain clusters as is schematically shown in Figure 4, which is closely related with the direction of magnetic field [21,37]. The Monte Carlo method was also used to simulate the distribution of magnetic nanoparticles around the fiber (see Figure 4c,d) [38]. When the magnetic field strength was 0 mT, the magnetic nanoparticles were randomly distributed around the fiber. However, when the magnetic field intensity was 30 mT, nanochain clusters parallel to the magnetic field were formed on both sides of the fiber, where the local RI increased. The RI of the MF around the structure changed with the magnetic field direction as well. In other words, the ERI of the SPM was affected by both the magnetic field strength and direction, which is the basic principle for vector-magnetic-field sensing.

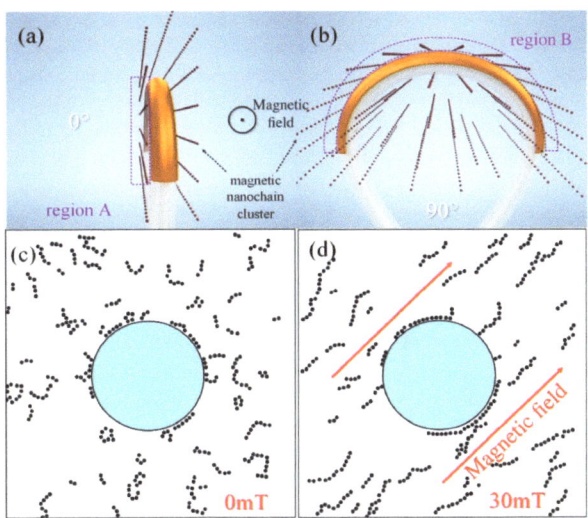

Figure 4. Schematic of nanochain clusters of magnetic nanoparticles under different magnetic field directions: (**a**,**b**); distribution of magnetic nanoparticles around optical fibers simulated with Monte Carlo method: (**c**) H = 0 mT, (**d**) H = 30 mT.

3. Experimental Details and Discussion

Figure 5 shows the experimental setup for investigating the sensing properties of the as-fabricated vector-magnetic-field sensor. The light source was tungsten halogen with a wavelength range from 360 nm to 2400 nm (HL-2000, Ocean Insight, Orlando, FL, USA). The sensor was fixed to a rotating platform placed in the center of the electromagnet. The magnetic field intensity was adjusted by changing the power supply current.

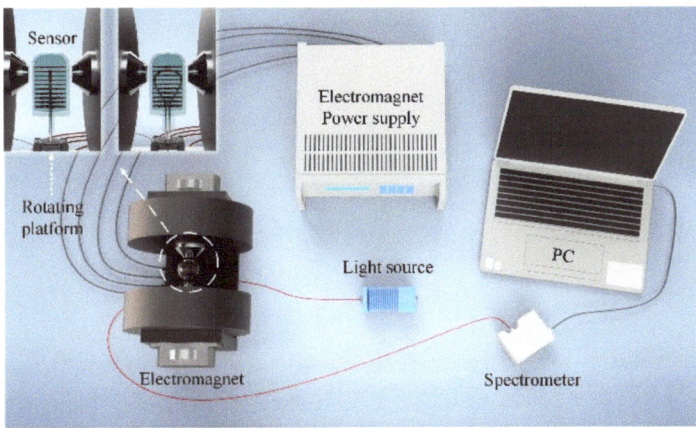

Figure 5. Experimental setup for investigating the sensing properties of the vector-magnetic-field sensor.

To verify the vector-magnetic-field-sensing characteristics, the magnetic field intensity was fixed at 5 mT and the magnetic field direction was rotated from 0° to 360° with a step of 5°. The angles 0° and 90° indicated that the plane of the bent-ring was parallel to and perpendicular to the magnetic field direction, respectively (see Figure 4 for the definition of direction). Figure 6 shows the typical transmission spectra of the as-fabricated three sensors at different magnetic field directions. Considering the spectral periodicity of the magnetic field direction ranged from 0° to 360°, Figure 7 only shows the spectra for the magnetic field direction in the range 0°–90° and 90°–180°. When the sensor's angle (the bent-ring with respect to the magnetic field direction) was rotated from 0° to 90° or 180° to 270°, all dips redshifted. On the contrary, when the included angle was rotated from 90° to 180° or 270° to 360°, all dips blueshifted.

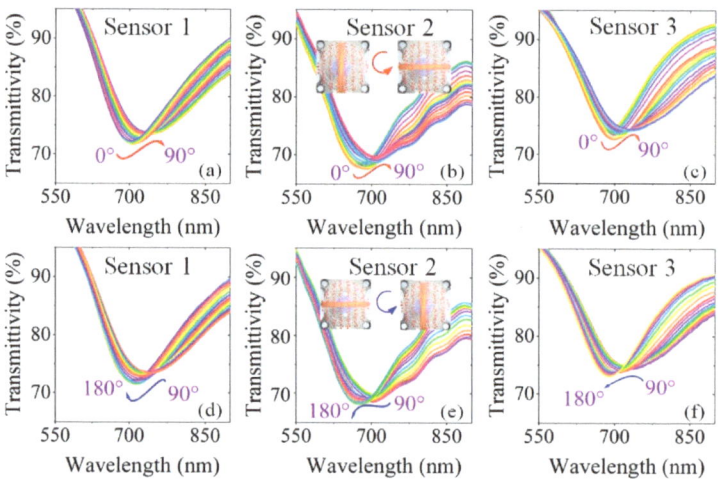

Figure 6. Transmittance spectral response to magnetic field direction under magnetic field intensity of 5 mT for sensor 1 (**a**,**d**), sensor 2 (**b**,**e**) and sensor 3 (**c**,**f**).

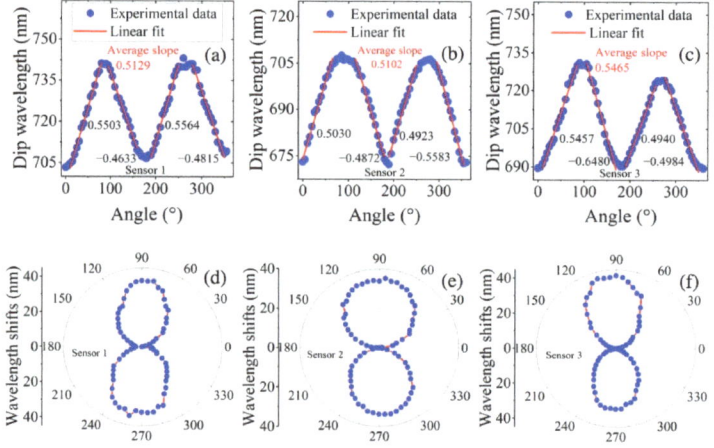

Figure 7. (**a**–**c**) Dip wavelength as a function of magnetic field orientation angle and the linear fitting of experimental data for sensor 1, sensor 2 and sensor 3; (**d**–**f**) shift of dip wavelength with magnetic field direction for sensor 1, sensor 2 and sensor 3 in polar coordinate system.

The above phenomenon can be explained by the uneven distribution of the magnetic nanochain clusters around the fiber, as shown in Figure 4. For the bent SPR structure, the SPM propagated along the outside cladding gold film. The ERI of the SPM mainly depends on the surrounding environment around the gold film. So, the CM–WGM–SPM coupling wavelength also shifted with the change in the surrounding environment. When the bending MMF plane was rotated to be parallel to the direction of the external magnetic field (see Figure 4a), the magnetic nanochain clusters were mainly distributed on both sides of the sensor (Region A in Figure 5) and the concentration of MF in the region above the sensor (Region B in Figure 4, i.e., the outside area of the bent part) decreased. Then, the ERI of the SPM decreased synchronically. When the bending MMF plane was rotated to be perpendicular to the direction of the external magnetic field (see Figure 4b), the magnetic nanochain clusters were mainly distributed in the region above the sensor (Region B in Figure 4) and the MF concentration therein increased. With the increase in the MF concentration, the RI increased, which resulted in the increase in the ERI of the SPM. These results were in good agreement with the experimental phenomena and the theoretical prediction from Equation (2).

To be clearer, the variation in the dip wavelength with the magnetic field direction is explicitly depicted in Figure 7. The dip wavelength varied periodically with the magnetic field direction. At $0°$, $90°$, $180°$ and $270°$, the variation in the dip wavelength with direction was slight, while the variation was very remarkable at around $45°$, $135°$, $225°$ and $315°$. As the magnetic field direction changed from $0°$ to $90°$, the magnetic nanochain clusters moved from both sides of the sensor (Region A in Figure 4) to the region above the sensor (Region B in Figure 4) and the RI changed significantly, so the directional sensitivity was very high (at around $45°$ or $225°$). For the cases that the magnetic nanochain clusters were mainly distributed in the regions near both the sides or above the sensor, the RI changed slightly, so the directional sensitivity was low (at around $0°$ or $90°$). The sensitivity corresponding to the magnetic field direction in the ranges of $0°\sim90°$, $90°\sim180°$, $180°\sim270°$ and $270°\sim360°$ was obtained through linear fitting of the experimental data, and the average highest direction sensitivity was 0.5465 nm/$°$ (see Figure 7c). In a polar coordinate system, the response of the dip wavelength shift with respect to the orientation angle exhibited an "8" shape (see Figure 7d–f). The directional response of the three sensors was very similar, but the shape of the wavelength drift with respect to the magnetic field direction for sensor 2 was more standardized (symmetrical) than those for sensor 1 and sensor 3. Thus, sensor 2 could measure the magnetic field direction more accurately. The slight asymmetry for the wavelength shift with respect to the magnetic field direction for sensors 1 and 3 may be assigned to the uniformity of the magnetic field produced by the electromagnet and the slight offset of the sensor position during the rotation, which led to the low mirror symmetry for sensor 1 and sensor 3.

Then, the magnetic field direction was fixed at $0°$ and $90°$. The magnetic field intensity increased with a step of 1 mT. The response of the sensors to the magnetic field intensity is characterized in Figure 8. When the magnetic field direction was at $0°$ and $90°$, the dips of sensors redshifted and blueshifted with the increase in the magnetic field intensity, respectively. At $0°$, the variation in the dip wavelength with respect to the magnetic field intensity was slight, while the variation was very remarkable at $90°$. At the $0°$ and $180°$ directions, the bending MMF plane was parallel to the magnetic field direction. As the magnetic nanochain clusters were mainly distributed on both sides of the sensor, the number of magnetic nanoparticles near the gold film at the outside region of the bent fiber was small. Therefore, even if the magnetic field intensity increased, the MF concentration near the gold film changed slightly, as did the ERI of the SPM. Thus, there was only a slight drift in the dip wavelength at this direction. Contrarily, at the $90°$ and $270°$ directions, the bending MMF plane was perpendicular to the magnetic field direction. The magnetic nanochain clusters were mainly distributed in the region above the sensor, where the number of magnetic nanoparticles was large. When the magnetic field intensity changed, the MF concentration changed obviously, as did the ERI of the SPM. So, the sensor had a good magnetic field response at 90°. The highest sensitivity was obtained at 9.749 nm/mT at 90° for sensor 1 (see Figure 8). In brief, the experimental results showed that the proposed sensing probe could detect not only the magnetic field intensity but also the magnetic field direction.

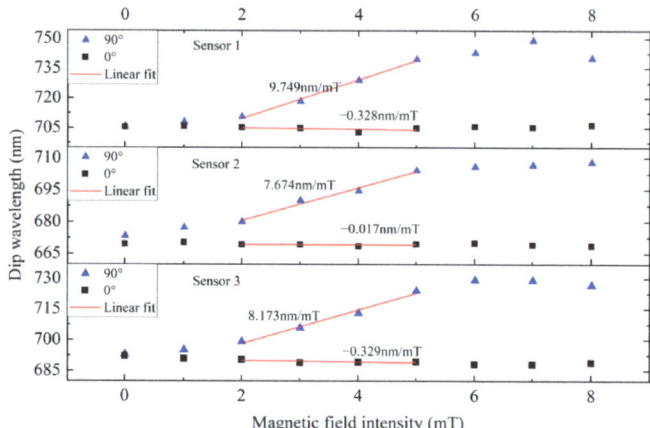

Figure 8. Shift of dip wavelength with magnetic field intensity at 0° and 90°.

For comparison, the sensing structure, fabrication method and sensing performance of various vector-magnetic-field sensors are listed in Table 1. Compared with other structures, the proposed vector-sensing structure in this work had the highest sensitivity and a better mechanical strength without destroying the integrity of the fiber. On the other hand, we also noticed that the detection range of our as-fabricated device was relatively small. Thus, the proposed structure may be more appropriate for applications in fields where high sensitivity is required but the magnetic field varies in a small range, such as biomedicine, medical diagnostics, mineral searches and archaeological excavations [39,40].

Finally, we would like to point out that the ambient temperature was kept constant in this work. However, the RI of the MF is also closely related to temperature. The thermo-optical coefficient of the MF was around -2.4×10^{-4} °C^{-1} [41]. So, temperature calibration or compensation may be necessary for applications in a harsh environment and/or high-precision applications. In addition, under a vibrating environment, the magnetic nanoparticle agglomeration will be destroyed to certain extent even though the magnetic field is unchanged. Then, the RI of the MF is affected by the vibration as well. Thus, the sensing performance will be influenced by the vibration. Therefore, anti-vibration design is necessary for the case when a high precision is required.

Table 1. Sensing performance of various optical fiber vector-magnetic-field sensors.

Sensing Structure	Fabrication Method	Detecting Range	Maximum Sensitivity	Reference
STS + lateral-offset	Offset splicing	0–16 mT	222.0 pm/mT	[21]
Tilted fiber grating	UV inscribed	0–15 mT	260 pm/mT	[42]
SMF fused with capillary	Tapered	0–110 mT	112 pm/mT	[43]
SPF-SNS	Side-polished	0–30 mT	2370 pm/mT	[44]
TFBG + gold film	Gold plating	0–18 mT	1800 pm/mT	[37]
FP + lateral-offset	Offset splicing	0–9 mT	4.63 pm/mT	[45]
D-shape fiber + gold film	Side polished + gold plating	0–23 mT	598.7 pm/Oe (5987 pm/mT)	[46]
Wedge-shape + gold film	Tip polished + gold plating	0–22 mT	1100 pm/mT	[47]
Bent MMF + gold film	Bent + gold plating	0–8 mT	**9749 pm/mT**	This work

4. Conclusions

In conclusion, a gold-clad bending MMF structure fixed in a vessel filled with MF was used for vector-magnetic-field sensing. Coupling among the CM, WGM and SPM could be realized simultaneously, which resulted in the resonant dip wavelength being closely related to the environmental RI. The magnitude and direction of the magnetic field could be measured by detecting the shift of the dip wavelength. The sensitivity in terms of direction and intensity were 546.5 pm/° and 9749 pm/mT, respectively. The sensor has the advantages of high sensitivity, easy fabrication and good integrity.

Author Contributions: Conceptualization, writing—original draft preparation, W.L.; methodology, validation, S.P.; software, Z.H.; investigation, J.W. (Jia Wang) and Y.F.; data curation, C.Z. and J.W. (Jingyue Wang) All authors have read and agreed to the published version of the manuscript.

Funding: This work was supported by the National Natural Science Foundation of China (Grant Nos. 62075130, 61675132), and the Shanghai "Shuguang Program" (Grant No. 16SG40).

Data Availability Statement: The data that support the findings of this study are available from the corresponding author upon reasonable request.

Conflicts of Interest: The authors declare no conflict of interest.

References

1. Zu, P.; Chan, C.C.; Lew, W.S.; Jin, Y.; Zhang, Y.; Liew, H.F.; Chen, L.H.; Wong, W.C.; Dong, X. Magneto-optical fiber sensor based on magnetic fluid. *Opt. Lett.* **2012**, *37*, 398–400. [CrossRef] [PubMed]
2. Zhao, Y.; Lv, R.; Li, H.; Wang, Q. Simulation and experimental measurement of magnetic fluid transmission characteristics subjected to the magnetic field. *IEEE Trans. Magn.* **2014**, *50*, 4600107. [CrossRef]
3. Zhao, Y.; Lv, R.; Zhang, Y.; Wang, Q. Novel optical devices based on the transmission properties of magnetic fluid and their characteristics. *Opt. Laser. Vol.* **2012**, *50*, 1177–1184. [CrossRef]
4. Zhang, C.; Pu, S.; Hao, Z.; Wang, B.; Yuan, M.; Zhang, Y. Magnetic field sensing based on whispering gallery mode with nanostructured magnetic fluid-infiltrated photonic crystal fiber. *Nanomaterials* **2022**, *12*, 862. [CrossRef]
5. Zhang, Y.; Pu, S.; Li, Y.; Hao, Z.; Li, D.; Yan, S.; Yuan, M.; Zhang, C. Magnetic field and temperature dual-parameter sensor based on nonadiabatic tapered microfiber cascaded with FBG. *IEEE Access* **2022**, *10*, 15478–15486. [CrossRef]
6. Deng, M.; Huang, C.; Liu, D.; Jin, W.; Zhu, T. All fiber magnetic field sensor with ferrofluid-filled tapered microstructured optical fiber interferometer. *Opt. Express* **2015**, *23*, 20668–20674. [CrossRef]
7. Dong, S.; Pu, S.; Wang, H. Magnetic field sensing based on magnetic-fluid-clad fiber-optic structure with taper-like and lateral-offset fusion splicing. *Opt. Express* **2014**, *22*, 19108–19116. [CrossRef]
8. Luo, L.; Pu, S.; Tang, J.; Zeng, X.; Lahoubi, M. Reflective all-fiber magnetic field sensor based on microfiber and magnetic fluid. *Opt Express* **2015**, *23*, 18133–18142. [CrossRef]
9. Xia, J.; Wang, F.; Luo, H.; Wang, Q.; Xiong, S. A magnetic field sensor based on a magnetic fluid-filled FP-FBG structure. *Sensors* **2016**, *16*, 620. [CrossRef]
10. Zhao, Y.; Lv, R.; Ying, Y.; Wang, Q. Hollow-core photonic crystal fiber Fabry–Perot sensor for magnetic field measurement based on magnetic fluid. *Opt. Laser Technol.* **2012**, *44*, 899–902. [CrossRef]
11. Candiani, A.; Margulis, W.; Sterner, C.; Konstantaki, M.; Pissadakis, S. Phase-shifted Bragg microstructured optical fiber gratings utilizing infiltrated ferrofluids. *Opt. Lett.* **2011**, *36*, 2548–2550. [CrossRef] [PubMed]
12. Candiani, A.; Konstantaki, M.; Margulis, W.; Pissadakis, S. A spectrally tunable microstructured optical fibre Bragg grating utilizing an infiltrated ferrofluid. *Opt. Express* **2010**, *18*, 24654–24660. [CrossRef] [PubMed]
13. Liu, T.; Chen, Y.; Han, Q.; Lv, X. Magnetic field sensor based on U-bent single-mode fiber and magnetic fluid. *IEEE Photonics J.* **2014**, *6*, 5300307. [CrossRef]
14. Fang, Y.L.; Huang, Y.H.; Kuo, C.Y.; Chiang, C.C. U-bend fiber optical sensor for magnetic field sensing. *Opt. Quantum Electron.* **2019**, *51*, 36. [CrossRef]
15. Chen, Y.; Han, Q.; Liu, T.; Lan, X.; Xiao, H. Optical fiber magnetic field sensor based on single-mode-multimode-single-mode structure and magnetic fluid. *Opt. Lett.* **2013**, *38*, 3999–4001. [CrossRef]
16. Wang, H.; Pu, S.; Wang, N.; Dong, S.; Huang, J. Magnetic field sensing based on singlemode-multimode-singlemode fiber structures using magnetic fluids as cladding. *Opt. Lett.* **2013**, *39*, 3765–3768. [CrossRef]
17. Mao, L.; Pu, S.; Su, D.; Wang, Z.; Zeng, X.; Lahoubi, M. Magnetic field sensor based on cascaded microfiber coupler with magnetic fluid. *J. Appl. Phys.* **2016**, *120*, 093102. [CrossRef]
18. Yuan, M.; Pu, S.; Li, D.; Li, Y.; Hao, Z.; Zhang, Y.; Zhang, C.; Yan, S. Extremely high sensitivity magnetic field sensing based on birefringence-induced dispersion turning point characteristics of microfiber coupler. *Results Phys.* **2021**, *29*, 104743. [CrossRef]

19. Zu, P.; Chan, C.C.; Siang, L.W.; Jin, Y.; Zhang, Y.; Fen, L.H.; Chen, L.; Dong, X. Magneto-optic fiber Sagnac modulator based on magnetic fluids. *Opt. Lett.* **2011**, *36*, 1425–1427. [CrossRef]
20. Lin, Q.; Xiao, Y.; Hu, Y.; Yan, F.; Hu, S.; Chen, Y.; Liu, G.; Chen, Y.; Luo, Y.; Chen, Z. Half-side gold-coated hetero-core fiber for highly sensitive measurement of a vector magnetic field. *Opt. Lett.* **2020**, *45*, 4746–4749. [CrossRef]
21. Yin, J.; Yan, P.; Chen, H.; Yu, L.; Jiang, J.; Zhang, M.; Ruan, S. All-fiber-optic vector magnetometer based on anisotropic magnetism-manipulation of ferromagnetism nanoparticles. *Appl. Phys. Lett.* **2017**, *110*, 5187. [CrossRef]
22. Candiani, A.; Konstantaki, M.; Margulis, W.; Pissadakis, S. Optofluidic magnetometer developed in a microstructured optical fiber. *Opt. Lett.* **2012**, *37*, 4467–4469. [CrossRef] [PubMed]
23. Zhang, X.; Peng, W. Bent fiber interferometer. *J. Lightwave Technol.* **2015**, *33*, 3351–3356. [CrossRef]
24. Chiang, C.C.; Chao, J.C. Whispering gallery mode based optical fiber sensor for measuring concentration of salt solution. *J. Nanomater.* **2013**, *2013*, 1–4. [CrossRef]
25. Nam, S.H.; Yin, S. High-temperature sensing using whispering gallery mode resonance in bent optical fibers. *IEEE Photonics Technol. Lett.* **2005**, *17*, 2391–2393. [CrossRef]
26. He, J.; Liao, C.; Yang, K.; Liu, S.; Yin, G.; Sun, B.; Zhou, J.; Zhao, J.; Wang, Y. High-sensitivity temperature sensor based on a coated single-mode fiber loop. *J. Lightwave Technol.* **2015**, *33*, 4019–4026. [CrossRef]
27. Homola, J. *Surface Plasmon Resonance Based Sensors*; Springer: Berlin/Heidelberg, Germany, 2006.
28. McDonagh, C.; Burke, C.S.; MacCraith, B.D. Optical chemical sensors. *Chem. Rev.* **2015**, *108*, 400–422. [CrossRef]
29. Homola, J. On the sensitivity of surface plasmon resonance sensors with spectral interrogation. *Sens. Actuators B* **1997**, *41*, 207–211. [CrossRef]
30. Slavík, R.; Homola, J.; Ctyroký, J. Single-mode optical fiber surface plasmon resonance sensor. *Sens. Actuators B* **1999**, *54*, 74–79. [CrossRef]
31. Liu, T.; Chen, Y.; Han, Q.; Liu, F.; Yao, Y. Sensor based on macrobent fiber Bragg grating structure for simultaneous measurement of refractive index and temperature. *Appl. Opt.* **2016**, *55*, 791–795. [CrossRef]
32. Chen, L.; Zhang, W.; Wang, L.; Zhou, Q.; Sieg, J.; Zhao, D.; Wang, B.; Yan, T.; Wang, S. Fiber refractive index sensor based on dual polarized Mach-Zehnder interference caused by a single-mode fiber loop. *Appl. Opt.* **2016**, *55*, 63–69. [CrossRef]
33. Li, Y.; Pu, S.; Hao, Z.; Yan, S.; Zhang, Y.; Lahoubi, M. Vector magnetic field sensor based on U-bent single-mode fiber and magnetic fluid. *Opt. Express* **2021**, *29*, 5236–5246. [CrossRef] [PubMed]
34. Dyshlyuk, A.V.; Vitrik, O.B.; Kulchin, Y.N.; Mitsai, E.V.; Cherepakhin, A.B.; Branger, C.; Brisset, H.; Iordache, T.V.; Sarbu, A. Numerical and experimental investigation of surface plasmon resonance excitation using whispering gallery modes in bent metal-clad single-mode optical fiber. *J. Lightwave Technol.* **2017**, *35*, 5425–5431. [CrossRef]
35. Zhang, Y.; Liang, P.; Wang, Y.; Liu, Z.; Wei, Y.; Zhu, Z.; Zhao, E.; Yang, J.; Yuan, L. Cascaded distributed multichannel fiber SPR sensor based on gold film thickness adjustment approach. *Sens. Actuators A* **2017**, *267*, 526–531. [CrossRef]
36. Haddouche, I.; Cherbi, L.; Ferhat, M.L. Analytical modelization of a fiber optic-based surface plasmon resonance sensor. *Opt. Commun.* **2017**, *402*, 618–623. [CrossRef]
37. Zhang, Z.; Guo, T.; Zhang, X.; Xu, J.; Xie, W.; Nie, M.; Wu, Q.; Guan, B.; Albert, J. Plasmonic fiber-optic vector magnetometer. *Appl. Phys. Lett.* **2016**, *108*, 101105. [CrossRef]
38. Lv, R.; Zhao, Y.; Li, H.; Hu, H.F. Theoretical analysis and experimental measurement of birefringence properties in magnetic fluid subjected to magnetic field. *IEEE Trans. Magn.* **2015**, *51*, 1–5. [CrossRef]
39. Nabighian, M.N.; Grauch, V.; Hansen, R.O.; Lafehr, T.R.; Pearson, W.C.; Phillips, J.D. The historical development of the magnetic method in exploration. *Geophysics* **2005**, *70*, 33–61. [CrossRef]
40. Savukov, I.; Karaulanov, T. Magnetic-resonance imaging of the human brain with an atomic magnetometer. *Appl. Phys. Lett.* **2013**, *103*, 115–120. [CrossRef]
41. Pu, S.; Chen, X.; Chen, Y.; Liao, W.; Chen, L.; Xia, Y. Measurement of the refractive index of a magnetic fluid by the retroreflection on the fiber-optic end face. *Appl. Phys. Lett.* **2015**, *86*, 611. [CrossRef]
42. Lu, T.; Sun, Y.; Moreno, Y.; Yan, Z.; Wang, C.; Sun, Q.; Wan, H.; Liu, D.; Zhang, L. Vector magnetic field sensor based on excessively tilted fiber grating assistant with magnetic fluids. In Proceedings of the 2018 Asia Communications and Photonics Conference, Hangzhou, China, 26–29 October 2018.
43. Cui, J.; Qi, D.; Tian, H.; Li, H. Vector optical fiber magnetometer based on capillaries filled with magnetic fluid. *Appl. Opt.* **2019**, *58*, 2754. [CrossRef] [PubMed]
44. Li, Y.; Pu, S.; Zhao, Y.; Zhang, R.; Jia, Z.; Yao, J.; Hao, Z.; Han, Z.; Li, D.; Li, X. All-fiber-optic vector magnetic field sensor based on side-polished fiber and magnetic fluid. *Opt. Express* **2019**, *27*, 35182–35188. [CrossRef] [PubMed]
45. Zhang, J.; Jiang, Y.; Zhang, X.; Chen, H.; Guo, Z.; Wang, W. Magnetic field vector sensor based on a cascaded structure of core-offset Fabry–Perot interferometer and a magnetic fluid infiltrated glass capillary. *J. Opt. Soc. Am. B* **2021**, *38*, 102902–102906. [CrossRef]
46. Jiang, Z.; Dong, J.; Hu, S.; Zhang, Y.; Chen, Y.; Luo, Y.; Zhu, W.; Qiu, W.; Lu, H.; Guan, H.; et al. High-sensitivity vector magnetic field sensor based on side-polished fiber plasmon and ferrofluid. *Opt. Lett.* **2018**, *43*, 4743–4746. [CrossRef] [PubMed]
47. Hao, Z.; Pu, S.; Wang, J.; Liu, W.; Zhang, C.; Fan, Y.; Lahoubi, M. Dual-channel temperature-compensated vector magnetic field sensor based on lab-on-a-fiber-tip. *Opt. Express* **2022**, *30*, 25208–25218. [CrossRef] [PubMed]

Electrospun PA66/Graphene Fiber Films and Application on Flexible Triboelectric Nanogenerators

Qiupeng Wu [1,2], Zhiheng Yu [3], Fengli Huang [2,*] and Jinmei Gu [2]

[1] College of Mechanical Engineering, Zhejiang University of Technology, Hangzhou 310014, China; 2112002087@zjut.edu.cn
[2] Key Laboratory of Advanced Manufacturing Technology of Jiaxing City, Jiaxing University, Jiaxing 341000, China; jmgu1218@zjxu.edu.cn
[3] College of Mechanical and Electrical Engineering, Jiaxing Nanhu University, Jiaxing 314000, China; yuzhiheng@jxnhu.edu.cn
* Correspondence: hfl@mail.zjxu.edu.cn or hfl@zjxu.edu.cn

Abstract: Triboelectric nanogenerators (TENGs) are considered to be the most promising energy supply equipment for wearable devices, due to their excellent portability and good mechanical properties. Nevertheless, low power generation efficiency, high fabrication difficulty, and poor wearability hinder their application in the wearable field. In this work, PA66/graphene fiber films with 0, 1 wt%, 1.5 wt%, 2 wt%, 2.5 wt% graphene and PVDF films were prepared by electrospinning. Meanwhile, TENGs were prepared with PA66/graphene fiber films, PVDF films and plain weave conductive cloth, which were used as the positive friction layer, negative friction layer and the flexible substrate, respectively. The results demonstrated that TENGs prepared by PA66/graphene fiber films with 2 wt% graphene showed the best performance, and that the maximum open circuit voltage and short circuit current of TENGs could reach 180 V and 7.8 µA, respectively, and that the power density was 2.67 W/m^2 when the external load was 113 MΩ. This is why the PA66/graphene film produced a more subtle secondary network with the addition of graphene, used as a charge capture site to increase its surface charge. Additionally, all the layered structures of TENGs were composed of breathable electrospun films and plain conductive cloth, with water vapor transmittance (WVT) of 9.6 Kgm^{-2}d^{-1}, reflecting excellent wearing comfort. The study showed that TENGs, based on all electrospinning, have great potential in the field of wearable energy supply devices.

Keywords: triboelectric nanogenerators; PA66/graphene fiber films; PVDF films; performance; wearing comfort

1. Introduction

With the development of Internet of Things technology (IoT), portable, flexible wearable electronic devices with various functions have been widely considered. In the IoT society, various devices offer many possibilities through data transmission and processing [1–4], such as real-time health monitoring, environmental monitoring, as well as biomedical assistance and human–machine interfaces [5–9]. It has brought about great changes in our lives. However, the flexibility and miniaturization of these electronic devices put forward higher requirements for power supply [10–12]. Triboelectric nanogenerators (TENGs), based on the combined action of friction generation and electrostatic induction, convert mechanical energy wasted in daily life into electrical energy required to drive tiny flexible electronic devices [13–15]. It has the advantages of high efficiency, wearable and environmental protection and suitability for integration into many flexible electronic devices.

All electrospun nanofiber films are widely used in TENGs design. The electrospun fibrous films have remarkable properties, such as inherent rough structure, large specific surface area and hierarchical porous structure [16,17], which could effectively increase the friction performance and air permeability, and the introduction of functional materials by electrospinning could affect the chemical and electrical properties of the resulting composites [18], optimizing the overall performance and TENGs' durability. Based on the previous studies, a series of effective microstructured electrospun films was prepared using different materials, such as polyvinylidene fluoride (PVDF) [19,20]. Meng [21] et al. successfully fabricated TENGs based on an all-electrospun silk fibroin/carbon nanotube (CNT) film as a substrate, with a power density of 317.4 $\mu W/cm^2$ under hand patching. Incorporating functional materials into polymer nanofibers could improve electrical output performance. Currently, most research focuses on improving output performance, such as selection of friction materials [22–24], surface modification [25–27] and micropatterning [28–30]. However, the above experiments all used dense metal films as electrodes. When worn, the breathability is reduced, making TENGs deficient in ideal breathability and wearing comfort, so that TENGs are short of the ideal air permeability and wearing comfort. Additionally, previous studies focused on materials or complex structures to a certain extent, ignoring the comfort of wearing. In fact, the multi-scale micro-nano structures in the friction layer structure could not only improve the output performance of TENGs, but also improve their mechanical properties and wearing comfort.

In this work, a method for all-electrospun fiber-based TENGs, consisting of two parts, including a PA66/graphene nanofiber-film-rubbed positive base layer and PVDF-rubbed negative electrode layer, was proposed. On the one hand, a PA66/graphene nanofiber film with multi-layer micro-nano structure could not only increase the effective friction area of contact, but also graphene dispersed in the PA66 fiber could be used as a charge capture site, which could increase the surface charge on the film surface, greatly improving its triboelectric properties [19]. In addition, all functional layers of the nanofiber-film-based triboelectric nanogenerators (TENGs) were constructed from nanofiber networks with a layered structure, ensuring air and moisture permeability. TENGs exhibited good electrical output performance as well as excellent flexibility and wear resistance. The maximum output voltage and current of the TENGs fabricated could reach 180 V and 7.8 μA, respectively, corresponding to a power density of 2.67 W/m^2. Water vapor transmission rate could be up to 9.6 $Kgm^{-2}d^{-1}$. In addition, the TENGs prepared could be placed anywhere on the body, detecting various human movements, and the mechanical energy that is wasted in life could be obtained. This energy can instead be used to power wearable electronic devices in daily life.

2. Materials and Methods

2.1. Materials

Polyamide66 (PA66) was bought from Aladdin Pharmaceuticals (Shanghai, China). Graphene was sold by Shenzhen Turing Artificial Intelligence Black Technology Co., Ltd. (Shenzhen, China). Polyvinylidene Difluoride (Mw 600000, PVDF) was produced by Arkema (Paris, France). Formic Acid and Polyethylene Oxide (PEO) were bought from Macklin (Shanghai, China). N, N-Dimethylformamide (DMF) and Acetone were produced by Sinopharm (Beijing, China). Plain weave conductive fabric was produced by Saintyear Electronic (125 ± 25 um, Hangzhou, China).

2.2. Fabrication of PA66/Graphene Fiber Film Layer

Firstly, 0.5 wt% PEO was added to 10 mL of Formic Acid and a magnetic stirrer was used to stir for 1 h to increase the viscosity of the solution, so that it could be well electrospun. Secondly, 15 wt% PA66 particles were put into Formic Acid/PEO solution, which were fully dissolved via a magnetic stirrer for 4 h to obtain 15 wt% PA66 solutions. Finally, different mass fractions of graphene (1 wt%, 1.5 wt%, 2 wt%, 2.5 wt%) were put into the prepared PA66 solution, respectively, and stirred to fully mix. Further, ultrasonic dispersion was applied for 6 h to make the graphene evenly dispersed in the solution.

The electrospinning process was performed with a supply voltage of 20 kV, a working distance between the spray nozzle and the substrate was 15 cm and the PA66/graphene solution was injected into a 10 mL syringe with a blunt needle (0.6 mm inner diameter).

The injection rate of the syringe was 0.1 mL/h. PA66/graphene nanofiber films were collected on plain weave conductive cloth (Ni-Cu) fabric electrodes, which were prepared at 25 °C and relative humidity of 50 ± 5%. The thickness of PA66/graphene fiber film was 80 ± 10 μm, measured by Talysurf Profiler (Dektak-XT-10 th, Bruker, MA, USA).

2.3. Fabrication of PVDF Nanofiber Films

Firstly, 3 mL Acetone was added into 7 mL DMF to obtain a DMF/Acetone solution in a 7-to-3 volume ratio. Secondly, 10 wt% PVDF powder was added into the mixed solution and magnetically stirred for 4 h under heating in a 60 °C water bath. Finally, the 10 wt% clear and transparent PVDF solution was obtained.

The electrospinning process was performed with a supply voltage of 12 kV, while the working distance between the spray nozzle and the substrate was 10 cm. The advance rate of the syringe was 0.4 mL/h. The PVDF nanofiber membrane was collected on the conductive fabric electrode, which was also collected under the same conditions, mentioned in Section 2.2. However, when dried under vacuum, the drying time of the PVDF was 80 °C for 4 h, 60 °C for 3 h and 50 °C for 1 h based on the previous experiment. The thickness of PVDF nanofiber films was 130 ± 15 um.

2.4. Assembly of TENG-Based Nanofiber Films

Based on the electrospinning process, the composite fabric of the two friction layers of TENGs was prepared. Firstly, the double-sided conductive tape was used as the electrode lead, which was tightly pasted on the back of the conductive fabric substrate where the different films were located. Then, the required size could be trimmed, such as 4 cm × 4 cm. Finally, these breathable films could be pasted on the corresponding positions of daily clothes, collecting the mechanical energy generated by wasted human movement.

2.5. Measurement of the TENGs

The surface morphologies of PA66 films, PA66/graphene films and PVDF films were characterized by the Field Emission Scanning Electron Microscopy (Hitachi S-4800, Tokyo, Japan). X-ray Diffraction (XRD, HAOYUAN DX-2700BH, Dandong, China) was used to study the elements and material structure of the electrospun fiber films. The I_{SC} and V_{OC} were measured by Keithley System Electrometer (6517b, Tektronix, Shanghai, China).

3. Results and Discussion

3.1. Preparation of the TENGs

The main concept of the design in this work was to use common fabric materials as the basic building unit to construct a wearable TENGs platform, so as to achieve effective harvesting of low-frequency, ubiquitous and easily wasted human biomechanical energy, such as walking and running. The all-electrospinning-based power-generating textile had multiple secondary structures and it could be better integrated with personal clothing, powering wearable electronic devices. These power-generating fabrics with excellent mechanical properties exhibited excellent breathability and cutability. It was an ideal device to make energy harvesters and sensors of a flexible wearable electronic system integrated, which guaranteed the integration of TENGs and wearable electronic devices with clothing.

A detailed preparation process for the PA66/graphene nanofiber film and PVDF film are schematically illustrated in Figure 1. The prepared organic solution flowed out slowly at a constant speed under the action of a syringe pump. Under the action of tens of thousands of volts of electrostatic field, the droplet at the needle of the organic solution could form a Taylor cone and the solvent in the sprayed droplet could evaporate in the air, becoming nano-scale filaments [31]. Finally, it was collected by the cylinder connected to the negative pole.

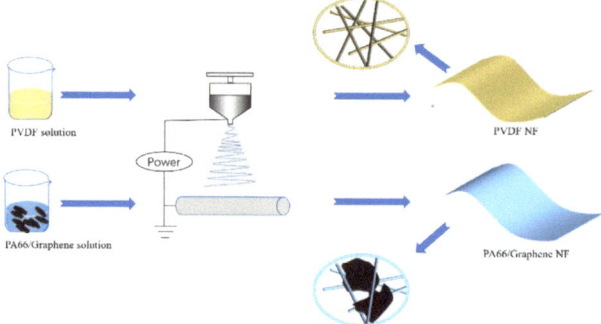

Figure 1. Schematic diagram of electrospinning film.

The prepared TENGs were composed of three functional components, including a PA66/graphene fiber film and PVDF film with two tribological polarities for contact electrification, and a plain weave conductive cloth for charge conduction. Based on previous studies [30], constructing a nanostructured triboelectric film with high-effective surface area was an effective method to improve the friction performance of TENGs. Electrospinning was widely used to fabricate fibrous films with surface structures. The PA66/graphene and PVDF films prepared by electrospinning were both thin and flexible. Due to the addition of graphene, the film was gray black, shown in Figure 2a. With the addition of graphene, the conductivity in the spinning solution increases; meanwhile, the viscosity in the spinning solution also increases. This results in an increase in the conductivity of the solution and the charge density, leading to reducing jet stability during spinning. This promotes fiber splitting, which results in the formation of a secondary fiber network (Figure 2b).The increase in solution viscosity could inhibit this phenomenon, so when too much graphene is added, the solution viscosity will be too high and the secondary network will be reduced [31].The generation of the secondary network can make the two friction layers have more contact area when they are in contact, promoting more charge flow when the TENGs works and improving the output performance. On the other hand, graphene dispersed between nanofibers could serve as charge-trapping sites, which could store more charge during the contact separation process of TENGs, enhancing the triboelectric surface charge and increasing the triboelectric properties of TENGs. For prepared TENGs, plain weave conductive cloth used instead of metal electrodes, shown in Figure 2d, greatly enhancing the air permeability and wearing comfort of electronic devices.

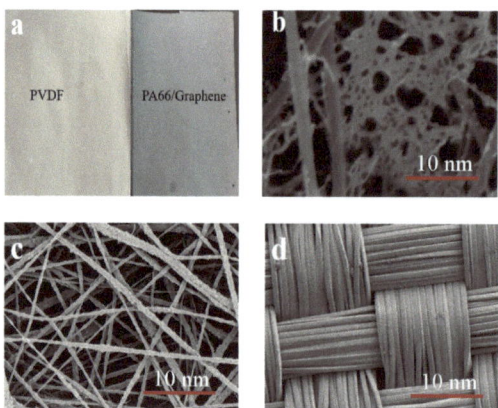

Figure 2. (**a**) Electrospun PVDF and PA66/graphene films; (**b**) SEM image of PA66/graphene film; (**c**) SEM image of PVDF film; (**d**) SEM image of plain weave conductive fabric.

3.2. Effects of Graphene with Different Mass Fractions on Output Performance

To study the elements in the films, the XRD patterns of PA66/graphene thin-film layers with different mass fractions of graphene were observed, shown in Figure 3. The two peaks (2θ = 19° and 2θ = 23.5°) correspond to the characteristic diffraction peak of an α-crystal form and were, therefore, obtained. What is more, the crystal form of the substance in the PA66 film was consistent with that of the PA66 particles. Small peaks around 45 and 50 degrees appeared and this is the reason that the PA66/graphene with different graphene mass fractions was fixed pinned with the adhesive tape, leading to the introduction of impurities. The results showed that the addition of graphene could not change the peak position of PA66 and from the diffraction peaks of the XRD pattern, it can be determined that the crystalline form of the material in the PA66 film is consistent with the PA66 particles, so it can be concluded that the material structure of the PA66 material in the film prepared by spinning has not changed.

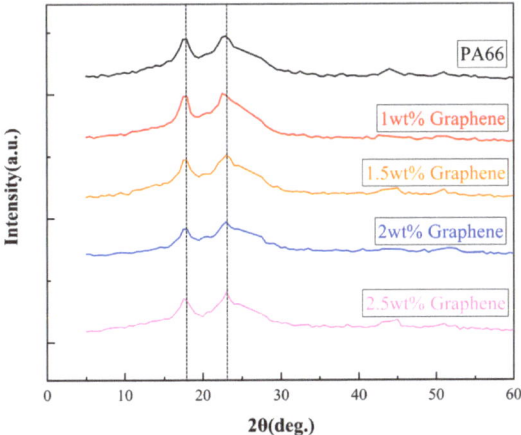

Figure 3. XRD patterns of PA66/graphene with different graphene mass fractions.

To study the effect of graphene doping on the output performance, contact-separation-mode TENGs based on PA66/graphene and PVDF films were designed. The two films were glued to two 3D-printed slabs with bosses (Figure 4a,b). The upper and bottom plates are connected by four studs. Between the two plates, four springs are placed over the studs to provide resilience (Figure 4c). This device can make the two friction layers contact under the slap and quickly return to the original position when the external force is removed. At the same time, the external force on contact can be controlled by replacing springs with different elastic coefficients. Connecting the lead-out conductive tape to the Keithley 6517b test leads (Figure 4d), the resulting voltage and current signals were collected by the Keithley 6517b for analysis. The synergistic effect of contact charging and electrostatic induction is shown in Figure 5a, which laid the theoretical foundation for TENGs as an energy harvester [32]. The triboelectric pair started with a separation attitude; no charge was generated at this time and there was no potential difference between the two triboelectric layers. When the TENGs were stimulated by an external force, the two triboelectric layers approached each other until they were in contact, and the two nano-friction layers would slide relative to each other at the microscopic level, due to the different electron affinities in the two friction materials, and the two friction layers would generate opposite charges. When the two friction layers were separated, a potential difference would be formed between the two friction layers to promote the flow of free electrons from the PVDF film to the PA66/graphene film through the external load, generating an instantaneous current. As the external force was continuously unloaded, this potential difference would continue to increase, and when it was fully restored to its original position, the potential difference would reach its maximum value. As the external force approached the PA66/graphene film again, the original electrostatic balance was broken again, causing the electrons in the PA66/graphene film to run back to the PVDF film, resulting in an opposite instantaneous current. When the two sides were in full contact, the potential difference became zero again. This continuous contact and separation process generated pulses of voltage and currents of opposite polarity, converting mechanical energy into electrical energy.

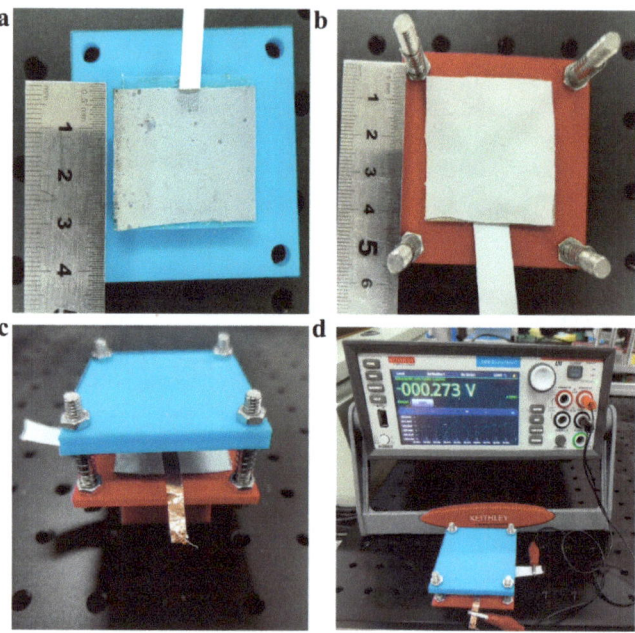

Figure 4. Testing flow diagram of the contact-separated TENGs. (**a**) PA66/graphene pasted on the upper plate; (**b**) PVDF film pasted on the bottom plate; (**c**) installation diagram after assembly; (**d**) signal testing diagram.

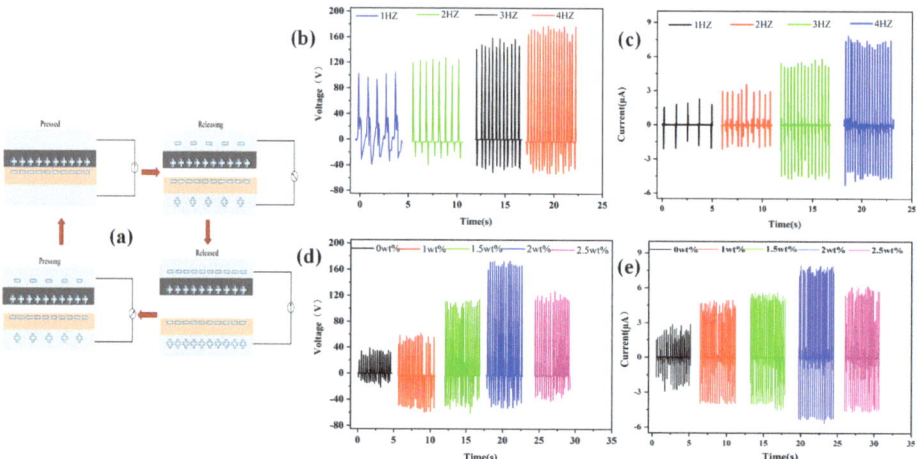

Figure 5. Output performance of nanofiber-film-based TENGs. (**a**) Working principle of the TENGs; (**b**) open-circuit voltage outputted of TENGs at different frequencies; (**c**) short-circuit current output of TENGs at different frequencies; (**d**,**e**) effects of the thickness of PA66/graphene.

According to previous studies [33–35], when the change in the distance between the two friction layers was much smaller than the area of the friction layer, the open-circuit voltage and short-circuit current of the TENGs could be calculated as follows:

$$V_{OC}^{TENG} = \frac{Qsc(x)}{C(x)} = \frac{\sigma x(t)}{\varepsilon_0} \quad (1)$$

$$I_{SC\,max}^{TENG} = \frac{S\sigma v(t)}{d_0} = \frac{2\pi S\sigma f x_{max}}{d_0} \quad (2)$$

where $C(x)$ was the capacitance between the two electrodes at different time-varying displacements x; ε_0 was the vacuum dielectric constant; S was the surface area of the friction layer; d_0 was the effective thickness of the dielectric layer; σ was the triboelectric surface charge density; $Q_{SC}(x)$ was the amount of short-circuit charge transfer at displacement x; and f was the frequency. Therefore, according to Equations (1) and (2), we could observe that the open-circuit voltage of TENGs was independent of frequency f and increased only with larger displacement between the two contact surfaces, and the short-circuit current was proportional to the frequency f, so the measured maximum short-circuit current increased with the frequency f, shown in Figure 5b,c. The results showed that TENGs still had good output performance at low frequencies and could collect low-frequency energy. Compared to the short-circuit current (I_{SC}) and open-circuit voltage (V_{OC}) of the TENGs and an external force exerted with a frequency of 4 Hz, a higher circuit signal and output power could be obtained.

3.3. Effects of the Dielectric Constant of the Electrospun Film on Output Performance

With the increase in graphene, the dielectric constant of the electrospun film also had an effect on the output performance. The enhanced dielectric constant of the triboelectric layer could improve the ability to transfer charge density and the surface potential also had a large impact on the performance of TENGs. A higher dielectric constant could use the electrospun PA66/graphene film as a friction layer to store more charge. However, an excessively large dielectric constant would generate a larger dielectric loss, which leads to the leakage of charge. Due to the changes in frequency and external strain during human motion, it was necessary to explore their effects on the output performance of wearable devices. As shown in Figure 3, the PA66/graphene fiber films of graphene with different mass fractions were used as the positive friction layer and PVDF was tested as the TENGs of the friction negative layer (both applied 4 Hz, 30 N external force), shown in Figure 5d,e. In the range of graphene mass fraction of 0–2 wt%, the short-circuit current and open-circuit voltage signals of TENGs increased with the increase in graphene mass fraction, while the maximum value could reach 180 V and 7.8 μA, respectively, when the graphene content was 2 wt%. Compared with the fiber film without graphene added, the modified V_{OC} and I_{SC} were both increased by about four-times. Therefore, the TENGs made of PA66/graphene fiber film with 2 wt% graphene content were the most suitable experimental sample. Further, the piezoelectric properties of PVDF films were tested by the method reported by Yin [36], so the highest output was only 0.8V, meaning, in this paper, the piezoelectric effect of PVDF films was not considered.

3.4. Effects of Load Resistance on Output of Voltage and Current

To study the effects of load resistance on output of the voltage and current, the output performance of nanofiber-film-based TENGs was researched, shown in Figure 6. As the load resistance increased, the voltage tended to increase, while the current tended to decrease due to ohmic losses. The output current and output voltage peak values could be reached 180 V and 7 μA, respectively. According to the formula $P = I^2R$, the power of TENGs could be calculated. At 113 MΩ resistance, the maximum power density could reach 2.67 W/m². The generated output power was sufficient to drive some low-energy electronic devices, solving the problem of sustainable power supply for wearable systems in a stable and reliable manner effectively.

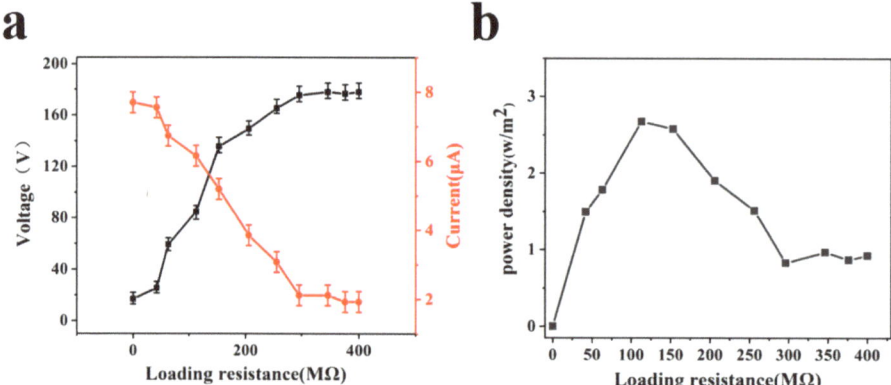

Figure 6. Relationship of load resistance on output of voltage and current. (**a**) Relationship of load resistance on output of voltage; (**b**) relationship of load resistance on output of current.

Due to the striking features of the porous structure, the fabricated power generation fabric had excellent water vapor permeability, which can be visually seen in Figure 7a. As we can see, the composite fabric was covered over a beaker filled with hot water. The abundant water vapour could move freely through the fabric without any restraint. To quantitatively demonstrate the breathability, the fabric was tested for water vapor transmission rate (WVTR) according to the ASTM E96 inverted cup standard at a relative humidity of 50% and a temperature of 38 °C. The WVTR value was calculated as follows.

$$WVT \text{ rate} = \frac{m_1 - m_2}{S} \times 24 \qquad (3)$$

where m_1 was the test cup mass before the test; m_2 was the test cup mass after the test; S was the test cup mouth area; and the obtained WVT rate was 9.6 Kgm^{-2}d^{-1}, which could confirm that the all-electrospun textile film had excellent air permeability. In other words, electrospun nanofibers had excellent wearing comfort, ensuring human thermal comfort and wide personal practical applications after their integration into wearable electronic devices.

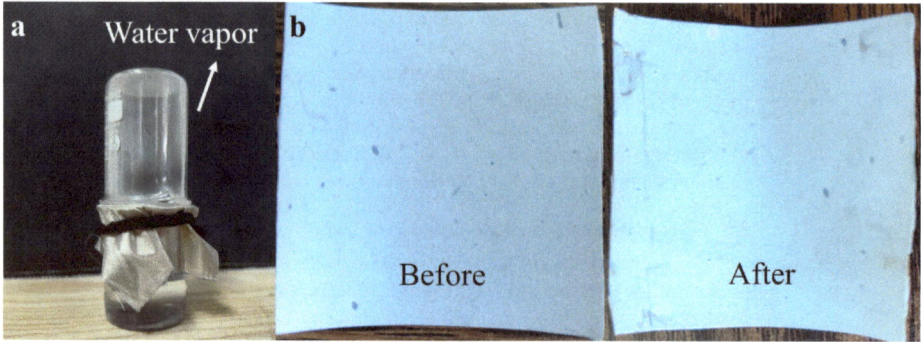

Figure 7. (**a**) Water vapor transmission rate test (WVT); (**b**) comparison before and after a thousand times of bending.

3.5. Test of Reliability and Application of TENGs

To explore the durability and stability of TENGs, the fabric current outputted after bending the two friction layers for 1000 cycles was tested and compared. It could be clearly seen that there was no significant mechanical damage to the surface, shown in Figure 7b. The outputted current did not present a significant reduction. It was indicated that the nanofiber films spun prepared had good mechanical properties.

To accurately evaluate the practical application of TENGs, a series of tests was carried out and the combined light chain composed of 110 LEDs in series could be lit by using the TENGs (Figure 8a), which showed that it was fully capable of outputting power to meet some low-power compatible electronic devices and it had the advantage of providing reliable power supply for wearable systems.

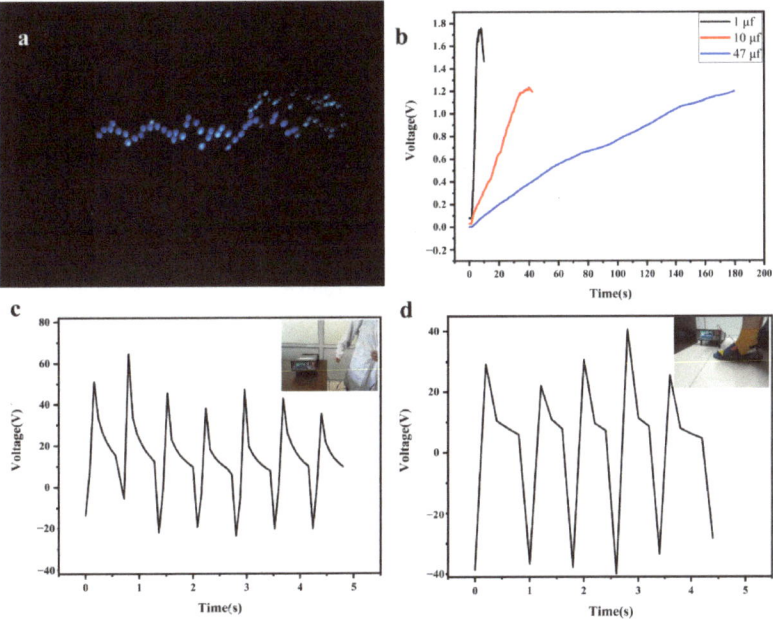

Figure 8. (**a**) 110 commercial led tandem light chains lit by TENGs; (**b**) three commercial capacitors to 1.2 V, respectively, charged by TENGs; (**c**) mechanical energy of the swing arm movement collected by TENGs; (**d**) mechanical energy from walking motions collected by TENGs.

To demonstrate the charging capability of TENGs, a circuit composed of a commercial capacitor and bridge rectifier was designed to test its charging efficiency. The charge in the capacitors (1 μF, 10 μF, 47 μF) to 1.2V with 4 s, 36 s, 180 s, respectively, is shown in Figure 8b.

To further demonstrate the application of TENGs based on all-electrospun nanofilms, they could be easily combined with textiles to harvest the mechanical energy generated from the human body, As shown in Figure 8c,d, two samples were attached to the clothing. The movement of the human body will lead to the contact and separation of the two friction layers, thereby converting the mechanical energy of the human body into electrical energy. In these experiments, the highest open-circuit voltages could reach 60 V and 40 V. The results showed that it was fully capable of collecting and supplying energy in daily life, and it not only functions as a power source, but also had the ability to collect external stimuli to generate signals. In the future, it may have the ability to collect human body posture, pulse, heartbeat and other human health information while supplying power.

4. Conclusions

In this work, an electrospun PA66/graphene nanofiber film with a secondary nano-network structure, as a tribo-cathode layer, was fabricated, and it was assembled with the electrospun PVDF film as the tribo-anode layer to form all-electrospun film-based TENGs. The rough surface of the two friction layers not only increased the effective contact area, but also improved the breathability. Furthermore, the porous fiber network in each functional layer guaranteed the air permeability of the TENGs. Compared with the pure PA66 fiber film, the introduced graphene increased the effective contact area of the PA66 nanofiber film and graphene could act as a charge trapping site, enhancing the electrical properties of the TENGs. The fabricated TENGs exhibited high Voc, Isc and peak power density of 180 V, 7.8 μA and 2.67 W/m^2, respectively. In addition, TENGs maintained excellent performance after 1000 cycles under cyclic pressure, keeping its advantages in durability and stability. TENGs with good comfort and high flexibility could obtain biomechanical energy from various human motions. There are great prospects in the system integration of wearable electronic devices.

Author Contributions: Conceptualization, F.H.; methodology, F.H., Z.Y. and Q.W.; validation, Z.Y. and Q.W.; formal analysis, Z.Y., J.G. and Q.W.; investigation, F.H., Z.Y. and Q.W.; resources, F.H.; data curation, F.H., Z.Y. and Q.W.; writing—original draft preparation, Z.Y. and Q.W.; writing—review and editing, Z.Y. and Q.W.; supervision, F.H. and Z.Y.; project administration, F.H.; funding acquisition, F.H. and Z.Y. All authors have read and agreed to the published version of the manuscript.

Funding: This research was funded by the Basic Public Welfare Research Program of Zhejiang Province (LGG20E050021) and the Science and Technology Bureau of Jiaxing City (2022AY10006).

Institutional Review Board Statement: Not applicable.

Informed Consent Statement: Not applicable.

Data Availability Statement: The data that support the findings of this study have not been made available but can be obtained from the author upon request.

Acknowledgments: The support of equipment and technology provided by College of Information Science and Engineering at Jiaxing University is gratefully acknowledged.

Conflicts of Interest: The authors declare no conflict of interest.

References

1. Ahmed, A.; Zhang, S.L.; Hassan, I.; Saadatnia, Z.; Zi, Y.; Zu, J.; Wang, Z.L. A washable, stretchable, and self-powered human-machine interfacing Triboelectric nanogenerator for wireless communications and soft robotics pressure sensor arrays. *Extreme Mech. Lett.* **2017**, *13*, 25–35. [CrossRef]
2. Chi, Q.P.; Yan, H.R.; Zhang, C.; Pang, Z.B.; Xu, L.D. A Reconfigurable Smart Sensor Interface for Industrial WSN in IoT Environment. *IEEE Trans. Ind. Inform.* **2014**, *10*, 1417–1425. [CrossRef]
3. Khan, Y.; Garg, M.; Gui, Q.; Schadt, M.; Gaikwad, A.; Han, D.; Yamamoto, N.A.D.; Hart, P.; Welte, R.; Wilson, W.; et al. Flexible Hybrid Electronics: Direct Interfacing of Soft and Hard Electronics for Wearable Health Monitoring. *Adv. Funct. Mater.* **2016**, *26*, 8764–8775. [CrossRef]
4. Sundaram, S.; Kellnhofer, P.; Li, Y.; Zhu, J.Y.; Torralba, A.; Matusik, W. Learning the signatures of the human grasp using a scalable tactile glove. *Nature* **2019**, *569*, 698–702. [CrossRef]
5. Gao, Y.; Yu, L.; Yeo, J.C.; Lim, C.T. Flexible Hybrid Sensors for Health Monitoring: Materials and Mechanisms to Render Wearability. *Adv. Mater.* **2020**, *32*, e1902133. [CrossRef]
6. Steinhauer, S. Gas Sensors Based on Copper Oxide Nanomaterials: A Review. *Chemosensors* **2021**, *9*, 51. [CrossRef]
7. Wang, Z.L. Triboelectric Nanogenerator (TENG)—Sparking an Energy and Sensor Revolution. *Adv. Energy Mater.* **2020**, *10*, 200137. [CrossRef]
8. Xiao, X.; Chen, G.; Libanori, A.; Chen, J. Wearable Triboelectric Nanogenerators for Therapeutics. *Trends Chem.* **2021**, *3*, 279–290. [CrossRef]
9. Zhu, M.L.; He, T.Y.Y.; Lee, C.K. Technologies toward next generation human machine interfaces: From machine learning enhanced tactile sensing to neuromorphic sensory systems. *Appl. Phys. Rev.* **2020**, *7*, 031305. [CrossRef]
10. Dong, K.; Peng, X.; An, J.; Wang, A.C.; Luo, J.; Sun, B.; Wang, J.; Wang, Z.L. Shape adaptable and highly resilient 3D braided triboelectric nanogenerators as e-textiles for power and sensing. *Nat. Commun.* **2020**, *11*, 2868. [CrossRef]

11. Guan, X.; Xu, B.; Wu, M.; Jing, T.; Yang, Y.; Gao, Y. Breathable, washable and wearable woven-structured triboelectric nanogenerators utilizing electrospun nanofibers for biomechanical energy harvesting and self-powered sensing. *Nano Energy* **2021**, *80*, 105549. [CrossRef]
12. Luo, J.; Gao, W.; Wang, Z.L. The Triboelectric Nanogenerator as an Innovative Technology toward Intelligent Sports. *Adv. Mater.* **2021**, *33*, e2004178. [CrossRef]
13. Chen, C.; Chen, L.; Wu, Z.; Guo, H.; Yu, W.; Du, Z.; Wang, Z.L. 3D double-faced interlock fabric triboelectric nanogenerator for bio-motion energy harvesting and as self-powered stretching and 3D tactile sensors. *Mater. Today* **2020**, *32*, 84–93. [CrossRef]
14. Dong, K.; Wu, Z.; Deng, J.; Wang, A.C.; Zou, H.; Chen, C.; Hu, D.; Gu, B.; Sun, B.; Wang, Z.L. A Stretchable Yarn Embedded Triboelectric Nanogenerator as Electronic Skin for Biomechanical Energy Harvesting and Multifunctional Pressure Sensing. *Adv. Mater.* **2018**, *30*, e1804944. [CrossRef] [PubMed]
15. Wang, N.; Wang, X.X.; Yan, K.; Song, W.; Fan, Z.; Yu, M.; Long, Y.Z. Anisotropic Triboelectric Nanogenerator Based on Ordered Electrospinning. *ACS Appl. Mater. Interfaces* **2020**, *12*, 46205–46211. [CrossRef]
16. Ding, Y.; Hou, H.; Zhao, Y.; Zhu, Z.; Fong, H. Electrospun polyimide nanofibers and their applications. *Prog. Polym. Sci.* **2016**, *61*, 67–103. [CrossRef]
17. Jiang, S.; Chen, Y.; Duan, G.; Mei, C.; Greiner, A.; Agarwal, S. Electrospun nanofiber reinforced composites: A review. *Polym. Chem.* **2018**, *9*, 2685–2720. [CrossRef]
18. Kaspar, P.; Sobola, D.; Castkova, K.; Dallaev, R.; Stastna, E.; Sedlak, P.; Knapek, A.; Trcka, T.; Holcman, V. Case Study of Polyvinylidene Fluoride Doping by Carbon Nanotubes. *Materials* **2021**, *14*, 1428. [CrossRef] [PubMed]
19. Bhatta, T.; Maharjan, P.; Cho, H.; Park, C.; Yoon, S.H.; Sharma, S.; Salauddin, M.; Rahman, M.T.; Rana, S.M.S.; Park, J.Y. High-performance triboelectric nanogenerator based on MXene functionalized polyvinylidene fluoride composite nanofibers. *Nano Energy* **2021**, *81*, 105670. [CrossRef]
20. Choi, G.-J.; Baek, S.-H.; Lee, S.-S.; Khan, F.; Kim, J.H.; Park, I.-K. Performance enhancement of triboelectric nanogenerators based on polyvinylidene fluoride/graphene quantum dot composite nanofibers. *J. Alloys Compd.* **2019**, *797*, 945–951. [CrossRef]
21. Su, M.; Kim, B. Silk Fibroin-Carbon Nanotube Composites based Fiber Substrated Wearable Triboelectric Nanogenerator. *ACS Appl. Nano Mater.* **2020**, *3*, 9759–9770. [CrossRef]
22. Yu, A.; Zhu, Y.; Wang, W.; Zhai, J. Progress in Triboelectric Materials: Toward High Performance and Widespread Applications. *Adv. Funct. Mater.* **2019**, *29*, 1900098. [CrossRef]
23. Zhao, Z.; Zhou, L.; Li, S.; Liu, D.; Li, Y.; Gao, Y.; Liu, Y.; Dai, Y.; Wang, J.; Wang, Z.L. Selection rules of triboelectric materials for direct-current triboelectric nanogenerator. *Nat. Commun.* **2021**, *12*, 4686. [CrossRef] [PubMed]
24. Zou, H.; Guo, L.; Xue, H.; Zhang, Y.; Shen, X.; Liu, X.; Wang, P.; He, X.; Dai, G.; Jiang, P.; et al. Quantifying and understanding the triboelectric series of inorganic non-metallic materials. *Nat. Commun.* **2020**, *11*, 2093. [CrossRef]
25. Feng, Y.; Zheng, Y.; Ma, S.; Wang, D.; Zhou, F.; Liu, W. High output polypropylene nanowire array triboelectric nanogenerator through surface structural control and chemical modification. *Nano Energy* **2016**, *19*, 48–57. [CrossRef]
26. Menge, H.G.; Kim, J.O.; Park, Y.T. Enhanced Triboelectric Performance of Modified PDMS Nanocomposite Multilayered Nanogenerators. *Materials* **2020**, *13*, 4156. [CrossRef]
27. Zhang, X.-S.; Han, M.-D.; Wang, R.-X.; Meng, B.; Zhu, F.-Y.; Sun, X.-M.; Hu, W.; Wang, W.; Li, Z.-H.; Zhang, H.-X. High-performance triboelectric nanogenerator with enhanced energy density based on single-step fluorocarbon plasma treatment. *Nano Energy* **2014**, *4*, 123–131. [CrossRef]
28. Gong, J.; Xu, B.; Tao, X. Breath Figure Micromolding Approach for Regulating the Microstructures of Polymeric Films for Triboelectric Nanogenerators. *ACS Appl. Mater. Interfaces* **2017**, *9*, 4988–4997. [CrossRef]
29. Huang, J.; Fu, X.; Liu, G.; Xu, S.; Li, X.; Zhang, C.; Jiang, L. Micro/nano-structures-enhanced triboelectric nanogenerators by femtosecond laser direct writing. *Nano Energy* **2019**, *62*, 638–644. [CrossRef]
30. Li, L.; Liu, S.; Tao, X.; Song, J. Triboelectric performances of self-powered, ultra-flexible and large-area poly(dimethylsiloxane)/Ag-coated chinlon composites with a sandpaper-assisted surface microstructure. *J. Mater. Sci.* **2019**, *54*, 7823–7833. [CrossRef]
31. Reneker, D.H.; Yarin, A.L.; Fong, H.; Koombhongse, S. Bending instability of electrically charged liquid jets of polymer solutions in electrospinning. *J. Appl. Phys.* **2000**, *87*, 4531–4547. [CrossRef]
32. Fan, F.-R.; Tian, Z.-Q.; Lin Wang, Z. Flexible triboelectric generator. *Nano Energy* **2012**, *1*, 328–334. [CrossRef]
33. Niu, S.; Wang, S.; Lin, L.; Liu, Y.; Zhou, Y.S.; Hu, Y.; Wang, Z.L. Theoretical study of contact-mode triboelectric nanogenerators as an effective power source. *Energy Environ. Sci.* **2013**, *6*, 3576. [CrossRef]
34. Yu, Z.H.; Zhang, T.C.; Li, K.F.; Huang, F.L.; Tang, C.L. Preparation of Bimodal Silver Nanoparticle Ink Base on Liquid Phase Reduction Method. *Nanomaterials* **2022**, *12*, 560. [CrossRef] [PubMed]
35. Yu, Z.H.; Xu, Y.; Tian, X. Graphene oxide loaded silver synergistically improved antibacterial performance of bone scaffold. *AIP Adv.* **2022**, *12*, 015024. [CrossRef]
36. Yin, J.Y.; Boaretti, C.; Lorenzetti, A.; Martucci, A.; Roso, M.; Modesti, M. Effects of Solvent and Electrospinning Parameters on the Morphology and Piezoelectric Properties of PVDF Nanofibrous Membrane. *Nanomaterials* **2022**, *12*, 962. [CrossRef]

Article

Effects of Three-Dimensional Circular Truncated Cone Microstructures on the Performance of Flexible Pressure Sensors

Weikan Jin [1,2], Zhiheng Yu [3], Guohong Hu [2], Hui Zhang [2,*], Fengli Huang [2,*] and Jinmei Gu [2]

1. School of Mechanical Engineering and Automation, Zhejiang Sci-Tech University, Hangzhou 310018, China; jinweikan@163.com
2. Key Laboratory of Advanced Manufacturing Technology of Jiaxing City, Jiaxing University, Jiaxing 341000, China; 2112002281@zjut.edu.cn (G.H.); jmgu1218@zjxu.edu.cn (J.G.)
3. College of Mechanical and Electrical Engineering, Jiaxing Nanhu University, Jiaxing 314000, China; yuzhiheng@jxnhu.edu.cn
* Correspondence: zhanghsx@163.com (H.Z.); hfl@mail.zjxu.edu.cn (F.H.)

Citation: Jin, W.; Yu, Z.; Hu, G.; Zhang, H.; Huang, F.; Gu, J. Effects of Three-Dimensional Circular Truncated Cone Microstructures on the Performance of Flexible Pressure Sensors. *Materials* 2022, *15*, 4708. https://doi.org/10.3390/ma15134708

Academic Editors: Jijun Feng and Shengli Pu

Received: 30 May 2022
Accepted: 2 July 2022
Published: 5 July 2022

Publisher's Note: MDPI stays neutral with regard to jurisdictional claims in published maps and institutional affiliations.

Copyright: © 2022 by the authors. Licensee MDPI, Basel, Switzerland. This article is an open access article distributed under the terms and conditions of the Creative Commons Attribution (CC BY) license (https://creativecommons.org/licenses/by/4.0/).

Abstract: Three-dimensional microstructures play a key role in the fabrication of flexible electronic products. However, the development of flexible electronics is limited in further applications due to low positioning accuracy, the complex process, and low production efficiency. In this study, a novel method for fabricating three-dimensional circular truncated cone microstructures via low-frequency ultrasonic resonance printing is proposed. Simultaneously, to simplify the manufacturing process of flexible sensors, the microstructure and printed interdigital electrodes were fabricated into an integrated structure, and a flexible pressure sensor with microstructures was fabricated. Additionally, the effects of flexible pressure sensors with and without microstructures on performance were studied. The results show that the overall performance of the designed sensor with microstructures could be effectively improved by 69%. Moreover, the sensitivity of the flexible pressure sensor with microstructures was 0.042 kPa^{-1} in the working range of pressure from 2.5 to 10 kPa, and the sensitivity was as low as 0.013 kPa^{-1} within the pressure range of 10 to 30 kPa. Meanwhile, the sensor showed a fast response time, which was 112 ms. The stability remained good after the 100 cycles of testing. The performance was better than that of the flexible sensor fabricated by the traditional inverted mold method. This lays a foundation for the development of flexible electronic technology in the future.

Keywords: microstructures; low-frequency ultrasonic resonance printing; flexible pressure sensor; sensitivity; response time

1. Introduction

Flexible electronic devices play an increasingly significant role in devices and systems of electronic skin [1–3], human–computer interactions [4], and physiological signal monitoring [5]. Flexible pressure sensors are an important part of flexible electronic devices, which can sense a small range of pressure, and convert it into electrical signals [6]. The sensing principles of flexible pressure sensors include capacitive sensing [7], piezoresistive sensing [8], piezoelectric sensing [9], and triboelectric sensing [10]. Flexible pressure sensors manufactured by different sensing principles have different application scenarios and characteristics [11]. At present, flexible pressure sensors have become a research hotspot because of their simple structure, simple tests, low preparation cost, and high sensitivity [12].

To improve the linear range and sensitivity of flexible pressure sensors, research into the microstructure has been the main research direction. Flexible pressure sensors with a micro-column structure and a pyramid structure are common geometric structures.

Zhang [13] prepared a flexible pressure sensor with a height-adjustable micro-column array. This sensor had a high shear force sensitivity (4.48 kPa^{-1}) and could accurately measure the shear force. Sally [14] et al. designed and developed an innovative and sensitive flexible sensor for efficient potentiometric monitoring of Ni (II) ions, which was constructed from highly porous activated flexible carbon cloth decorated with nitrogen and spherical porous carbon nanoparticles derived from low-cost cotton doped with polypyrrole nanoparticles via a simple carbonization–activation process followed by dip-coating in a membrane cocktail containing 2D Ni-MOF nanosheets as an electroactive material. The detection limit of the sensor was 2.7×10^{-6} with the pH range of 2–8. Nour [15] et al. developed a method of synthesizing polyaniline–polypyrrole composite materials with a network morphology, and prepared a thin layer of polypyrrole. Sally [16] et al. prepared a nanocomposite, and developed new sensitive and selective modified carbon paste electrodes (CPE), which showed good discriminating capability toward ClO-HCl with regard to a number of interfering materials. Choi [17] prepared PDMS films with pyramid microstructures, and spin-coated conductive active materials on their surfaces. The flexible pressure sensor manufactured by this method had the advantages of a large measurement range (10–100 kPa) and a fast response time (210 ms). To improve the performance of the sensor elements, microstructures such as spherical microstructures [18], bionic microstructures [19], folded microstructures [20], porous microstructures [8], and so on, are also being studied. Therefore, a reasonably designed microstructure is an effective method to fabricate highly sensitive flexible pressure sensors. Most of the research on the microstructure of flexible pressure sensors has mainly focused on the active material layer, and little research has focused on the fabrication of electrode layer microstructures. To facilitate their application, the manufacturing of flexible pressure sensors must have the characteristics of miniaturization and array. The preparation methods of the microstructure have the characteristics of diversity and comprehensiveness because of the complexity of the materials. Sun [21] coated a conductive polymer on a sandpaper template, obtaining a layered spinel microstructure. Ji [22] placed the cured polymer in water to dissolve NaCl to prepare an elastomer with a porous structure by using the template etching method. Chai [23] added Ca^{2+} as crosslinking agent to a graphene oxide sol to convert it into printable gel ink, and prepared a porous conductive network structure by using 3D printing technology. Because these methods need many processing steps and a complex operation, it is difficult to realize large-area repetitive manufacturing of microstructures without an efficient production process.

In view of the existing problems, in this work, we propose a new low-frequency ultrasonic resonant printing technology for high-precision and high-efficiency fabrication of microstructures. A flexible pressure sensor with microstructured electrodes was designed by using low-frequency ultrasonic resonant printing technology. The fabrication method of the sensor is introduced in this study. The surface structure of each part of the sensor was characterized, and the related tests were carried out after packaging. The effects of the microstructured electrode on the performance of the sensor were compared. The test results showed that the electrode with a microstructure could effectively improve the performance of the sensor, and the sensor had a fast response time and good repetition stability. It has great application potential in the field of flexible electronic devices.

2. Materials and Methods

2.1. Materials

Multi-walled carbon nanotubes (MWCNTS, outer diameter, 110~190 nm; length, 5~9 μm; Chengdu Zhongke Times Nano Energy Tech Co., Ltd., Chengdu, China), polydimethylsiloxane (PDMS, Sylgard 184, Nanjing Danpei Chemical Co., Ltd., Nanjing, China), isopropyl alcohol (IPA, Shanghai Macklin Biochemical Co., Ltd., Shanghai, China), silicone precision film (Model BD film KRR-200, Hangzhou Guizhi New Material Technology Co., Ltd., Hangzhou, China), silver nanoparticle ink (solid content, 20~30 wt%; 10~15 cp; 30 nm; Xi'an Qiyue Biotechnology Co., Ltd., Xi'an, China).

2.2. Structure and Printing Mechanism of the Flexible Pressure Sensor

The bottom layer of the flexible pressure sensor was a layer of flexible PET film. The top layer of the flexible PET film was an electrode, which was composed of an interdigital electrode and microstructures. To ensure the sensor had high sensitivity, the flexible film for the uppermost package of the flexible pressure sensor was made of PDMS material, which had good elasticity. The lower layer of the PDMS film was a rough active material layer prepared from CNT/PDMS composites, which had excellent electrical conductivity. CNT/PDMS composites are widely used in pressure sensors. A schematic diagram of the structure and working principles of the sensor is shown in Figure 1.

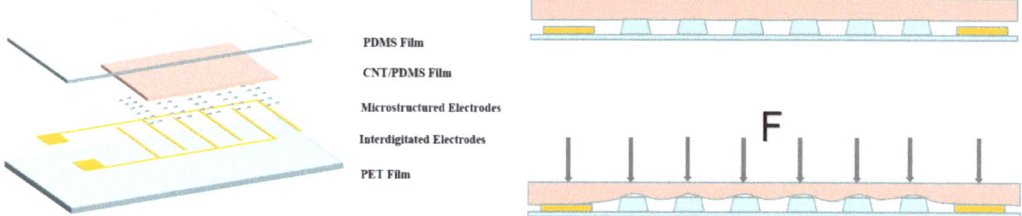

Figure 1. Diagram of the structure and working principles of the flexible pressure sensor.

In the initial state without external pressure, the sensor had the highest contact resistance, because the surface of the active material layer was only in contact with the microstructured electrodes, the contact area was minimal, and the active material layer was not deformed. Under the application of external pressure, compression of the active material layer led to the complete connection of adjacent CNTs in the rough CNT/PDMS film to form a conductive network, which increased the electrical conductivity of the composite. Secondly, after the external pressure was applied, the active material layer gradually contacted the interdigitated electrodes, resulting in an increase in the contact area to increase the conductive path and reduce the resistance. The electrode with a microstructure improved the performance of the sensor in two aspects. Firstly, the active material layer and the interdigital electrode were separated without an external force, which increased the initial resistance and improved the sensitivity of the sensor. Secondly, the microstructured electrodes allowed the active material layer to deform more when pressure was applied; thereby, the performance of the sensor was improved.

3. Fabrication of Flexible Pressure Sensors

3.1. Fabrication of Microstructures

The preparation of the flexible pressure sensor mainly included the preparation of the rough CNT/PDMS composite active material layer and the preparation of the electrode layer with microstructures. The rough CNT/PDMS composite active material layer was made of PDMS and MWCNT. Sandpaper was used as a template. This material was prepared by doping MWCNT with PDMS. It was prepared by ultrasonication, coating, curing, stripping, and other processes. The preparation process is shown in Figure 2.

The detailed preparation process was as follows. Firstly, MWCNT and isopropanol were mixed at a weight ratio of 50:1, and ultrasonicated for 1 h. Secondly, the solution was added to PDMS-A, and the solution was placed on a magnetic mixer to stir evenly for 4 h to ensure the MWCNT and PDMS were fully mixed. Thirdly, the solution was heated in an oil bath. It was stirred at 80 °C for 2 h to fully volatilize the isopropanol in the solution. Fourthly, a curing agent (PDMS-B) was added and stirred evenly. Then the mixture was put into a vacuum drying oven and dried at room temperature for 1 h to remove bubbles, then CNT/PDMS composites were prepared. Fifthly, the mixture was uniformly coated onto sandpaper by a rotary coater. It was then cured in an oven at 60 °C for 2 h. Finally, it was stripped to obtain a rough CNT/PDMS film.

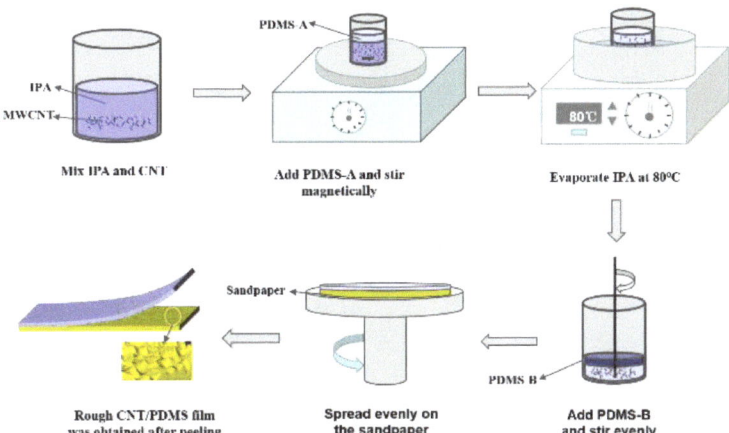

Figure 2. Flowchart of CNT/PDMS film preparation.

The microstructure array of the electrode layer and the interdigital electrode were prepared by SonoPlot GIX Microplotter II printing equipment. The equipment was mainly composed of a high-precision mobile operation platform, a real-time monitoring CCD industrial camera, a capillary glass needle, a distributor, a voltage source, a vacuum adsorption controller, a heating controller, and a control unit. The equipment is shown in Figure 3. Its working principle was that the liquid at the tip of the glass tube needle contacted the surface of the substrate to form a "liquid bridge" to connect with the substrate. At the same time, the distribution voltage was applied to make the glass tube vibrate, which ensured the continuous flow of liquid from the tip.

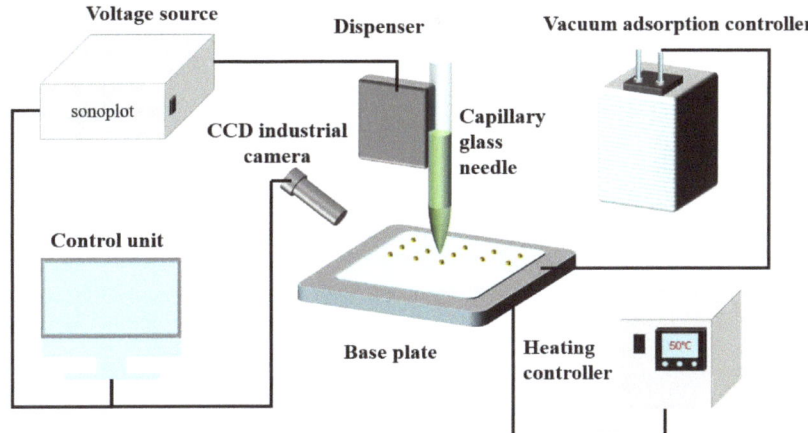

Figure 3. Structural diagram of the printing equipment.

The preparation process was as follows. Firstly, the configured silver nanoparticle ink was sucked into the capillary needle tube by the capillary effect. Secondly, the vacuum adsorption device was opened to adsorb the PET substrate on the platform. Thirdly, the capillary needle tube was slowly brought close to the PET substrate by the control unit, so that the liquid at its tip contacted the substrate surface to form a liquid bridge, which connected to the substrate. The system controlled the output distribution voltage of the voltage source to make the needle tube vibrate, and then the control unit controlled the movement of the capillary needle to carry out the patterning process. Finally, the sample was placed into a vacuum drying oven and dried under a vacuum at 60 °C for 12 h to complete the fabrication of the electrode layer. The prepared sample is shown in Figure 4.

Figure 4. Physical layout of the electrode layer.

3.2. Morphological Characterization of the Microstructures

To study the surface morphology of the prepared microstructures and the active material layer, the microstructure's surface morphology was characterized by emission scanning electron microscopy, and the results are shown in Figure 5.

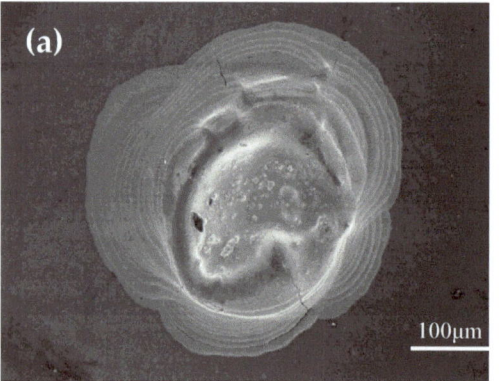
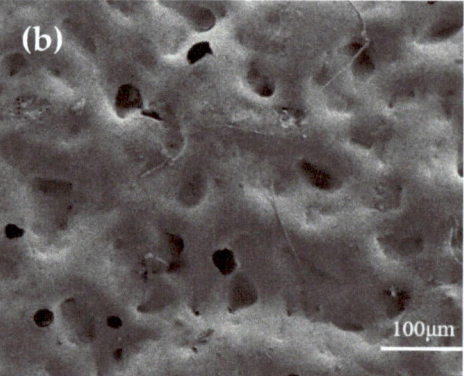

Figure 5. SEM images of the microstructure and CNT/PDMS film. (**a**) Surface morphology of the microstructure (**b**) Surface morphology of the thin CNT/PDMS films.

As seen in Figure 5a, microstructures could be printed by the printing equipment, which was layer-by-layer printing. The ink did not appear to escape when the microstructure was being printed, because one thin film of silver nanoparticle ink was deposited in a picoliter volume, so the film dried in a very short time at room temperature.

In Figure 5b, it can be seen that the surface microstructures of the CNT/PDMS films prepared by sandpaper had irregular pore structures. However, the rough surface could improve the sensitivity of the flexible pressure sensors, because it could deform more easily than a smooth one under the same pressure.

4. Results

4.1. The Effects of Microstructures on Flexible Pressure Sensors' Sensitivity

To evaluate the sensing performance of the sensor, the flexible pressure sensor was connected to the electrical sensing analyzer through conductive copper wires, and it was placed on the pressure testing machine for real-time sensing recording. The testing device is shown in Figure 6. The electronic universal material testing machine set the force loading mode on the computer, and monitored the applied force in real time through the force sensor. External force was applied to the sensor through a circular loader with a diameter of 10 mm.

Figure 6. Sensitivity testing device used for the sensor.

To study the effects of the electrode with microstructures on the sensor, two flexible pressure sensors with and without microstructured electrodes were prepared and tested. The formula for calculating sensitivity (S) is as follows:

$$S = [(R - R_0)/R_0]/\Delta p \qquad (1)$$

where R_0 is the initial resistance of the device without pressure, R is the resistance after applying pressure, and Δp is the change in the pressure value. The sensitivity of the flexible pressure sensor with and without microstructures under different pressures is shown in Figure 7. It was demonstrated that the sensitivity of the flexible pressure sensor with microstructures was significantly better than that of the one without microstructures. The sensitivity with microstructures was 0.042 kPa^{-1} under pressure from 2.5 to 10 kPa, and the sensitivity was 0.013 kPa^{-1} under pressure from 10 to 30 kPa, which was lower by more than 69%. The performance was better than that of the flexible sensor fabricated by the traditional inverted mold method. This was due to the difference in the contact area when the sensor worked.

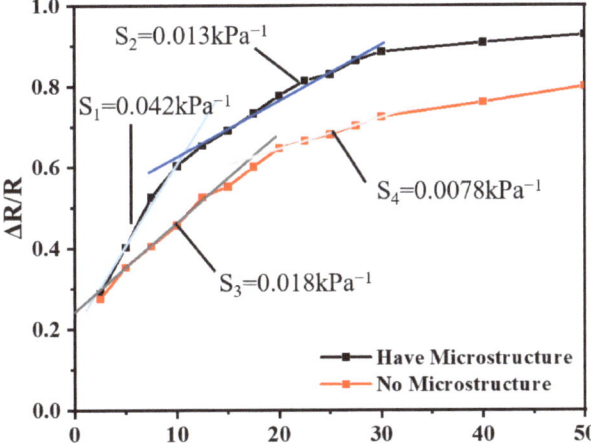

Figure 7. Relationship between the rate of change in the resistance and pressure of sensors with and without microstructured electrodes.

4.2. The Effects of Microstructures on Flexible Pressure Sensors' Response Time

To further explore the response characteristics of the flexible pressure sensor with microstructures, the response time and repeatability of the sensor were studied. The results of the response time of sensor when loaded with pressure are shown in Figure 8. The test demonstrated that the response time was 112 ms, which was better than the sensor fabricated by casting [17].

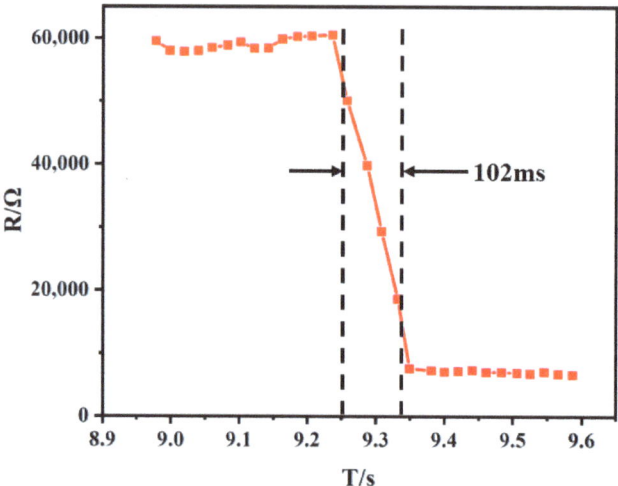

Figure 8. The curve of the sensor's response time.

4.3. The Stability of Flexible Pressure Sensors

To test the stability of the sensor, the sensor was tested for 100 cycles under pressures of 10 kPa, 20 kPa, and 30 kPa, as displayed in Figure 9. The results showed that the relative resistance change rate of the flexible pressure sensors was almost unchanged and became steady. Therefore, the flexible pressure sensors fabricated by low-frequency ultrasonic resonance printing had good stability.

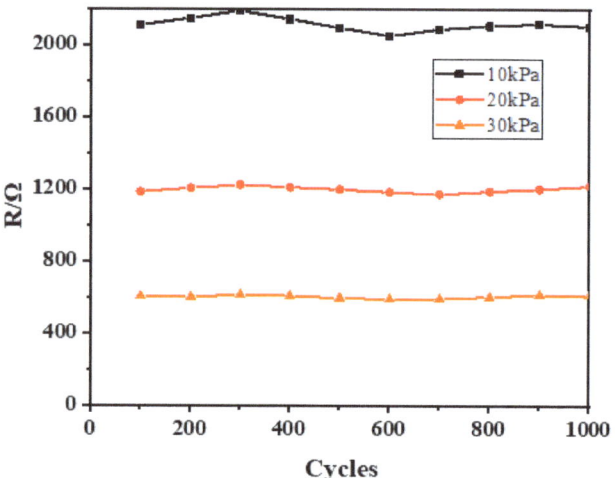

Figure 9. Results of the sensor repeatability test.

4.4. Economic Feasibility of Sensor Category Expansion

According to the results of printing the three-dimensional circular truncated cone microstructure described above, low-frequency ultrasonic resonance printing is an economical and convenient method to fabricate three-dimensional microstructures, and thus obtaining 3D microstructures could greatly simplify the fabrication process of flexible sensors, reducing the fabrication costs. However, it could print microstructures of various shapes only through the microstructure design software which came with the low-frequency ultrasonic resonance printing device. Therefore, low-frequency ultrasonic resonance printing technology could be applied for the fabrication of various flexible sensor devices, promoting the development of flexible electronic technology.

5. Conclusions

In this work, a low-frequency ultrasonic resonance printing technique was proposed to print the microstructures, and a flexible pressure sensor with microstructured electrodes was fabricated. Microstructured thin films were prepared on sandpaper by spin-coating based on CNT/PDMS composites, and the microstructures and the electrode were printed by low-frequency ultrasonic resonance printing. The performance of the flexible pressure sensor with and without microstructures was tested. It was demonstrated that the performance of the flexible pressure sensor with microstructures improved effectively. What is more, the flexible pressure sensor with microstructures could withstand a wide range of pressure from 0 to 30 kPa, and it had a fast response time and higher sensitivity (lower than 112 ms and 0.013 kPa^{-1}, respectively).

Author Contributions: Conceptualization, F.H.; methodology, F.H., Z.Y. and W.J.; validation, Z.Y. and W.J.; formal analysis, Z.Y., H.Z., J.G., G.H. and W.J.; investigation, F.H., Z.Y. and W.J.; resources, F.H.; data curation, F.H., Z.Y., G.H. and W.J.; writing—original draft preparation, W.J., Z.Y., J.G. and H.Z.; writing—review and editing, Z.Y., W.J. and H.Z.; supervision, F.H. and Z.Y.; project administration, F.H.; funding acquisition, F.H. and Z.Y. All authors have read and agreed to the published version of the manuscript.

Funding: This research was funded by the Basic Public Welfare Research Program of Zhejiang Province (Grant No. LGG20E050021), and the Science and Technology Bureau of Jiaxing City (Grant No. 2022AY10006).

Institutional Review Board Statement: Not applicable.

Informed Consent Statement: Not applicable.

Data Availability Statement: The data that support the findings of this study have not been made available but can be obtained from the author upon request.

Acknowledgments: Additional financial support was provided by the Top-level Talent Project of Zhejiang Province. The support of the equipment and technology provided by the College of Information Science and Engineering at Jiaxing University is gratefully acknowledged.

Conflicts of Interest: The authors declare that there are no conflict of interest regarding the publication of this study.

References

1. Han, M.; Lee, J.; Kim, J.K.; An, H.K.; Kang, S.W.; Jung, D. Highly sensitive and flexible wearable pressure sensor with dielectric elastomer and carbon nanotube electrodes. *Sens. Act. A. Phys.* **2020**, *305*, 111941. [CrossRef]
2. Yu, Z.H.; Zhang, T.C.; Li, K.F.; Huang, F.L.; Tang, C.L. Preparation of Bimodal Silver Nanoparticle Ink Base on Liquid Phase Reduction Method. *Nanomaterials* **2022**, *12*, 560. [CrossRef] [PubMed]
3. Yu, Z.H.; Xu, Y.; Tian, X. Graphene oxide loaded silver synergistically improved antibacterial performance of bone scaffold. *AIP Adv.* **2022**, *12*, 015024. [CrossRef]
4. Zhang, T.C.; Yu, Z.H.; Huang, F.L.; Tang, C.L.; Yang, C. Method of multi-layer near-field electrohydraulic printing and sintering of nano-silver ink prepared by liquid phase reduction. *AIP Adv.* **2021**, *11*, 085220. [CrossRef]
5. Jiang, D.; Shi, B.; Ouyang, H.; Fan, Y.; Wang, Z.L.; Li, Z. Emerging Implantable Energy Harvesters and Self-Powered Implantable Medical Electronics. *ACS Nano* **2020**, *14*, 6436–6448. [CrossRef]

6. Ozgur, A.; Asli, A.; Joshua, G.; Conor, W. A Highly Sensitive Capacitive-Based Soft Pressure Sensor Based on a Conductive Fabric and a Microporous Dielectric Layer. *Adv. Mater. Technol.* **2018**, *3*, 1700237. [CrossRef]
7. Yu, Z.H.; Huang, F.L.; Zhang, T.C.; Tang, C.L.; Cui, X.H.; Yang, C. Effects of Different Thermal Sintering Temperatures on Pattern Resistivity of Printed Silver Ink with Multiple Particle Sizes. *AIP Adv.* **2021**, *11*, 115116. [CrossRef]
8. Qin, R.; Hu, M.; Li, X.; Yan, L.; Wu, C.; Liu, J.; Gao, H.; Shan, G.; Huang, W. A highly sensitive piezoresistive sensor based on MXene and polyvinyl butyral with a wide detection limit and low power consumption. *Nanoscale* **2020**, *34*, 17715–17724. [CrossRef]
9. Gogurla, N.; Roy, B.; Park, J.-Y.; Kim, S. Skin-contact actuated single-electrode protein triboelectric nanogenerator and strain sensor for biomechanical energy harvesting and motion sensing—ScienceDirect. *Nano Energy* **2019**, *62*, 674–681. [CrossRef]
10. Ha, M.; Lim, S.; Cho, S.; Lee, Y.; Na, S.; Baig, C.; Ko, H. Skin-Inspired Hierarchical Polymer Architectures with Gradient Stiffness for Spacer-Free, Ultrathin, and Highly Sensitive Triboelectric Sensors. *ACS Nano* **2018**, *4*, 3964–3974. [CrossRef]
11. Liu, X.; Wei, Y.; Qiu, Y. Advanced Flexible Skin-Like Pressure and Strain Sensors for Human Health Monitoring. *Micromachines* **2021**, *12*, 695. [CrossRef] [PubMed]
12. Chen, W.; Yan, X. Progress in achieving high-performance piezoresistive and capacitiveflexible pressure sensors: A review. *J. Mater. Sci. Technol.* **2020**, *43*, 177–190. [CrossRef]
13. Zhang, X.; Hu, Y.; Gu, H.; Zhu, P.; Jiang, W.; Zhang, G.; Sun, R.; Wong, C.P. A Highly Sensitive and Cost-Effective Flexible Pressure Sensor with Micropillar Arrays Fabricated by Novel Metal-Assisted Chemical Etching for Wearable Electronics. *Adv. Mater. Technol.* **2019**, *4*, 367–378. [CrossRef]
14. Sally, E.; Nour, F.; Hyunchul, O. Design and Fabrication of Novel Flexible Sensor Based on 2D Ni-MOF Nanosheets as a Preliminary Step Toward Wearable Sensor for Onsite Ni (II) ions Detection in Biological and Environmental Samples. *Anal. Chim. Acta* **2022**, *1197*, 339518. [CrossRef]
15. Nour, F.; Kurt, E. Polyaniline-polypyrrole Composites with Enhanced Hydrogen Storage Capacities. *Macromol. Rapid Commun.* **2013**, *34*, 931–937. [CrossRef]
16. Sally, E.; Nour, F.; Omara, M.; Hager, M. Cost-effective and Green Synthesized Electroactive Nanocomposite for High Selective Potentiometric Determination of Clomipramine Hydrochloride. *Microchem. J.* **2019**, *151*, 104222. [CrossRef]
17. Choi, H.B.; Oh, J.; Kim, Y.; Pyatykh, M.; Chang, Y.J.; Ryu, S.; Park, S. Transparent Pressure Sensor with High Linearity Over a Wide Pressure Range for 3D Touch Screen Applications. *ACS Appl. Mater. Interfaces* **2020**, *12*, 16691–16699. [CrossRef]
18. Zhang, Y.; Hu, Y.; Zhu, P.; Han, F.; Zhu, Y.; Sun, R.; Wong, C.P. Flexible and Highly Sensitive Pressure Sensor Based on Microdome-Patterned PDMS Forming with Assistance of Colloid Self-Assembly and Replica Technique for Wearable Electronics. *ACS Appl. Mater. Interfaces* **2017**, *9*, 35968–35976. [CrossRef]
19. Nie, P.; Wang, R.; Xu, X.; Cheng, Y.; Wang, X.; Shi, L.; Sun, J. High-Performance Piezoresistive Electronic Skin with Bionic Hierarchical Microstructure and Microcracks. *ACS Appl. Mater. Interfaces* **2017**, *9*, 14911–14919. [CrossRef]
20. Yan, J.; Ma, Y.; Li, X.; Zhang, C.; Cao, M.; Chen, W.; Luo, S.; Zhu, M.; Gao, Y. Flexible and high-sensitivity piezoresistive sensor based on MXene composite with wrinkle structure. *Ceram. Int.* **2020**, *46*, 23592–23598. [CrossRef]
21. Sun, Q.J.; Zhao, X.H.; Zhou, Y.; Yeung, C.C.; Wu, W.; Shishir, V.; Xu, Z.X.; Jonathan, J.W.; Li, W.J.; Vellaisamy, A.L.R. Fingertip-kin-inspired Highly Sensitive and Multifunctional Sensor with Hierarchically Structured Conductive Graphite/Polydimethylsiloxane Foams. *Adv. Funct. Mater.* **2019**, *29*, 1808829. [CrossRef]
22. Ji, B.; Zhou, Q.; Wu, J.; Gao, Y.; Wen, W.; Zhou, B. Synergistic Optimization toward the Sensitivity and Linearity of Flexible Pressure Sensor via Double Conductive Layer and Porous Microdome Array. *ACS Appl. Mater. Interfaces* **2020**, *29*, 31021–31035. [CrossRef] [PubMed]
23. Jiang, Y.; Xu, Z.; Huang, T.; Liu, Y.; Guo, F.; Xi, J.; Gao, W.; Gao, C. Direct 3D Printing of Ultralight Graphene Oxide Aerogel Microlattices. *Adv. Funct. Mater.* **2018**, *28*, 1707024. [CrossRef]

Article

BTO-Coupled CIGS Solar Cells with High Performances

Congmeng Li [1,2], Haitian Luo [1,2], Hongwei Gu [1,2] and Hui Li [1,2,*]

1. Institute of Electrical Engineering Chinese Academy of Sciences, Beijing 100190, China
2. University of Chinese Academy of Sciences, Beijing 100049, China
* Correspondence: lihui2021@iphy.ac.cn; Tel.: +86-82547286

Abstract: In order to improve the power conversion efficiency (PCE) of Cu(In,Ga)Se$_2$ (CIGS) solar cells, a BaTiO$_3$ (BTO) layer was inserted into the Cu(In,Ga)Se$_2$. The performances of the BTO-coupled CIGS solar cells with structures of Mo/CIGS/CdS/i-ZnO/AZO, Mo/BTO/CIGS/CdS/i-ZnO/AZO, Mo/CIGS/BTO/CdS/i-ZnO/AZO, Mo/CIGS/CdS/BTO/i-ZnO/AZO, Mo/CIGS/BTO/i-ZnO/AZO, Mo/CIGS/CdS/BTO/AZO, and Mo/ CIGS/CdS(5 nm)/BTO(5 nm)/i-ZnO/AZO were systematically studied via the SCAPS-1D software. It was found that the power conversion efficiency (PCE) of a BTO-coupled CIGS solar cell with a device configuration of Mo/CIGS/CdS/BTO/AZO was 24.53%, and its open-circuit voltage was 931.70 mV. The working mechanism for the BTO-coupled CIGS solar cells with different device structures was proposed. Our results provide a novel strategy for improving the PCE of solar cells by combining a ferroelectric material into the *p-n* junction materials.

Keywords: thin-film solar cells; CIGS solar cells; ferroelectric materials; BaTiO$_3$; SCAPS simulation

Citation: Li, C.; Luo, H.; Gu, H.; Li, H. BTO-Coupled CIGS Solar Cells with High Performances. *Materials* **2022**, *15*, 5883. https://doi.org/10.3390/ma15175883

Academic Editors: Jijun Feng and Shengli Pu

Received: 15 April 2022
Accepted: 17 May 2022
Published: 25 August 2022

Publisher's Note: MDPI stays neutral with regard to jurisdictional claims in published maps and institutional affiliations.

Copyright: © 2022 by the authors. Licensee MDPI, Basel, Switzerland. This article is an open access article distributed under the terms and conditions of the Creative Commons Attribution (CC BY) license (https:// creativecommons.org/licenses/by/ 4.0/).

1. Introduction

Solar cells show great potential for solving the urgent energy and environmental crisis. The mainstream solar cells in the market are Si solar cells, which have more than 90% of the market share [1]. Compared with Si solar cells, thin-film solar cells, such as Cu(In,Ga)(Se)$_2$ (CIGS), CdTe, and perovskite solar cells, show superior advantages in weak light effect and fabrication of flexible devices, rendering their potential applications in building integrated photovoltaic, portable application, and so on [1]. Among thin-film solar cells, CIGS solar cells have attracted intensive attention due to their high record power conversion efficiency (PCE) of 23.35% [2], easily tunable bandgap (E_g) in the range of 1.04–1.68 eV, great potential applications in perovskite tandem solar cells, low cost of $0.34 W^{-1}, low-temperature coefficient of −0.32%/K, and high stability [1]. However, the average PCE of a practical CIGS module is only 13–15%, which is substantially lower than the theoretical PCE of ~33.2% according to the Shockley–Queisser (SQ) limit [3,4]. The low practical PCEs are mainly ascribed to the open-circuit voltage (V_{oc}) loss ($V_{oc,\,loss} = E_g/q − V_{oc}$). The $V_{oc,\,loss}$ mainly stems from the non-radiative recombination induced by the interface and bulk defects.

Many strategies have been applied to passivate the bulk and interface recombination defects, which is beneficial to the improvement of the V_{oc} and PCE. Nevertheless, the practical PCEs of solar cells with ideal passivation still have limitations in surpassing the theoretical PCE based on the SQ limit [5]. In contrast, the ferroelectric-coupled photovoltaic (PV) device, where a ferroelectric depolarization field, aroused by the ferroelectric materials, is coupled into the common *p-n* junction, shows an extremely high PCE because of the anomalously high built-in electric field of ~10^5 V/cm [6–8]. The ferroelectric-coupled PV device has great potential to exceed the theoretical PCE. For this, ferroelectric materials have been widely combined into organic PV devices and perovskite solar cells. For instance, a 10 nm ferroelectric polymer film was inserted into the polymeric organic solar cell [9]. The PCE of the solar cell was increased from 1–2% to 4–5%. The improved PCE is due to the enhanced electron transport capability, which is due to the large and permanent internal depolarization field of the ferroelectric layer. Similarly, an ultrathin BaTiO$_3$ (BTO) layer was inserted into the TiO$_2$ and perovskite layers in the perovskite solar cell to reduce the charge recombination. The BTO layer retarded the charge recombination and improved the carrier extraction rate at the interface, thus increasing the PCE (from 16.13% to 17.87%) [10]. These results confirm that the ferroelectric-coupled solar cells show great potential for the PCE improvement by reducing the $V_{oc, loss}$.

Various ferroelectric materials, such as PVDF-TrFE [8], Pb(Zr,Ti)O$_3$ [11], BiFeO$_3$, Bi$_2$FeCrO$_6$, BTO, LiNbO$_3$ [12], Be$_x$Cd$_y$Zn$_{1-x-y}$O, PbTiO$_3$, and Pb$_{(1-x)}$La$_x$ZrTiO$_3$ [13], have been applied to enhance the internal electric field of a solar cell by the combination of the ferroelectric depolarization field with the p-n junction field. Among these ferroelectric materials, BTO shows superior properties of a wide bandgap (E_g ~ 3.4 eV), room-temperature ferroelectricity (T_c ~ 120 °C), substantial remnant polarization (P_r = 0.5 C/m^2), environmental-friendly advantage, easy fabrication, excellent stability, and so on [14–20]. The internal electric field of a solar cell is expected to be significantly enhanced by the insertion of a BTO ferroelectric layer into the device. The spontaneous polarization and domain structure of the ferroelectric BTO layer is used to improve the built-in electric field. The improved electric field can enhance the carrier separation and transportation and reduce the carrier recombination, which is beneficial to the improvement of the V_{oc}, short-circuit current density (J_{sc}), and, thus, the PCE.

Compared with other experiments, theoretical simulation provides a feasible approach to elucidate the performance and working mechanism of BTO-coupled solar cells clearly, which is helpful to the design and fabrication of practical devices. SCAPS-1D is a widely applied simulation tool to study the performance of a solar cell [21,22]. Lots of information, including the static energy band diagram, current density-voltage (J-V) curve, external quantum efficiency (EQE) curve, capacitance-voltage (C-V) curve, capacitance-frequency (C-f), carrier transportation, and recombination currents from bulk and interface defects, are easily obtained via the simulation via the SCAPS-1D software. For example, the performances of the novel Cu$_2$BaSnS$_4$ solar cells were well studied by the SCAPS-1D software [23,24]. The SCAPS-1D simulation indicates that an added different Back Surface Field (BSF) layer in Cu$_2$BaSnSSe$_3$ solar cells could increase the V_{oc}, which is consistent with the experimental results [25]. The software is also applied to simulate and modify CIGS [21], Cu$_2$ZnSnS$_4$ (CZTS) [26], and perovskite solar cells [27] to study the effect of the materials and thickness of the window layer [28], buffer layer [29], and the electrode on the device performances.

SCAPS-1D is a one-dimensional solar cell simulation software. It is based on solving three non-linear differential equations: Poisson's equation, the continuity equation for free electrons, and the continuity equation for free holes, as shown in Equations (1)–(3) [30]:

$$\frac{d}{dx}\left(\varepsilon(x)\frac{d\varphi}{dx}\right) = q[p(x) - n(x) + N_D^+(x) - N_A^-(x) + p_t(x) - n_t(x)] \quad (1)$$

$$-\frac{1}{q}\frac{dJ_n}{dx} + R_n(x) - G(x) = 0 \quad (2)$$

$$\frac{1}{q}\frac{dJ_p}{dx} + R_p(x) - G(x) = 0 \quad (3)$$

where ψ, n, p, n_t, p_t, N_D^+, and N_A^- are the electrostatic potential, free electrons density, free holes density, electrons distribution, holes distribution, ionized donors' concentration, and ionized acceptors concentration, respectively. $R_n(x)$, $R_p(x)$, $G(x)$, J_n, J_p, ε, and q are the electrons recombination rate, holes recombination rate, generation rate, electron current density, hole current density, permittivity, and charge of the electron, respectively [24].

In this study, a BTO ferroelectric layer was incorporated into the CIGS solar cell. The impact of the BTO layer on the performance of the CIGS solar cell was systematically investigated via the SCAPS-1D software. BTO-coupled CIGS solar cells with various device structures were simulated. Our results show that the optimized device architecture is: Mo/CIGS/CdS/BTO (5 nm)/AZO with a PCE of 24.53% and a V_{oc} of 931.70 mV. The combination of a ferroelectric layer into a common *p-n* junction solar cell is helpful for the improvement of the device's performance. Our work provides a novel device structure for the PCE improvement of the CIGS solar cell by the incorporation of a ferroelectric layer into the device structure. This strategy is also applicable to other solar cells.

2. Methodology

In order to elucidate the impact of a BTO layer on the performance of the CIGS solar cell, four device configurations were proposed and simulated: soda-lime glass (SLG)/Mo/BTO/CIGS/CdS/i-ZnO/Al-doped zinc oxide (AZO)/Au (Figure 1a), SLG/Mo/CIGS/BTO/CdS/i-ZnO/AZO/Au (Figure 1b), SLG/Mo/CIGS/CdS/BTO/i-ZnO/AZO/Au (Figure 1c), and SLG/Mo/CIGS/CdS/BTO/AZO/Au (Figure 1d). For comparison, the performance of the CIGS solar cell with a typical device configuration of SLG/Mo/CIGS/CdS/i-ZnO/AZO/Au (Figure 2) was also simulated. Note that the BTO was easily polarized under an external field, which shows that the depolarization field is induced by the *p-n* junction field during the simulation.

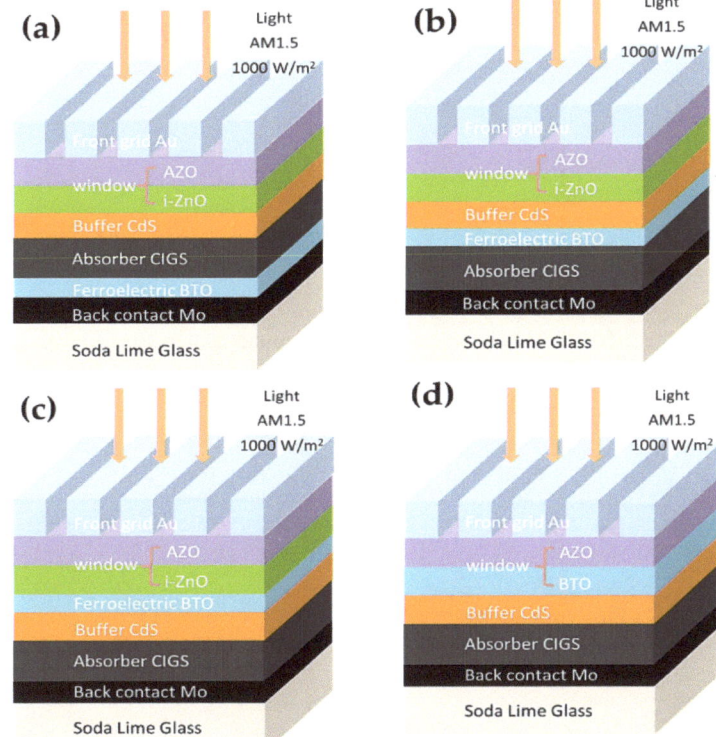

Figure 1. Device architectures for simulation: (**a**) SLG/Mo/BTO/CIGS/CdS/i-ZnO/AZO/Au, (**b**) SLG/Mo/CIGS/BTO/CdS/i-ZnO/AZO/Au, (**c**) SLG/Mo/CIGS/CdS/BTO/i-ZnO/AZO/Au, (**d**) SLG/Mo /CIGS/CdS/BTO/AZO/Au.

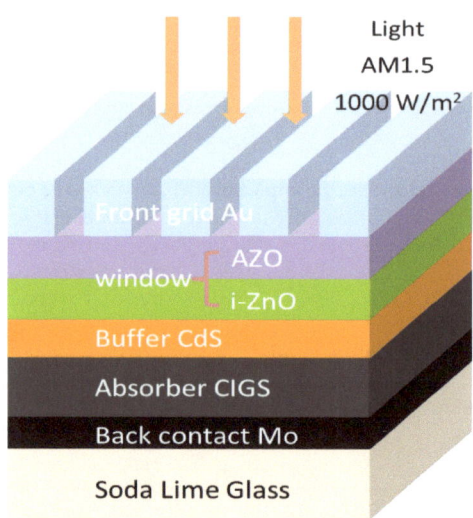

Figure 2. Device architecture of the simulated CIGS solar cell.

The parameters for different layers in the simulated devices are listed in Table 1. The J-V curves, EQE curves, J_{sc}, V_{oc}, fill factor (FF), and PCEs for the simulated devices were obtained under the condition of AM1.5 G, an incident solar power (P) of 100 mW/cm^2, and a temperature of 300 K. In addition, the band diagram and C-V curves were also obtained.

Table 1. Input simulation parameters for different layers in CIGS and BTO-coupled CIGS solar cells [31–33].

Parameters	p-CIGS	n-CdS	i-ZnO	AZO	BTO
	Absorber	Buffer	Window		Ferroelectric
Thickness (nm)	2000	5–50	20	30	5–100
Bandgap E_g (eV)	1.04–1.68	2.45	3.30	3.37	3.40
Electron affinity χ (eV)	4.6	4.4	4.3	4.3	4.5
Relative dielectric permittivity ε_r	13.6	10.0	9.0	9.0	290.0
Effective conduction band density N_c (cm^{-3})	6.8×10^{17}	1.3×10^{18}	3×10^{18}	1×10^{20}	4×10^{18}
Effective valence band density N_v (cm^{-3})	1.5×10^{19}	9.1×10^{18}	1.7×10^{19}	3×10^{18}	9×10^{18}
Electron thermal velocity v_n (cm/s)	1×10^7	3.1×10^7	1×10^7	1×10^7	1×10^7
Hole thermal velocity v_p (cm/s)	1×10^7	1×10^7	1×10^7	1×10^7	1×10^7
Electron mobility μ_n (cm^2/(Vs))	100	72	100	100	50
Hole mobility μ_p (cm^2/(Vs))	12.5	20	31	31	20
Donor concentration N_D (cm^{-3})	0	5×10^{17}	1×10^{17}	1×10^{20}	5×10^{17}
Acceptor concentration N_A (cm^{-3})	2×10^{16}	0	0	0	0
Defect density (cm^{-2})	5×10^{13}	3×10^{13}	1×10^{16}	3×10^{16}	0

3. Results and Discussion

3.1. Impact of CIGS Bandgap on Performances of CIGS Solar Cells

It is well known that the PCE of a solar cell is closely related to the bandgap of the absorbing layer. Only photons with an energy $h\nu$ higher than the optical bandgap of the absorbing layer can be absorbed. The bandgap of CIGS is well modified in the region of 1.04–1.68 eV by the modification of the In and Ga atomic ratio. Therefore, it is especially important to clarify the impact of the bandgap of CIGS on the device's performance. For this, the CIGS solar cell with a typical structure of SLG/Mo/CIGS/CdS/i-ZnO/AZO/Au (Figure 2) is firstly investigated via the SCAPD-1D software. The program is developed at the department of Electronics and Information Systems of the University of Gent, Belgium. The version number we use is SCAPS 3.3.07, and it can be freely available. As shown in Figure 3, the V_{oc}, J_{sc}, FF, and PCE show a close relationship with the bandgap of CIGS. The V_{oc} linearly increases with the bandgap of CIGS (Figure 3a). It is well known that the V_{oc} can be calculated based on the following equations [34,35]:

$$V_{oc} = \frac{kT}{q} \ln\left(\frac{J_{sc}}{J_0} + 1\right) \quad (4)$$

$$J_0 \propto n_i^2 = N_C N_V \exp\left(-\frac{E_g}{kT}\right) I_0 = 1.5 \times 10^5 \exp\left(-\frac{E_g}{kT}\right) \quad (5)$$

where k, T, and J_0 are the Boltzmann constant (1.38×10^{-23} J/K), thermodynamic temperature, and reverse saturation current, respectively.

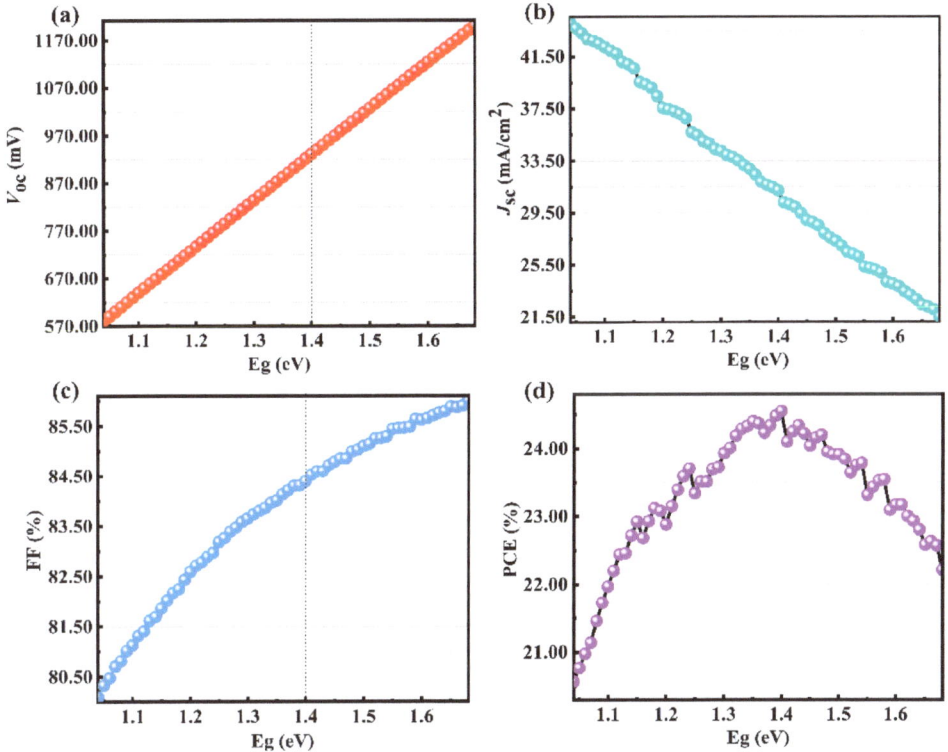

Figure 3. Dependence of (**a**) V_{oc}, (**b**) J_{sc}, (**c**) FF, and (**d**) PCE of CIGS solar cells on CIGS bandgap.

Therefore, the V_{oc} linearly increases with the bandgap of CIGS. On the contrary, J_{sc} linearly decreases with the bandgap of CIGS (Figure 3b). The short-circuit current (I_{sc}) is calculated according to the following equations [36]:

$$I_{sc} = -I_L \quad (6)$$

$$I_L = qAG(L_e + W + L_h) \quad (7)$$

Each photon reaching the solar cell surface with an energy greater than the bandgap of the absorption layer creates an electron-hole pair. Thus, the J_{sc} ($J_{sc} = I_{sc}$/device area) decreases with the increase of the CIGS bandgap. The FF shows a quasi-linearly relationship with the bandgap of CIGS (Figure 3c). Under an ideal condition, the FF is only related to the V_{oc} based on Equations (8) and (9) [37]:

$$FF_0 = \frac{v_{oc} - \ln(v_{oc} + 0.72)}{v_{oc} + 1} \tag{8}$$

$$v_{oc} = V_{oc}\frac{q}{kT} \tag{9}$$

However, in a practical device, the FF is also affected by the series resistance (R_s) and shunt resistance (R_{sh}), leading to the quasi-linearly increase of the FF with the bandgap of CIGS. The PCE increases with the bandgap of CIGS when the E_g is lower than 1.40 eV (Figure 3d). When the bandgap of CIGS is larger than 1.40 eV, the PCE decreases with the increase of E_g (Figure 3d). The PCE of a solar cell is closely related to the V_{oc}, J_{sc}, and FF based on the following equation [38]:

$$PCE = \frac{V_{oc} \times J_{sc} \times FF}{P_{in}} \times 100\% \tag{10}$$

Therefore, based on the relationship between V_{oc}, J_{sc}, and FF and the bandgap of CIGS, the dependence of PCE on the bandgap of CIGS is easily achieved, as shown in Figure 3d. The PCE of the CIGS solar cell firstly increases and then decreases (Figure 3d) with the bandgap of CIGS in the region of 1.04–1.68 eV. Interestingly, the PCE is quite high in the whole bandgap region of CIGS, with a value of 20.57–24.55%. The maximum PCE is 24.55% when the CIGS bandgap is 1.40 eV. Thus, the CIGS bandgap is set to 1.40 eV for the following simulation.

Figure 4a,b shows the energy band diagrams of the CIGS solar cell with a CIGS bandgap of 1.40 eV under dark and AM1.5 G light illumination. Under light illumination, a spike-like conduction band offset (CBO) is observed at the interfaces of CIGS/CdS (~0.2 eV) and CdS/i-ZnO (~0.1 eV), implying a low interface carrier recombination [24]. According to the simulated C-V result (Figure 4c), the internal electric field is ~1.00 eV, consistent with the V_{oc} of 931.70 mV obtained from the I-V curve (Table 2). The $V_{oc, loss}$ (~0.40 eV) is probably due to the small width of the deletion region, determined from the energy band diagram under light illumination (Figure 4b). The low J_{sc} mainly stems from the optical loss in the short wavelength (<418 nm) seen from the simulated EQE curve (Figure 4d). Thereby, the buffer layer, with a much wider bandgap than 2.45 eV of CdS, is required to reduce the optical loss and, thus, improve the PCE.

Table 2. The simulated device parameters for PV devices with different device architectures.

Device Architectures	V_{oc} (mV)	J_{sc} (mA/cm^2)	FF (%)	PCE (%)
Mo/CIGS/CdS/i-ZnO/AZO	931.70	31.21	84.40	24.55
Mo/BTO/CIGS/CdS/i-ZnO/AZO	83.60	13.51	18.05	0.20
Mo/CIGS/BTO/CdS/i-ZnO/AZO	930.10	29.52	84.25	23.13
Mo/CIGS/CdS/BTO/i-ZnO/AZO	931.60	31.13	84.40	24.48
Mo/CIGS/BTO/i-ZnO/AZO	931.60	31.13	84.34	24.46
Mo/CIGS/CdS/BTO/AZO	931.70	31.19	84.40	24.53
Mo/CIGS/CdS(5 nm)/BTO(5 nm)/i-ZnO/AZO	931.60	31.12	84.33	24.45

(These data are obtained under AM1.5 G, 1SUN, 1000 W/m^2 light illumination conditions.)

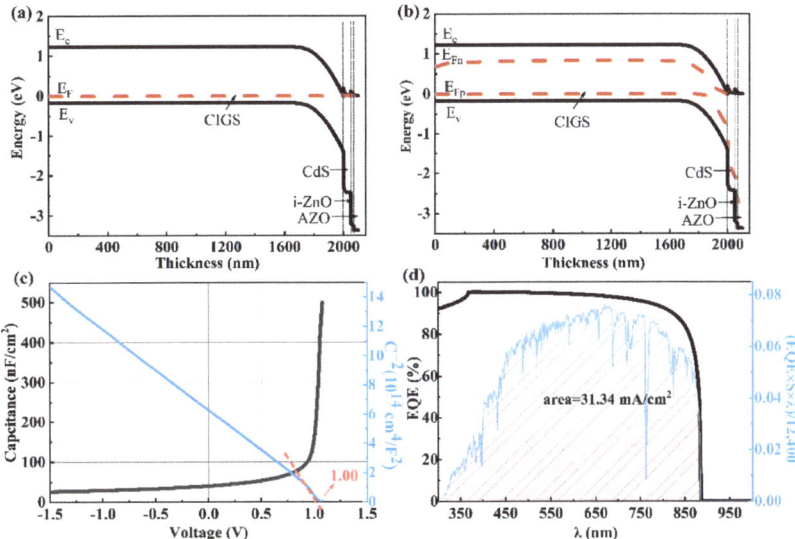

Figure 4. Energy band diagram of the CIGS (E_g = 1.4 eV) solar cell with a device structure of Mo/CIGS/CdS/i-ZnO/AZO under (**a**) dark and (**b**) AM1.5 G light illumination. (**c**) C-V and C^{-2}-V curves, (**d**) EQE curve and the J_{sc} value from the integral curve based on the expression $\frac{EQE \times S \times \lambda}{12,400}$, λ is a wavelength of sunlight; S is the standard AM1.5 G spectrum.

3.2. Impact of BTO Thickness on Performances of BTO-Coupled CIGS Devices

In our study, the BTO was inserted into the different locations of the CIGS solar cell to clarify the impact of BTO on the performance of the CIGS solar cell, as shown in Figure 1. The internal electric field was expected to be enhanced because the BTO ferroelectric layer has a large depolarization field (E_{dp}) after poling [39]:

$$E_{dp} = \frac{d\sigma_p}{\varepsilon_0 \, \varepsilon_{FE} L} \tag{11}$$

where σ_p is the polarization charge density, d is the thickness of the BTO thin film, L is the thickness of the semiconductor layer, and ε_{FE} is the relative dielectric constant of the BTO. The thickness of the BTO ferroelectric film should be very thin because the ultrathin ferroelectric layer has a large surface charge density. For this, the thickness of the BTO layer was set to be 5–100 nm. The simulated results show that the location of BTO has a significant effect on the V_{oc}, J_{sc}, FF, and PCE. The V_{oc}, J_{sc}, FF, and PCE decreased significantly when a BTO layer was inserted between the Mo and CIGS layers. The maximum PCE of the solar cell was quite low, only 0.20%, with a V_{oc}, J_{sc}, and FF of 0.08 V, 13.51 mA/cm^2, and 18.05% (Figure 5 and Table 2). The device performance of the Mo/BTO/CIGS/CdS/i-ZnO/AZO was poor and nearly independent of the BTO thickness (Figure 5). The reason for the low PCE was due to the formed BTO/CIGS back field, as seen from the energy band diagram (Figure 6a,b). The electric field between the BTO/CIGS showed an opposite direction to that of the CIGS/CdS because of the *n*-type of the CdS and BTO. The Fermi level difference between E_{fn} and E_{fp} had nearly the same value (Figure 6b). Thus, the internal electric field strength between the BTO/CIGS and CIGS/CdS showed little difference, which may be due to the similar *n*-type nature and bandgap of the BTO (E_g = 3.4 eV) and AZO (E_g = 3.3 eV) [39]. The internal electric field, calculated from the C-V curve (Figure 6c), was only 0.16 V [40], leading to a very low V_{oc}. The integrated J_{sc} from the EQE curve (Figure 6d) was 13.54 mA/cm^2, in good agreement with the J_{sc} of 13.51 mA/cm^2 obtained from the I-V result (Figure 5b and Table 2). The same capability in the separation and transportation of carriers in the opposite directions of the BTO/CIGS and CIGS/CdS/i-

ZnO/AZO led to the weak separation and transportation of light-induced carriers, resulting in the low J_{sc} (Figure 5b). The low V_{oc} resulted in a low FF. The low V_{oc}, J_{sc}, and FF resulted in the low PCE of the solar cell. Thereby, it is seen that the BTO located between the CIGS and Mo layers significantly reduced the device's performance.

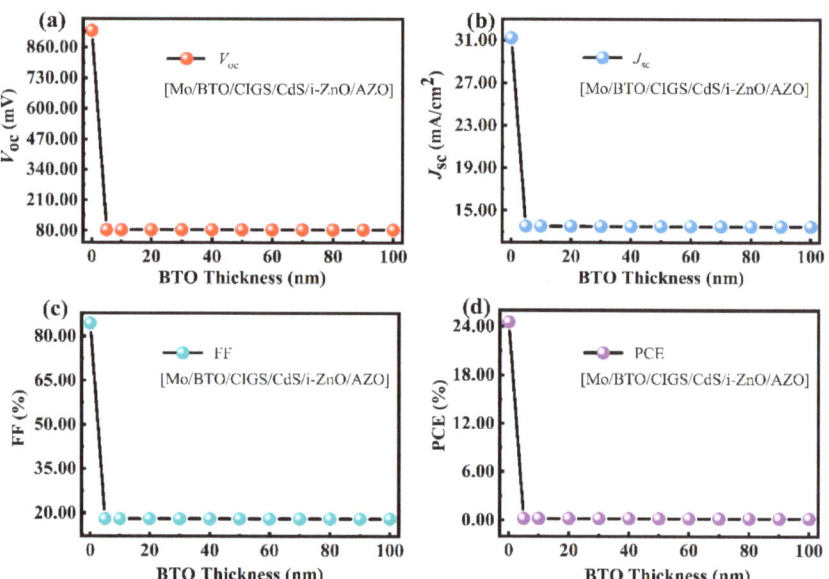

Figure 5. Dependence of (**a**) V_{oc}, (**b**) J_{sc}, (**c**) FF, and (**d**) PCE on the thickness of the BTO layer of the BTO-coupled CIGS solar cell with a device configuration of Mo/BTO/CIGS/CdS/i-ZnO/AZO.

Based on the simulated results, the performance of the CIGS solar cell is expected to be improved when the BTO is moved onto the top of the *p*-type CIGS because of the *n*-type and ferroelectric nature of the BTO. As shown in Figure 7, the performance of the device still shows little dependence on the thickness of the BTO layer. The V_{oc}, J_{sc}, FF, and PCE of the device with a device structure of Mo/CIGS/BTO/CdS/i-ZnO/AZO are greatly improved when the BTO is moved onto the top of CIGS (Figure 7). Importantly, the V_{oc} increases from 83.60 mV to 930.10 mV. The enhanced V_{oc} is attributed to the enhanced strength of the internal electric field. The strength of the internal electric field, calculated from the *C-V* curve, is 1.01 V (Figure 8c). The enhanced internal electric field is largely due to the change of the energy band diagram of Mo/CIGS/BTO/CdS/i-ZnO/AZO compared with that of Mo/BTO/CIGS/CdS/i-ZnO/AZO, as shown in Figure 8a,b. Clearly, the energy band diagram at the back contact is greatly changed, showing that the location of the BTO greatly affects the internal electric field. The energy band diagram shows a similar character to that of the CIGS solar cell with a device structure of Mo/CIGS/CdS/i-ZnO/AZO (Figure 4), indicating the enhanced internal electric field. The enhanced internal electric field improves the V_{oc} and the separation and transportation of the light-induced carriers, which contributes to the enhanced J_{sc}. The integrated J_{sc} from the EQE curve (Figure 8d) is 29.63 mA/cm^2, which is in good agreement with the J_{sc} of 29.52 mA/cm^2 obtained from the simulated *J-V* curve (Figure 7b and Table 2). The enhanced V_{oc} leads to the enhanced FF. The improved V_{oc}, J_{sc}, and FF thus enhance the PCE of the solar cell. The maximum PCE is 23.13% with a V_{oc}, J_{sc}, and FF of 930.10 mV, 29.52 mA/cm^2, and 84.25% (Figure 7 and Table 2) when the thickness of BTO is 5 nm.

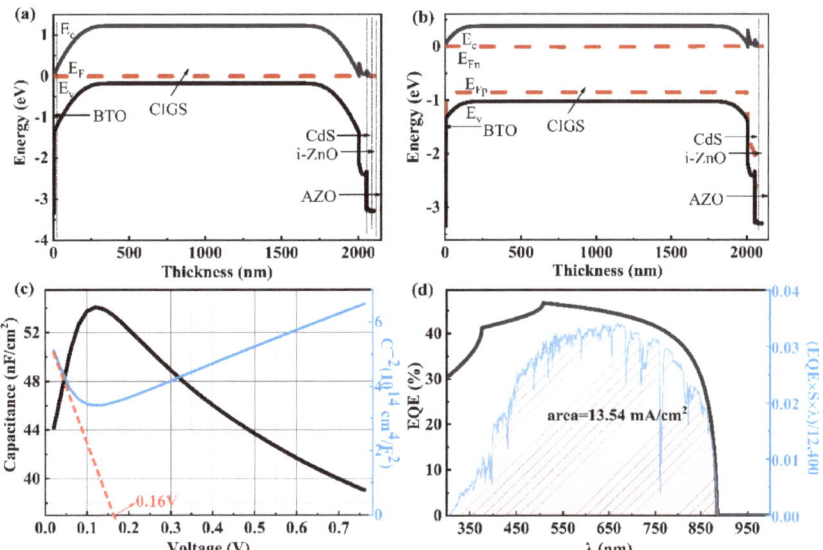

Figure 6. The energy band diagram, *C-V*, and EQE curves of the solar cell with a device structure of Mo/BTO/CIGS/CdS/i-ZnO/AZO and a CIGS bandgap of 1.40 eV: (**a**) under non-light conditions, (**b**) under the AM1.5 G, 1SUN, 1000 W/m² condition, (**c**) *C-V* curve and C^{-2}-*V* curve, (**d**) EQE curve and the J_{sc} value from the integral curve based on the expression $\frac{EQE \times S \times \lambda}{12,400}$.

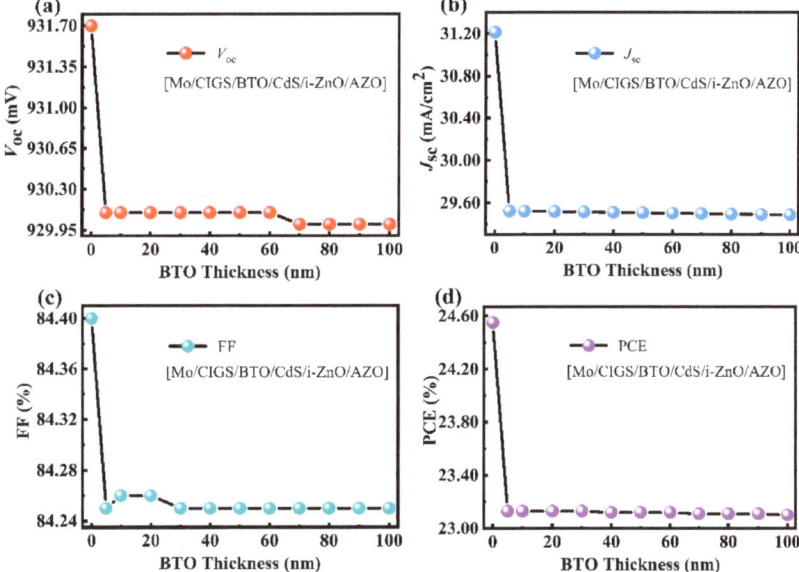

Figure 7. Dependence of (**a**) V_{oc}, (**b**) J_{sc}, (**c**) FF, and (**d**) PCE on the thickness of the BTO in the ferroelectric-coupled CIGS solar cell with a device structure of Mo/CIGS/BTO/CdS/i-ZnO/AZO.

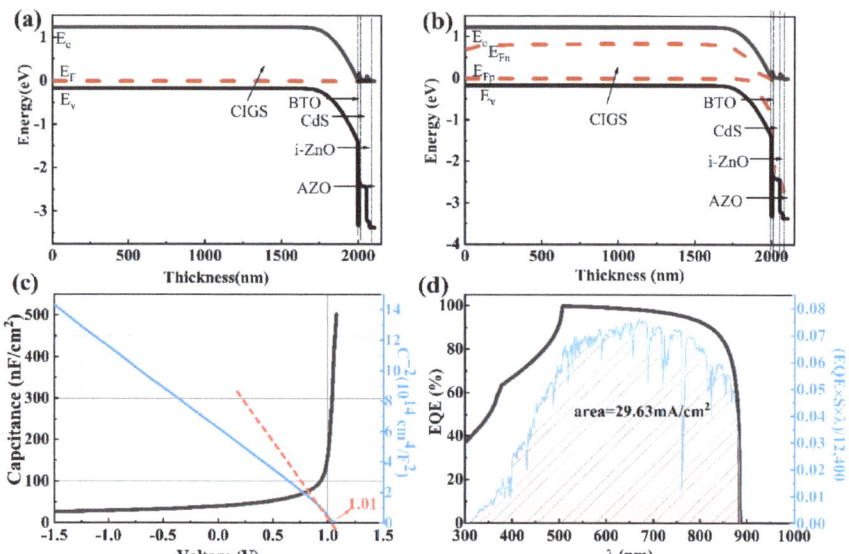

Figure 8. The energy band diagram, *C-V*, and EQE curves of the ferroelectric-coupled CIGS solar cell with a device structure of Mo/CIGS/BTO/CdS/i-ZnO/AZO and a CIGS bandgap of 1.40 eV: (**a**) under non-light conditions, (**b**) under the AM1.5 G, 1SUN, 1000 W/m² condition, (**c**) *C-V* curve and C^{-2}-*V* curve, (**d**) EQE curve and the J_{sc} value from the integral curve based on the expression $\frac{EQE \times S \times \lambda}{12,400}$.

As shown in Figure 9a, when the BTO thickness is 5–80 nm, the V_{oc} improves from 930 mV to 932 mV when the BTO layer is on the top of CdS. Although the V_{oc} decreases with the increase of the BTO thickness when the thickness of the BTO is larger than 80 nm, the V_{oc} is still higher than that of the device with an architecture of Mo/CIGS/BTO/CdS/i-ZnO/AZO. The decrease of the V_{oc} with the thickness of the BTO is due to the large surface charge density at the ultrathin ferroelectric layer [39]. As seen in Figure 10c, the internal electric field is 1.05 V, which is higher than that of 1.01 V for the device with a structure of Mo/CIGS/BTO/CdS/i-ZnO/AZO. The increased internal electric field results in the increased V_{oc}. The energy band diagram, especially the valence band, becomes much more suitable for the separation and transport of the light-induced carriers (Figure 10a,b) and thus leads to the enhanced J_{sc}. The J_{sc} linearly decreases with the BTO thickness (Figure 9b), which attributes to the optical absorption of the BTO. The maximum J_{sc} is 31.25 mA/cm² (Figure 10d). The FF shows a similar dependence of the BTO thickness with that of V_{oc} (Figure 9c), which can be well explained based on Equations (8) and (9). Thus, the PCE displays a similar relationship with that of J_{sc} between the BTO thickness (Figure 9d), which is well explained according to Equation (10). The maximum PCE of the solar cells is 24.48%, as listed in Table 2. Thereby, it is further confirmed that the BTO location greatly affects the performance of the ferroelectric-coupled CIGS solar cells. As shown in Table 2, the PCE, V_{oc}, and J_{sc} of the solar cell show nearly no change when the CdS is removed. Interestingly, the electric parameters and the energy band diagram of the devices also show nearly no change when the i-ZnO is removed from the device (Figures 11 and 12). The highest PCE of the device with a 5 nm BTO layer is increased to 24.53% (Table 2 and Figure 11d), which is mainly due to the much flatter valence band offset (Figure 12b) when compared to that of the ferroelectric-coupled CIGS solar cell with a device structure of Mo/CIGS/CdS/BTO/i-ZnO/AZO (Figure 10b). The maximum J_{sc} is 31.20 mA/cm² when the thickness of the BTO is 5 nm (Figure 11b). The J_{sc} is slightly higher than the J_{sc} of 31.12 mA/cm² for the device with a configuration of Mo/CIGS/CdS/BTO/i-ZnO/AZO. The slightly increased J_{sc} is owed to the removal of the parasitic optical absorption layer of i-ZnO.

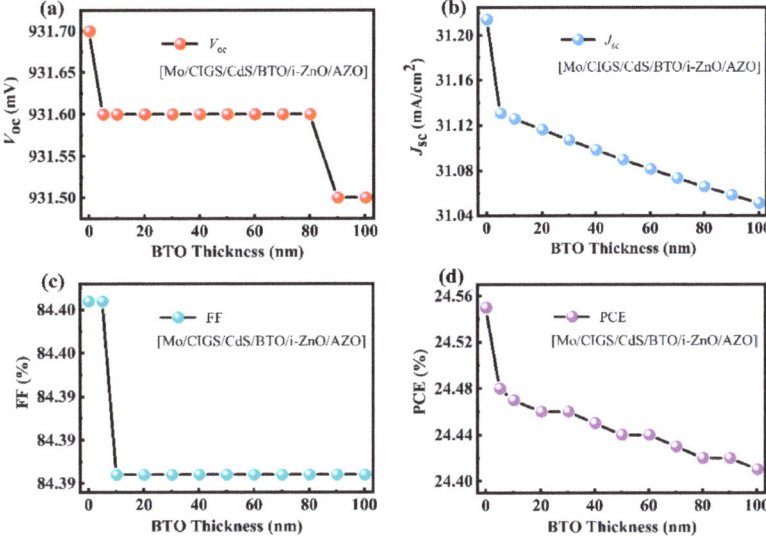

Figure 9. Effect of the BTO thickness on the (**a**) V_{oc}, (**b**) J_{sc}, (**c**) FF, and (**d**) PCE of the CIGS solar cell with a device structure of Mo/CIGS/CdS/BTO/i-ZnO/AZO.

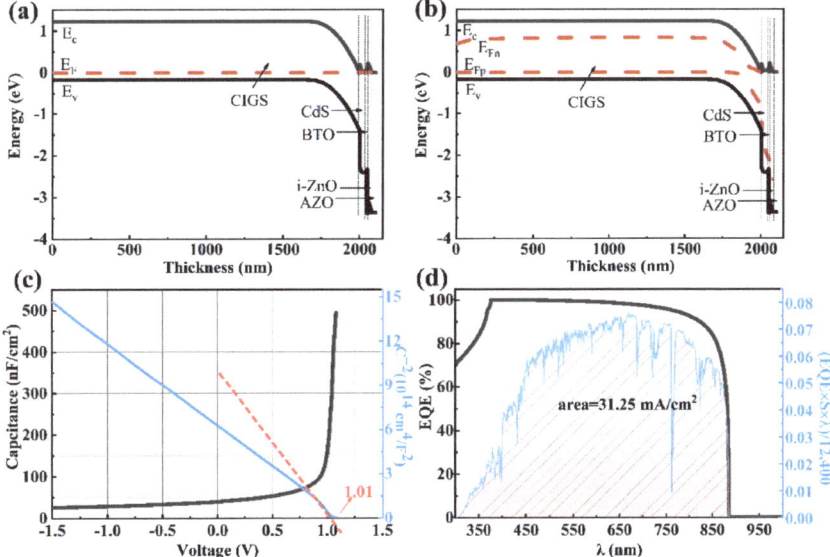

Figure 10. The energy band diagram, C-V, and EQE curve of the ferroelectric-coupled CIGS solar cell with a device structure of Mo/CIGS/CdS/BTO/i-ZnO/AZO: (**a**) under dark, (**b**) under the AM1.5 G light illumination, (**c**) C-V curve and C^{-2}-V curves, and (**d**) EQE curve and the J_{sc} value from the integral curve based on the expression $\frac{EQE \times S \times \lambda}{12,400}$.

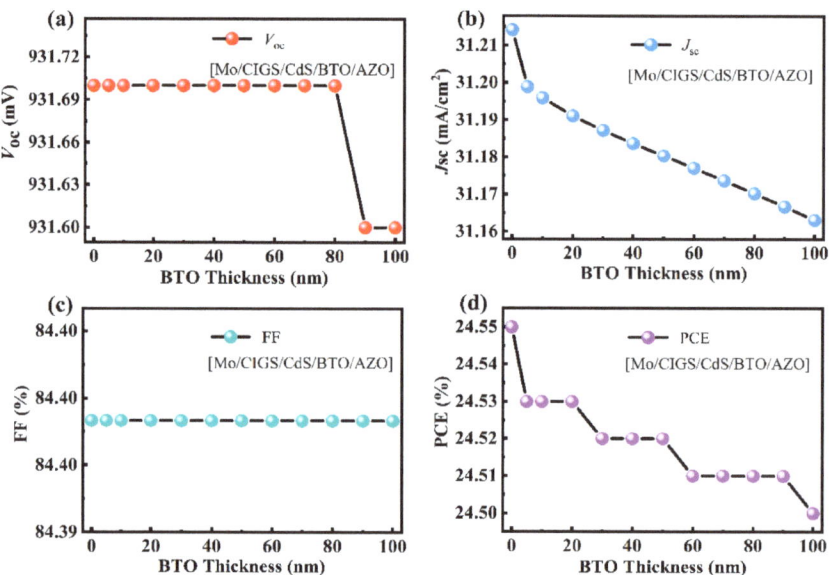

Figure 11. Effect of the BTO layer thickness on the electrical parameters of the ferroelectric-coupled CIGS solar cell with a device structure of Mo/CIGS/CdS/BTO/AZO: (**a**) V_{oc}, (**b**) J_{sc}, (**c**) FF, and (**d**) PCE.

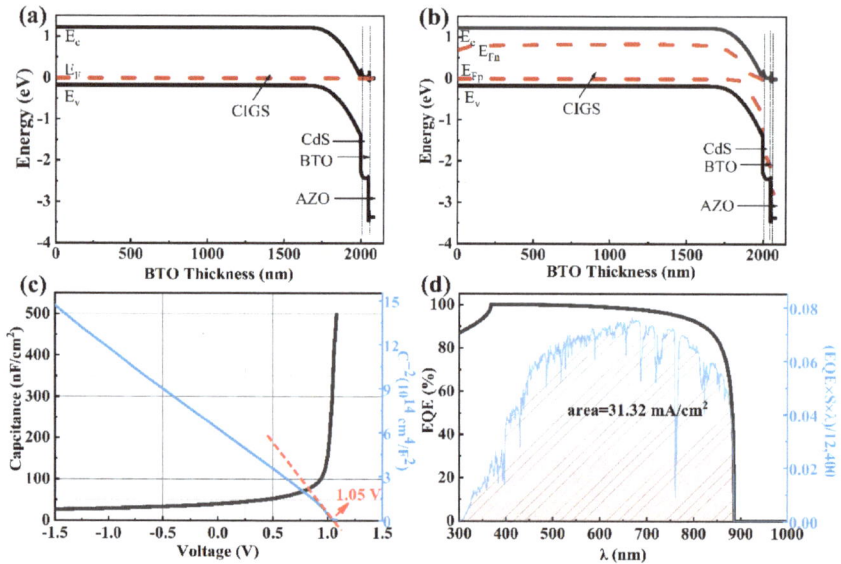

Figure 12. The energy band diagram, C-V, and EQE curves of the ferroelectric-coupled CIGS solar cell with a device structure of Mo/CIGS/CdS/BTO/AZO: (**a**) under non-light conditions, (**b**) under the AM1.5 G light illumination, (**c**) C-V and C^{-2}-V curves, (**d**) EQE curve and the J_{sc} value from the integral curve based on the expression $\frac{EQE \times S \times \lambda}{12,400}$.

Thereby, it is concluded that the ferroelectric-coupled CIGS solar cell is a promising solar cell. The J-V, corresponding EQE, and the integrated EQE curves for the devices with the highest simulated PCE are shown in Figure 13. The corresponding device parameters are listed in Table 2 in detail. The performance of this device is influenced little by the thickness of the CdS when the CdS thickness is 5–50 nm (Table 2), showing the potential low fabrication cost of the solar cells thanks to less material usage.

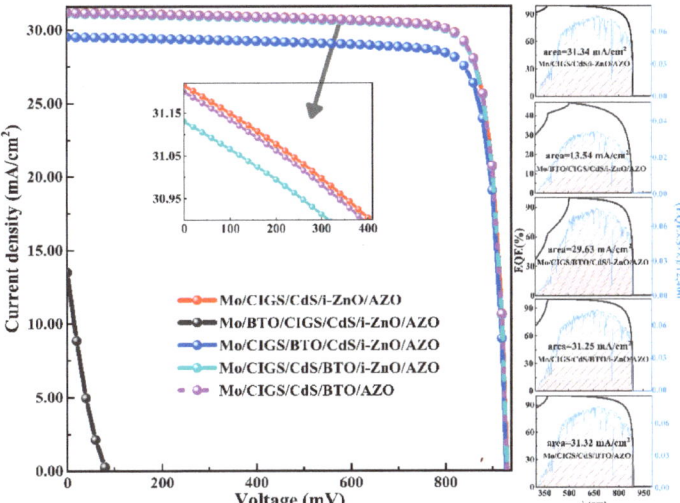

Figure 13. The *J-V*, corresponding EQE, and the integrated EQE curves for the different devices with the highest simulated PCE.

Thus, the optimized device configuration is Mo/CIGS/CdS/BTO/AZO with a 5 nm BTO. The thinner BTO is beneficial to the carries tunnel and provides a stronger ferroelectric field. The PCE for this device is 24.53% with a V_{oc} of 931.70 mV, which is comparable to the PCE of 24.55% for the CIGS solar cell with a typical device configuration of Mo/CIGS/CdS/i-ZnO/AZO (Figure 13 and Table 2). The PCE for the device with a device configuration of Mo/CIGS/CdS/BTO/i-ZnO/AZO is 24.48% with a V_{oc} of 931.60 mV.

The working mechanism for the CIGS solar cells with different device structures is shown in Figure 14. When the BTO is located between Mo and CIGS, a *p-n* junction field is formed between the BTO and CIGS. The *p-n* junction electric field between the BTO/CIGS has an opposite direction to that of the CIGS/CdS (Figure 14a), which leads to the low PCE of the solar cells. However, the *p-n* junction electric field between the BTO and CIGS shows the same direction as the CIGS/CdS *p-n* junction electric field (Figure 14b) when the BTO is between the CIGS and CdS. When the BTO is poled, the depolarization field also shows the same direction as that of the built-in *p-n* electric field. Therefore, the strength of the total field is enhanced greatly. After the BTO is moved to the top of the CdS, the CIGS/CdS *p-n* junction electric strength is not affected by the insertion of the BTO. When the BTO is poled, the depolarization field shows the same direction as the built-in *p-n* electric field (Figure 14c). The total electrical field is thus enhanced, leading to an enhanced performance of the device. For the device with a structure of SLG/Mo/CIGS/CdS/BTO/AZO/Au (Figure 14d), the CIGS/CdS *p-n* junction electric strength is not affected by the removal of i-ZnO; thus the coupled *p-n* junction electric field and depolarization electric field can be retained. Interestingly, the parasitic optical absorption is reduced by the removal of the i-ZnO, resulting in the slightly enhanced PCE of the device.

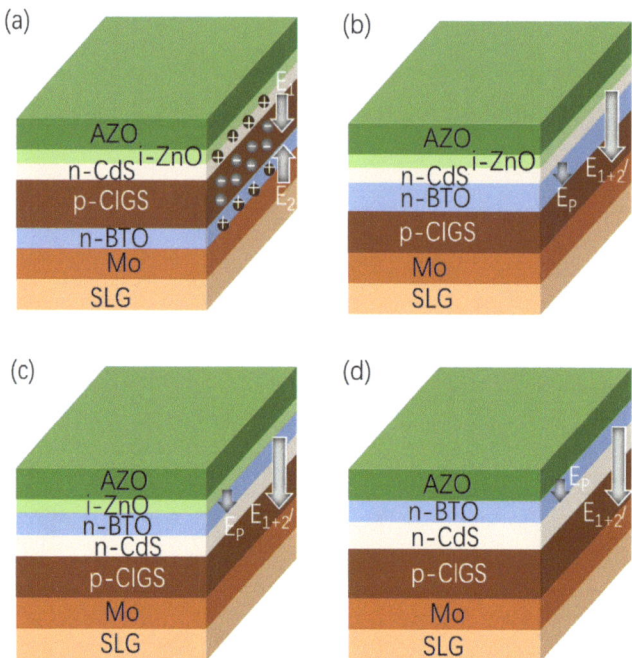

Figure 14. The different device structures of CIGS cells (not drawn to scale) and the schematic diagram of the electric field induced by the polarized BTO thin film and the field-assisted carrier separation. (**a**) SLG/Mo/BTO/CIGS/CdS/i-ZnO/AZO/Au, (**b**) SLG/Mo/CIGS/BTO/CdS/i-ZnO/AZO/Au, (**c**) SLG/Mo/CIGS/CdS/BTO/i-ZnOAZO/Au, and (**d**) SLG/Mo/CIGS/CdS/BTO/AZO/Au.

4. Conclusions

In this study, BTO-coupled CIGS solar cells with various structures have been systematically investigated via the SCAPS-1D software. The performance of the CIGS solar cell was closely related to the location of the BTO layer. The CIGS solar cell had the maximum PCE when the BTO layer was on the top of the *p-n* junction. The effect of the BTO thickness on the performance of the CIGS solar cell was also studied. The optimized device configuration was Mo/CIGS/CdS/BTO/AZO, with a 5 nm-thick BTO. The PCE of this device was 24.53% with an FF of 84.40%, a J_{sc} of 31.19 mA/cm^2, and a V_{oc} of 931.70 mV.

Author Contributions: Writing, original draft preparation, C.L. and H.L. (Haitian Luo); software, C.L. and H.L. (Haitian Luo); review and editing, C.L. and H.L. (Hui Li); project administration, H.L. (Hui Li) and H.G.; funding acquisition, H.L. (Hui Li) and H.G. All authors have read and agreed to the published version of the manuscript.

Funding: This research was funded by the National Key Research and Development Program of China, grant number 2019YFB1503500, the National Natural Science Foundation of China, grant number 11774082, the State Key Laboratory of Metastable Materials Science and Technology, grant number 201901, Fujian Key Laboratory of Photoelectric Functional Materials, grant number FJPFM-201902, and the Newton Advanced Fellowship, grant number 192097.

Institutional Review Board Statement: Not applicable.

Informed Consent Statement: Not applicable.

Data Availability Statement: Not applicable.

Acknowledgments: Acknowledged for providing help with the chart production are Qi Zou and Taiguang Li.

Conflicts of Interest: The authors declare no conflict of interest.

References

1. Li, H.; Zhang, W. Perovskite Tandem Solar Cells: From Fundamentals to Commercial Deployment. *Chem. Rev.* **2020**, *120*, 9835. [CrossRef] [PubMed]
2. Best Research-Cell Efficiencies, NREL Maintains a Chart of the Highest Confirmed Conversion Efficiencies for Research Cells for a Range of Photovoltaic Technologies, Plotted from 1976 to the Present. Available online: https://www.nrel.gov/pv/cell-efficiency.html (accessed on 13 October 2021).
3. Krause, M.; Nikolaeva, A.; Maiberg, M.; Jackson, P.; Hariskos, D.; Witte, W.; Marquez, J.A.; Levcenko, S.; Unold, T.; Scheer, R.; et al. Microscopic origins of performance losses in highly efficient Cu(In,Ga)Se$_2$ thin-film solar cells. *Nat. Commun.* **2020**, *11*, 4189. [CrossRef] [PubMed]
4. Guillemoles, J.; Kirchartz, T.; Cahen, D.; Rau, U. Reply to 'Ideal solar cell efficiencies'. *Nat. Photonics* **2021**, *15*, 165–166. [CrossRef]
5. Shockley, W.; Queisser, H.J. Detailed Balance Limit of Efficiency of P-N Junction Solar Cells. *J. Appl. Phys.* **1961**, *32*, 510. [CrossRef]
6. Zhu, L.; Wang, Z. Recent Progress in Piezo-Phototronic Effect Enhanced Solar Cells. *Adv. Funct. Mater.* **2019**, *29*, 1808214. [CrossRef]
7. Kim, K.; Bae, S.H.; Toh, C.T.; Kim, H.; Cho, J.H.; Whang, D.; Lee, T.W.; Ozyilmaz, B.; Ahn, J.H. Ultrathin organic solar cells with graphene doped by ferroelectric polarization. *ACS Appl. Mater. Interfaces* **2014**, *6*, 3299–3304. [CrossRef]
8. Asadi, K.; de Bruyn, P.; Blom, P.W.M.; de Leeuw, D.M. Origin of the efficiency enhancement in ferroelectric functionalized organic solar cells. *Appl. Phys. Lett.* **2011**, *2011*. *98*, 183301. [CrossRef]
9. Yuan, Y.; Reece, T.J.; Sharma, P.; Poddar, S.; Ducharme, S.; Gruverman, A.; Yang, Y.; Huang, J. Efficiency enhancement in organic solar cells with ferroelectric polymers. *Nat. Mater.* **2011**, *10*, 296–302. [CrossRef]
10. Qin, J.; Zhang, Z.; Shi, W.; Liu, Y.; Gao, H.; Mao, Y. Enhanced Performance of Perovskite Solar Cells by Using Ultrathin BaTiO$_3$ Interface Modification. *ACS Appl. Mater. Interfaces* **2018**, *10*, 36067–36074. [CrossRef]
11. Zheng, F.; Xin, Y.; Huang, W.; Zhang, J.; Wang, X.; Shen, M.; Dong, W.; Fang, L.; Bai, Y.; Shen, X.; et al. Above 1% efficiency of a ferroelectric solar cell based on the Pb(Zr,Ti)O$_3$ film. *J. Mater. Chem. A* **2014**, *2*, 1363–1368. [CrossRef]
12. Liu, X.; Zhang, Q.; Li, J.; Valanoor, N.; Tang, X.; Cao, G. Increase of power conversion efficiency in dye-sensitized solar cells through ferroelectric substrate induced charge transport enhancement. *Sci. Rep.* **2018**, *8*, 17389. [CrossRef] [PubMed]
13. Zhang, J.; Su, X.; Shen, M.; Dai, Z.; Zhang, L.; He, X.; Cheng, W.; Cao, M.; Zou, G. Enlarging photovoltaic effect: Combination of classic photoelectric and ferroelectric photovoltaic effects. *Sci. Rep.* **2013**, *3*, 2109. [CrossRef] [PubMed]
14. Choi, K.J.; Biegalski, M.; Li, Y.; Sharan, A.; Schubert, J.; Uecker, R.; Reiche, P.; Chen, Y.; Pan, X.; Gopalan, V.; et al. Enhancement of ferroelectricity in strained BaTiO$_3$ thin films. *Science* **2004**, *306*, 1005–1009. [CrossRef] [PubMed]
15. Acosta, M.; Novak, N.; Rojas, V.; Patel, S.; Vaish, R.; Koruza, J.; Rossetti, G.A.; Rodel, J. BaTiO$_3$-based piezoelectrics: Fundamentals, current status, and perspectives. *Appl. Phys. Rev.* **2017**, *4*, 041305. [CrossRef]
16. Jaffe, H. Piezoelectric Ceramics. *J. Am. Ceram. Soc.* **1958**, *41*, 494–498. [CrossRef]
17. Devi, L.G.; Nithya, P.M. Preparation, characterization and photocatalytic activity of BaTiF$_6$ and BaTiO$_3$: A comparative study. *J. Environ. Chem. Eng.* **2018**, *6*, 3565–3573. [CrossRef]
18. Kappadan, S.; Thomas, S.; Kalarikkal, N. BaTiO$_3$/ZnO heterostructured photocatalyst with improved efficiency in dye degradation. *Mater. Chem. Phys.* **2020**, *255*, 123583. [CrossRef]
19. Wu, H.; Lu, S.; Aoki, T.; Ponath, P.; Wang, J.; Young, C.; Ekerdt, J.G.; McCartney, M.R.; Smith, D. Direct Observation of Large Atomic Polar Displacements in Epitaxial Barium Titanate Thin Films. *Adv. Mater. Interfaces* **2020**, *7*, 2000555. [CrossRef]
20. Scholtz, L.; Sutta, P.; Calta, P.; Novak, P.; Solanska, M.; Mullerova, J. Investigation of barium titanate thin films as simple antireflection coatings for solar cells. *Appl. Surf. Sci.* **2018**, *461*, 249–254. [CrossRef]
21. Brammertz, G.; Kohl, T.; de Wild, J.; Buldu, D.G.; Birant, G.; Meuris, M.; Poortmans, J.; Vermang, B. Bias-Dependent Admittance Spectroscopy of Thin-Film Solar Cells: Experiment and Simulation. *IEEE J. Photovolt.* **2020**, *10*, 1102–1111. [CrossRef]
22. Khattak, Y.H.; Baig, F.; Toura, H.; Beg, S.; Soucase, B.M. Efficiency enhancement of Cu$_2$BaSnS$_4$ experimental thin-film solar cell by device modeling. *J. Mater. Sci.* **2019**, *54*, 14787–14796. [CrossRef]
23. Hameed, K.Y.; Faisal, B.; Hanae, T.; Mari, S.B.; Saira, B.; Kaim, K.N.A. Modelling of novel-structured copper barium tin sulphide thin film solar cells. *Bull. Mater. Sci.* **2019**, *42*, 231. [CrossRef]
24. Luo, H.; Zhang, Y.; Li, H. Effect of MoS2 interlayer on performances of copper-barium-tin-sulfur thin film solar cells via theoretical simulation. *Sol. Energy* **2021**, *223*, 384–397. [CrossRef]
25. Ghobadi, A.; Yousefi, M.; Minbashi, M.; Kordbacheh, A.H.A.; Abdolvahab, A.R.H.; Gorji, N.E. Simulating the effect of adding BSF layers on Cu$_2$BaSnSSe$_3$ thin film solar cells. *Opt. Mater.* **2020**, *107*, 109927. [CrossRef]
26. Houimi, A.; Gezgin, S.Y.; Mercimek, B.; Kilic, H.S. Numerical analysis of CZTS/n-Si solar cells using SCAPS-1D. A comparative study between experimental and calculated outputs. *Opt. Mater.* **2021**, *121*, 111544. [CrossRef]
27. Widianto, E.; Rosa, E.S.; Triyana, K.; Nursam, N.M.; Santoso, I. Performance analysis of carbon-based perovskite solar cells by graphene oxide as hole transport layer: Experimental and numerical simulation. *Opt. Mater.* **2021**, *121*, 111584. [CrossRef]
28. Sobayel, K.; Shahinuzzaman, M.; Amin, N.; Karim, M.R.; Dar, M.A.; Gul, R.; Alghoul, M.A.; Sopian, K.; Hasan, A.K.M.; Akhtaruzzaman, M. Efficiency enhancement of CIGS solar cell by WS2 as window layer through numerical modelling tool. *Sol. Energy* **2020**, *207*, 479–485. [CrossRef]

29. Chen, J.; Shen, H.; Zhai, Z.; Li, Y.; Li, S. Cd-free Cu(InGa)Se$_2$ solar cells with eco-friendly a-Si buffer layers. *Appl. Surf. Sci.* **2020**, *512*, 145729. [CrossRef]
30. Ihalane, E.; Atourki, L.; Kirou, H.; Ihlal, A.; Bouabid, K. Numerical study of thin films CIGS bilayer solar cells using SCAPS. *Mater. Today Proc.* **2016**, *3*, 2570–2577. [CrossRef]
31. Rezaei, N.; Procel, P.; Simor, M.; Vroon, Z.; Zeman, M.; Isabella, O. Interdigitated back-contacted structure: A different approach towards high-efficiency ultrathin copper indium gallium (di)selenide solar cells. *Prog. Photovolt.* **2020**, *28*, 899–908. [CrossRef]
32. Heriche, H.; Rouabah, Z.; Bouarissa, N. New ultra thin CIGS structure solar cells using SCAPS simulation program. *Int. J. Hydrogen Energy* **2017**, *42*, 9524–9532. [CrossRef]
33. Guirdjebaye, N.; Teyou Ngoupo, A.; Ouédraogo, S.; Mbopda Tcheum, G.L.; Ndjaka, J.M.B. Numerical analysis of CdS-CIGS interface configuration on the performances of Cu(In,Ga)Se$_2$ solar cells. *Chin. J. Phys.* **2020**, *67*, 230–237. [CrossRef]
34. Zhao, Y.; Yuan, S.; Kou, D.; Zhou, Z.; Wang, X.; Xiao, H.; Deng, Y.; Cui, C.; Chang, Q.; Wu, S. High Efficiency CIGS Solar Cells by Bulk Defect Passivation through Ag Substituting Strategy. *ACS Appl. Mater. Interfaces* **2020**, *12*, 12717–12726. [CrossRef] [PubMed]
35. Sites, J.R.; Mauk, P.H. Diode Quality Factor Determination for Thin-Film Solar-Cells. *Sol. Cells* **1989**, *27*, 411–417. [CrossRef]
36. Qu, J.; Zhang, L.; Wang, H.; Song, X.; Zhang, Y.; Yan, H. Simulation of double buffer layer on CIGS solar cell with SCAPS software. *Opt. Quantum Electron.* **2019**, *51*, 383. [CrossRef]
37. Hovel, H.J. *Semiconductors and Semimetals*; Elsevier: New York, NY, USA, 1975.
38. Goetzberger, A.; Hoffmann, V. *Photovoltaic Solar Energy Generation*; Springer: Berlin Heidelberg, Germany, 2005.
39. Maeda, K. Rhodium-doped barium titanate perovskite as a stable p-type semiconductor photocatalyst for hydrogen evolution under visible light. *ACS Appl. Mater. Interfaces* **2014**, *6*, 2167–2173. [CrossRef]
40. Zhao, Y.; Yuan, S.; Chang, Q.; Zhou, Z.; Kou, D.; Zhou, W.; Qi, Y.; Wu, S. Controllable Formation of Ordered Vacancy Compound for High Efficiency Solution Processed Cu(In,Ga)Se$_2$ Solar Cells. *Adv. Funct. Mater.* **2020**, *31*, 2007928. [CrossRef]

MDPI AG
Grosspeteranlage 5
4052 Basel
Switzerland
Tel.: +41 61 683 77 34

Materials Editorial Office
E-mail: materials@mdpi.com
www.mdpi.com/journal/materials

Disclaimer/Publisher's Note: The title and front matter of this reprint are at the discretion of the Guest Editors. The publisher is not responsible for their content or any associated concerns. The statements, opinions and data contained in all individual articles are solely those of the individual Editors and contributors and not of MDPI. MDPI disclaims responsibility for any injury to people or property resulting from any ideas, methods, instructions or products referred to in the content.

www.ingramcontent.com/pod-product-compliance
Lightning Source LLC
LaVergne TN
LVHW072353090526
838202LV00019B/2538